SUPERELECTROPHILES
AND THEIR CHEMISTRY

THE WILEY BICENTENNIAL–KNOWLEDGE FOR GENERATIONS

*E*ach generation has its unique needs and aspirations. When Charles Wiley first opened his small printing shop in lower Manhattan in 1807, it was a generation of boundless potential searching for an identity. And we were there, helping to define a new American literary tradition. Over half a century later, in the midst of the Second Industrial Revolution, it was a generation focused on building the future. Once again, we were there, supplying the critical scientific, technical, and engineering knowledge that helped frame the world. Throughout the 20th Century, and into the new millennium, nations began to reach out beyond their own borders and a new international community was born. Wiley was there, expanding its operations around the world to enable a global exchange of ideas, opinions, and know-how.

For 200 years, Wiley has been an integral part of each generation's journey, enabling the flow of information and understanding necessary to meet their needs and fulfill their aspirations. Today, bold new technologies are changing the way we live and learn. Wiley will be there, providing you the must-have knowledge you need to imagine new worlds, new possibilities, and new opportunities.

Generations come and go, but you can always count on Wiley to provide you the knowledge you need, when and where you need it!

WILLIAM J. PESCE
PRESIDENT AND CHIEF EXECUTIVE OFFICER

PETER BOOTH· WILEY
CHAIRMAN OF THE BOARD

SUPERELECTROPHILES AND THEIR CHEMISTRY

GEORGE A. OLAH

DOUGLAS A. KLUMPP

BICENTENNIAL
1807
WILEY
2007
BICENTENNIAL

WILEY-INTERSCIENCE

A John Wiley & Sons, Inc., Publication

Chemistry Library

Published by John Wiley & Sons, Inc., Hoboken, New Jersey.
Published simultaneously in Canada.

For general information on our other products and services or for technical support, please contact our Customer Care Department within the United States at (800) 762-2974, outside the United States at (317) 572-3993 or fax (317) 572-4002.

Wiley also publishes its books in a variety of electronic formats. Some content that appears in print may not be available in electronic formats. For more information about Wiley products, visit our web site at www.wiley.com.

Wiley Bicentennial Logo: Richard J. Pacifico

Library of Congress Cataloging-in-Publication Data:

I. Olah, George A. (George Andrew), 1927- II. Klumpp, Douglas A., 1964-
   Superelectrophiles and their chemistry / by George A. Olah and Douglas A. Klumpp.
       p. cm.
   Includes index.
   ISBN: 978-0-470-04961-7 (cloth)
1. Superelectrophiles. 2. Chemical affinity. 3. Organic
compounds–Synthesis.   I. Klumpp, Douglas A. II. Title.
   QD271.35.E54043 2007
   547′.2–dc22

                                                                    2007019597

Printed in the United States of America.

10 9 8 7 6 5 4 3 2 1

CONTENTS

PREFACE

Our book is about the emerging field of "Superelectrophiles and Their Reactions." It deals first with the differentiation of usual electrophiles from superelectrophiles, which show substantially increased reactivity. Ways to increase electrophilic strength, the classification into gitionic, vicinal, and distonic superelectrophiles, as well as the differentiation of superelectrophilic solvation from involvement of *de facto* dicationic doubly electron deficient intermediates are discussed. Methods of study including substituent and solvent effects as well as the role of electrophilic solvation in chemical reactions as studied by kinetic investigations, spectroscopic and gas-phase studies, and theoretical calculations are subsequently reviewed. Subsequently, studied superelectrophilic systems and their reactions are discussed with specific emphasis on involved gitionic, vicinal, and distonic superelectrophiles. A brief consideration of the significance of superelectrophilic chemistry and its future outlook concludes this book.

Results of substantial experimental and theoretical work of the field accumulated in recent years warrant a comprehensive review and discussion. This should be of general use to chemists not only with academic and research fields interest but also to advanced students. Because of relevance to potential significant practical applications (including the pharmaceutical and petrochemical fields), industrial chemists should also benefit from it.

We believe that continuing work will result in much further progress and practical applications. If our book will be of help toward this endeavor, our goal will be achieved.

GEORGE A. OLAH
DOUGLAS A. KLUMPP

1

GENERAL ASPECTS

Electrophiles (i.e., electron-deficient species) are of fundamental importance to chemistry. The concept of nucleophiles (lit. "nucleus seeking")
and electrophiles (lit. "electron seeking") was suggested by Ingold following similar views implied by Lapworth's description of "anionoid" and
"cationoid" reagents, Robinson's concepts, and Lewis's theory of bases
(electron donors) and acids (electron acceptors).[1]

   The realization of carbon electrophiles or carbocations dates back to
1901, with the reports of the ionization of triphenylmethyl alcohol in con-
centrated sulfuric acid and triphenylmethyl chloride with aluminum and
tin chlorides.[1b,2] These reactions gave deeply colored solutions, which
are now attributed to the formation of the π-conjugatively delocalized
triphenylmethyl cation. In later studies by Meerwein, Ingold, Hughes,
Whitmore, Roberts, Winstein, Schleyer, and others, using kinetic, stere-
ochemical, and varied experimental methods, carbocation electrophiles
were recognized as intermediates in reactions. It was Olah who discov-
ered in the early 1960s methods to prepare and study long-lived persistent
carbocations, for which he received the Nobel Prize in 1994. The topic
was well reviewed and there is no need for further discussion here.[3] Varied
diverse electrophilic reagents, functionalities, and intermediates have been
further studied in detail.[3] They were reviewed in preceding monographs,
which are referred to for the interested reader.[3b,d] With the advance of our

Superelectrophiles and Their Chemistry, by George A. Olah and Douglas A. Klumpp
Copyright © 2008 John Wiley & Sons, Inc.

structural and mechanistic understanding, it became clear that electrophilic reactivity is an important driving force in many chemical reactions.

Extensive efforts have been made to characterize nucleophile and electrophile strengths. Hammett first correlated[4a] the acidities of substituted benzoic acids (1)

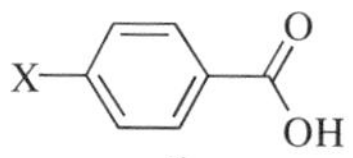

with the structures of the substituent groups and set up his equation as $\log k/k_o = \sigma p$ (where k_o is the rate or equilibrium constant for X=H, k is the rate or equilibrium constant for the substituted benzoic acid, p is a constant for the given reaction, and σ (Hammet's constant) is the value characteristic for the substituent).[4] In their linear free-energy studies, Swain and Scott characterized nucleophiles and electrophiles in kinetic experiments by comparing reaction rates according to the equation 1,

$$\log k_x / k_{H_2O} = s n_x \tag{1}$$

where s is the parameter characteristic for the electrophile and n_x is the parameter characteristic for the nucleophile.[5] More recently, Mayr and co-workers have conducted extensive kinetic studies in estimating the electrophilicities and nucleophilicities of a wide variety of reactants (Figure 1).[6] Using equation 2,

$$\log k(20°C) = s(N + E) \tag{2}$$

the rate constants k for nucleophile-electrophile reactions may be calculated from three parameters (N the nucleophilicity parameter, E the electrophilicity parameter, and s the nucleophile-dependent slope parameter). By analyzing pseudo-first order rate constants with various types of nucleophiles, the electrophilicities of many cationic and neutral species have been established. These electrophiles include dithiocarbenium ions, iminium ions, cationic organometallic complexes (such as propargyl cations with cobalt carbonyl stabilization, cationic palladium complexes, and others), and quinone methides. When experimentally observed, rate constants k are compared with those values of k predicted from the three parameter equation (eq 2), they are generally accurate to within a factor of 10 to 100, excluding reactions with bulky reagents or multi-centered reactions (like $S_N 2$). Moreover, the three-parameter equation may be used to

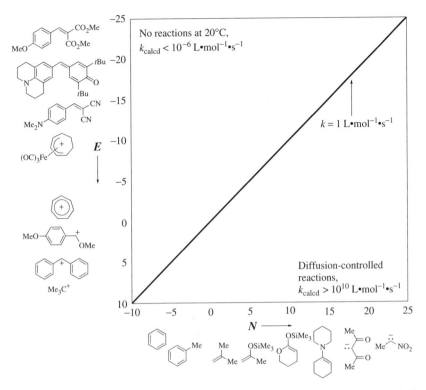

Figure 1. Estimated reaction rates, k_{calcd}, using nucleophilicity parameter N and electrophilicity parameter E.

describe numerous types of reactions involving electrophiles in such reactions as Michael additions, Mannich aminoalkylations, palladium-catalyzed allylations, Freidel-Crafts alkylations, and others.

In addition to the linear free energy studies discussed, there have been many attempts to estimate the thermodynamic stabilities of electrophilic species, such as carbocations.[7] The pK_{R+} values for carbocations reveal trends in relative stability and is defined as, according to the equilibrium established between the carbinol

$$pK_{R+} = \log([R^+]/ROH]) + H^+$$

and carbocation in acidic solution. The pK_{R+} value for triphenylmethyl cation is -6.63 and that of the tri(p-nitrophenyl)methyl cation is -16.27, which is consistent with the resonance destabilization of the cationic center by the nitrophenyl substituents. Gas-phase ionization techniques have also been used to provide thermodynamic data for a variety of electrophiles.[8]

It has been recognized that electrophiles must have sufficiently high reactivity in order to react with weak nucleophiles. This concept of electrophilic reactivity is well demonstrated in the carbonyl chemistry of aldehydes and ketones. The carbonyl group is a reactive electrophilic center when encountering strong nucleophiles like Grignard and organolithium reagents (eq 3). The rapidly formed alkoxide product is thermodynamically heavily favored. With weaker nucleophiles like water or alcohols, reaction rates are considerably slower and the equilibria often favor the starting carbonyl compounds. However, protonation or complexation of a carbonyl group increases its electrophilic reactivity, and weaker nucleophiles may then react with the resulting carboxonium ion (eq 4).

$$CH_3Li + \ H_3C \overset{\overset{\textstyle O^{\delta-}}{\|}}{\underset{CH_3}{C}}\!\!\!{}^{\delta+} \longrightarrow H_3C\overset{OLi}{\underset{CH_3}{\overset{|}{C}}}\!\!-CH_3 \qquad (3)$$

$$CH_3OH + \ H_3C\overset{\overset{\textstyle {}^+O{-}H}{\|}}{\underset{CH_3}{C}} \rightleftharpoons H_3C\overset{OH}{\underset{CH_3}{\overset{|}{C}}}\!\!-OCH_3 \qquad (4)$$

Examples of acid-catalyzed carbonyl chemistry are abundant in synthetic organic chemistry, biochemistry, industrial processes (such as in the synthesis of malachite green; eq 5), and in polymer chemistry (such as in the synthesis of bisphenols of derived epoxy and polycarbonate resins; eq 6).[9]

$$2 \ \overset{}{\underset{N(CH_3)_2}{\bigcirc}} + \ PhCHO \ \xrightarrow{HCl} \ \text{leucomalachite green} \qquad (5)$$

$$2 \ \overset{OH}{\bigcirc} + \ H_3C\overset{\overset{\textstyle O}{\|}}{\underset{}{C}}CH_3 \ \xrightarrow{H_2SO_4} \ \text{bisphenol A} \longrightarrow \ \textit{epoxy and polycarbonate resins} \qquad (6)$$

Without protonation of the carbonyl group, weak nucleophiles (*N,N*-dimethylaniline and phenol) would only react slowly or not at all with the carbonyl groups. Similarly, complexation with Lewis acids can enhance the electrophilic reactivities of carbonyl compounds. This occurs by decreasing participation (using Winstein's concept) of the neighboring oxygen into the developing carbocationic center and thus increasing the polarized character of the (complexed) carbonyl group and lowering the energy of the LUMO (as described by the frontier molecular orbital theory). It has also been shown that substitution by electron withdrawing groups can significantly increase the electrophilic reactivity of carbonyl groups and their related protonated carboxonium ions. Whereas protonated acetophenone (**2**) is unreactive towards benzene (a weak π-nucleophile), the carboxonium ion from 2,2,2-trifluoroacetophenone reacts to give the condensation product in high yield (eqs 7–8).[10]

$$
\underset{}{\text{(structure: acetophenone)}} \quad \xrightarrow[C_6H_6]{CF_3SO_3H} \quad \underset{\mathbf{2}}{\text{(structure: protonated acetophenone, } \overset{+}{OH})} \quad \xrightarrow{\quad\not\quad} \quad \text{unreactive to benzene} \qquad (7)
$$

$$
\underset{}{\text{(structure: trifluoroacetophenone, } CF_3)} \quad \xrightarrow[C_6H_6]{CF_3SO_3H} \quad \underset{\mathbf{3}}{\text{(structure, } \overset{+}{OH}, CF_3)} \quad \xrightarrow{C_6H_6} \quad \xrightarrow[-H_2O]{C_6H_6} \quad F_3C-\underset{C_6H_5}{\overset{C_6H_5}{\underset{|}{\overset{|}{C}}}}-C_6H_5 \qquad (8)
$$

The electron withdrawing inductive effects of the fluorine substituents render the carboxonium ion **3** more electrophilic than carboxonium ion **2**, and consequently it reacts with benzene. Thus, the electrophilic reactivity of the carbonyl group can be greatly enhanced by Brønsted or Lewis acid solvation and by substitution with electron withdrawing groups.

Although increased electrophilicity can lead to reactions with weak nucleophiles, highly electrophilic cations can exist as stable, long-lived species in solutions of low nucleophilicity.[11] Superacidic media are especially well suited for studies of such highly electrophilic species. Superacids and their chemistry, a topic extensively reviewed in a previous monograph,[11] have enabled the preparation and study of varied long-lived cationic electrophiles such as carbocations, acyl and carboxonium cations, and varied onium ions such as oxonium, sulfonium, halonium, nitronium, and azonium ions.[3d] Whereas these electrophilic species react instantaneously with many common electron donor solvents, the superacidic media is essentially an environment of very low nucleophilicity. For example, efficient routes to carbocation electrophiles include the ionization of alkyl fluorides in SbF_5 and the ionization of alcohols in magic acid, $FSO_3H\text{-}SbF_5$ (eqs 9–10).[12,13]

$$\underset{H_3C}{\overset{F}{\underset{}{\overset{}{C}}}}\overset{CH_3}{\underset{CH_3}{}} \xrightarrow{SbF_5} \underset{H_3C}{\overset{CH_3}{\underset{}{\overset{+}{C}}}}CH_3 \quad SbF_6^- \tag{9}$$

$$\underset{}{\overset{OH}{\underset{CH_3}{\bigcirc}}} \xrightarrow{FSO_3H\text{-}SbF_5} \underset{}{\overset{+}{\bigcirc}}\text{—}CH_3 \tag{10}$$

$$[FSO_3\text{-}(SbF_5)n]^-$$

In the superacids, the resulting counterions are weak nucleophiles: SbF_6^- or $[SbF_6^-(SbF_5)_n]^-$, $[FSO_3\text{-}(SbF_5)_n]^-$, etc.

Gillespie proposed the widely accepted definition of superacids as those being stronger than 100% H_2SO_4 for Brønsted acids (i.e., $H_0 \leq -12$).[11] Similarly, according to Olah, Lewis acids stronger than anhydrous $AlCl_3$ are considered superacidic. Brønsted superacids span the logarithmic Hammett acidity scale from $H_0 -12$ for anhydrous H_2SO_4, to -27 for FSO_3H-SbF_5 (9:1) and ca. -30 for HF-SbF_5(1:1) (Figure 2). The isolated ("naked") proton is unobtainable in solution chemistry, but by comparing gas phase data with superacid solution chemistry, its acidity has been estimated to be in the -50 to -60 H_0 range. Superacids can react with weak base-sites like the n-electrons of carbonyl and other groups, the π-electrons of unsaturated groups (alkenes, alkynes, and arenes), and even with σ-electrons of alkanes. Not only carbocationic but also other varied reactive electrophiles can be generated as long-lived species in the superacids, and these electrophiles can often be studied directly by spectroscopic methods. They can also participate in many superacid-catalyzed reactions.

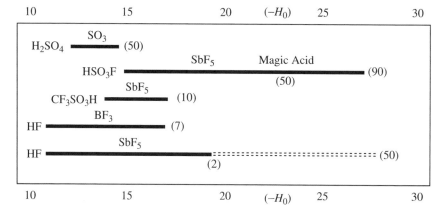

Figure 2. Acidity ranges for several common superacids. The solid bars are measured using indicators, while the broken bar is estimated by kinetics measurements; in (%) mol Lewis acid.

With respect to electrophiles and electron-deficient varied species, there have been suggestions of "non-coordinating solvents" and "non-coordinating anions." Of course, by definition anions are electron donors. In order to prepare highly electron-deficient species such as trialkylsilyl cations, claims were made for the use of non-nucleophilic or non-coordinating anions.[14] Even more surprisingly, some of these reactive ions have been prepared from toluene solution, referred to as a "non-nucleophilic solvent."[14] It should be mentioned that "the myth of non-coordinating anions" was discussed as early as 1973 by Rosenthal, who concluded that "it is clear that the notion of the non-coordinating anion should be put to rest alongside the notion of non-coordinating solvent."[15] Olah et al. subsequently discussed this point critically. As a result, the terms "least coordinating" and more correctly "weakly coordinating" anions were substituted.

Much effort has been made recently to find new weakly coordinating anions as counter ions for strong electrophiles and acids.[16] Like the conjugate bases of superacids, these weakly coordinating anions are generally characterized by the anionic charge being delocalized over the entirety of large anions, with no individual atom bearing a substantial part of the charge. Many of the most useful weakly coordinating anions have hydrogen and fluorine atoms on their periphery, thus avoiding the presence of strongly Lewis basic-sites. Among the most common low or weakly coordinating anions are borate anions, such as Meerwein's BF_4^- (which is of course the conjugate base of the important superacid $HF-BF_3$), Wittig's $B(C_6H_5)_4^-$, and the more recent $B(C_6F_5)_4^-$, and $[B(OTeF_5)_4]^-$, as well as Olah's SbF_6^- or $SbF_6^--(SbF_5)_n$. The weakly coordinating borates are of particular practical importance to the activity of the electrophilic, one-component Ziegler-Natta olefin polymerization catalysts.[17] These cationic, metallocene catalysts have been shown to have high activities as polymerization catalysts, due in large part to their electrophilic metal center and weakly coordinating anions. It has been shown that for a given type of catalytic site, the more weakly coordinating anions result in more active polymerization catalysts. Another class of less coordinating anions are Reed's 1-carba-*closo*-dodecaborate monoanions ($CB_{11}H_{12}^-$ and related halogenated analogs). In addition to being used as counter ions in active, cationic polymerization catalysts, these anions have also been shown to be useful in the preparation of salts of varied electrophilic cations.[18] These include four-coordinate Fe(III)porphyrins, crystalline salts of carbocations (eq 11), a weakly coordinated, highly crowded stannyl cation (eq 12) and highly stabilized silicenium cations (eq 13), and other salts, despite the high electrophilic reactivities of the cations. Reactive electrophilic salts

$$\text{(CH}_3\text{)}_3\text{CH} + \text{CH}_3[\text{CHB}_{11}(\text{CH}_3)_5\text{Br}_6] \xrightarrow[\substack{\text{CH}_2\text{Cl}_2 \\ -30°\text{C}}]{-\text{CH}_4} (\text{CH}_3)_3\text{C}^+ \ [\text{CHB}_{11}(\text{CH}_3)_5\text{Br}_6]^- \quad (11)$$

$$\underset{\substack{n\text{-Bu} \\ n\text{-Bu}}}{\overset{\substack{n\text{-Bu} \\ n\text{-Bu}}}{n\text{-Bu-Sn-Sn-}n\text{-Bu}}} + 2\,\text{CB}_{11}(\text{CH}_3)_{12}\bullet \longrightarrow \ \text{Sn}^+ \ [\text{CB}_{11}(\text{CH}_3)_{12}]^- \quad (12)$$

$$\underset{\text{Ar}}{\overset{\text{Ar}}{\text{Ar-Si}}}\!\!-\!\!\diagup\!\!\diagdown + (\text{CH}_3\text{CH}_2)_3\text{Si}[\text{CHB}_{11}(\text{CH}_3)_5\text{Br}_6] \xrightarrow{-(\text{CH}_3\text{CH}_2)_3\text{SiCH}_2\text{CH=CH}_2} \underset{\text{Ar}}{\overset{\text{Ar}}{\text{Ar-Si}^+}} \ [\text{CHB}_{11}(\text{CH}_3)_5\text{Br}_6]^- \quad (13)$$

Ar = mesityl

$$\underset{\text{Ph}}{\overset{\overset{\text{O}}{\|}}{\text{C}}}\!\!-\!\!\text{CH}_3 \xrightarrow[\substack{(\text{CH}_3\text{CH}_2)_3\text{SiH} \\ -78°\text{C}}]{(\text{C}_6\text{H}_5)_3\text{C}^+ \ ^-\text{B}(\text{C}_6\text{F}_5)_4} \underset{\underset{\mathbf{4}}{\text{Ph}}}{\overset{\overset{\text{OSi}(\text{CH}_2\text{CH}_3)_3}{|}}{\text{C}}}\!\!-\!\!\text{CH}_3 \ ^-\text{B}(\text{C}_6\text{F}_5)_4 + (\text{C}_6\text{H}_5)_3\text{CH} \quad (14)$$

such as the silylated carboxonium ion (**4**, eq 14) and the tris-mesityl silicenium ion have been prepared as tetra(pentafluorophenyl)borate salts. Recent reviews have been published on weakly coordinating anions, especially with respect to polymerization catalysts.[16,17]

Although monocationic carbon electrophiles have been involved in chemical reactions for many years, multiply charged organic electrophiles (dications, trications, etc.) have only been studied recently.[19] It was the high reactivity of dicationic vicinal and geminal electrophiles that led to the concept of superelectrophilic activation as proposed by Olah in the 1970s.[20] In 1973, Brouwer and Kiffen reported the results of superacid-catalyzed reactions between protonated aldehydes and ketones (carboxonium ions) and alkanes, as well as the reactions of the acetyl cation with alkanes (Scheme 1).[21] In solutions of HF–SbF$_5$ or HF–BF$_3$, reaction products are formed that are consistent with hydride transfer between the acetyl cation (generated *in situ* from acetic acid) and isobutane. Earlier studies by Olah and co-workers showed, however, that acetyl salts like acetyl hexafluoroantimonate do not abstract hydride from alkanes in aprotic solvents (SO$_2$, SO$_2$ClF, or CH$_2$Cl$_2$). In order to explain the enhanced reactive of the acetyl cation in superacid, Olah proposed the formation of a protosolvated, superelectrophilic intermediate (**5**).[20] Despite the fact that the acetyl cation has a positive charge, it has nonbonding oxygen electron pairs, which are capable of interacting with the superacidic media. In the limiting case, the hydrogen-bonded species (**7**) can lead toward the formation of the highly electrophilic, doubly electron deficient, dicationic

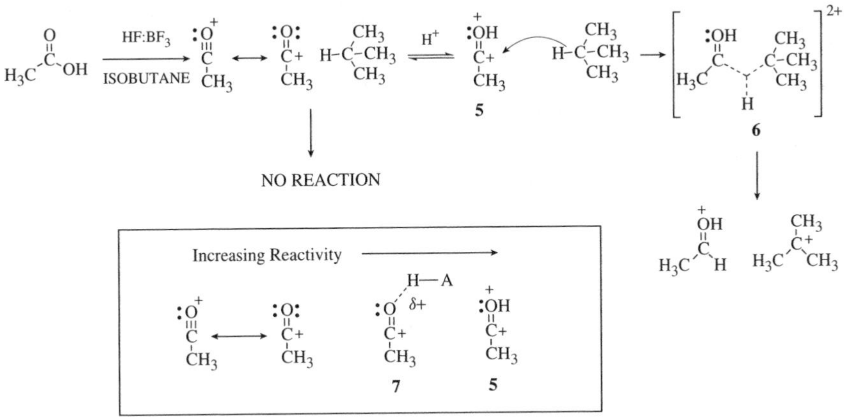

Scheme 1. Protosolvation of the acetyl cation and its reaction with isobutane.

species **5**. It is the superelectrophilic intermediate (**5** or **7**) that is capable of reacting with the C–H bond of the hydrocarbon substrate (isobutane). Subsequently, the complex **6** then leads to the formed hydride abstraction products. As described more thoroughly in Chapter 5, the proposed superelectrophilic activation is also supported by thermodynamic calculations.

About the same time, a similar type of activation was observed in the reactions of nitronium salts.[20] Nitronium salts (such as $NO_2^+BF_4^-$ or NO_2^+ PF_6^-) show little or no tendency to react with deactivated arenes or alkanes in aprotic media. However, in fluorosulfuric acid or HF-BF$_3$ solution, nitration takes place giving nitration products even nitromethane (eq 15).

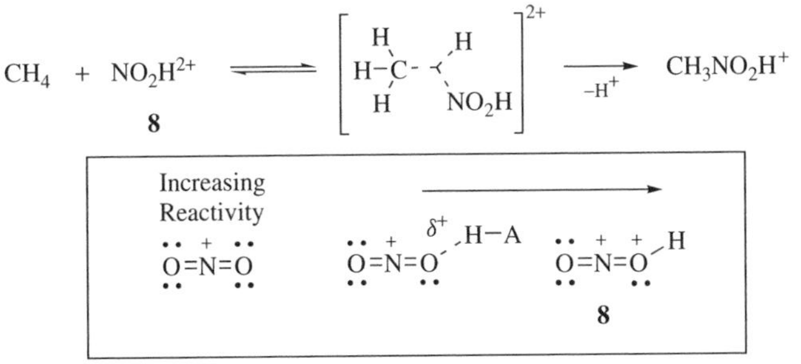

These results can be interpreted in terms of protosolvation of the nitronium ion. While the monocationic nitronium ion is a sufficiently polarizible electrophile to react with strong nucleophiles such as olefins and activated arenes, it is generally not reactive enough to react with weak nucleophiles including methane. Partial or complete protonation of the nitronium oxygen then leads to the superelectrophilic species **8**. The heightened electrophilic reactivity of **8** allows the reactions with deactivated aromatics and alkanes.

The protionitronium ion (**8**, NO_2H^{2+}) and the protioacetyl cation (**5**, CH_3COH^{2+}) were the first examples of superelectrophilic intermediates. Their electrophilic reactivities are much greater than that of the corresponding parent monocations.[22] As such, these superelectrophiles are capable of reacting with weaker nucleophiles than the nitronium (NO_2^+) and acetyl (CH_3CO^+) cations. A defining feature of these superelectrophiles (and those described subsequently) is the further complexation (solvation) of the monocationic electrophile by Brønsted or Lewis acids. As a result of this interaction, neighboring group participation with the electrophilic center is decreased and the resulting electrophiles are increasingly electron deficient and reactive.[22] In the limiting cases, multiply charged *de facto* dications (even multications) may result. As discussed in Chapter 7, superelectrophiles are distinguished from such distant dications in which the two charged groups are isolated. Distant onium dications exhibit chemistry no different than the monocationic onium ions. The term "superelectrophile" previously has been applied occasionally to a number of other chemical systems, including metal complexes in high oxidation states, electrophiles bearing multiple electron-withdrawing groups, and other highly reactive electrophiles. While these systems may exhibit unique chemistry, they are not superelectrophiles within the context of discussed acid-base interactions. Consequently, their chemistry will not be included in our discussions in this book.

Superelectrophilic intermediates have been categorized into two distinct groups: the *distonic* (distant) and the *gitonic* (close) superelectrophiles (Table 1).[22] Distonic superelectrophiles are defined as electrophiles in

Table 1. Classes and examples of superelectrophiles

| Gitonic Superelectrophiles | Distonic Superelectrophiles |

which the positive charge centers are separated by two or more carbon or hetero atoms, while gitonic superelectrophiles are characterized by the positive charge centers being in close proximity. Both types of superelectrophiles and their chemistry will be discussed in subsequent chapters.

Theoretical calculations have been done on varied superelctrophilic species. They are often found in deep potential energy wells on the energy surfaces, although others are higher lying minima. Moreover, the calculated gas-phase structures are often only kinetically stable species but with sizable energy barriers to proton loss or other fragmentations. These calculations have been verified by the observation of a number of superelectrophiles by gas-phase mass spectroscopy studies (*vide infra*). Reactions of superelctrophiles in the condensed phase frequently involve discreetly formed dicationic (or tricationic) species. However, as noted with the protioacetyl dication (CH_3COH^{2+}, **5**) and the protionitronium dication (NO_2H^{2+}, **8**), formation of effective dications may be the limiting case. Partial protonation or weaker donor-acceptor interaction with a Lewis acid, what we now call electrophilic solvation, can also activate electrophiles to produce superelectrophiles. Along these same lines, there has been kinetic evidence to suggest varying degrees of protonation in the transitions states involving superelectrophiles.[23] For many superelectrophiles, it has not been possible to de facto directly observe these species, even with fast spectroscopic methods. It has been proposed in several of these studies that superelectrophiles are formed in only low concentrations. As an explanation for these reactions, superelectrophilic transition states may be involved with no persistent intermediates.

Two types of interactions have been shown to be involved in superelectrophilic species. Superelectrophiles can be formed by the further interaction of a conventional cationic electrophile with Brønsted or Lewis acids (eq 16).[23] Such is the case with the further protonation (protosolvation) or Lewis acid coordination of suitable substitutents at the electron deficient site, as for example in carboxonium cations. The other involves further protonation or complexation formation of a second proximal onium ion site, which results in superelectrophilic activation (eq 17).[24]

$$(16)$$

$$(17)$$

10

Both types of dicationic species (**9–10**) exhibit the properties and reactivities of superelectrophiles.

It should be noted that it was Pauling who predicted the viability of a doubly charged molecular structure in the 1930s by suggesting the kinetic stability of the helium dimer dication (He_2^{2+}).[25] Despite the large estimated exothermic energy of dissociation (200 kcal/mol), theoretical calculations predicted a substantial energy barrier to dissociation (33.2 kcal/mol). The recent mass spectrometric observation of the helium dimer dication, He_2^{2+}, confirms Pauling's prediction.[25b] The kinetic stability of the helium dimer dication can be understood by the bonding interaction (He_2^{2+} is isoelectronic with the hydrogen molecule) offsetting the large electrostatic charge-charge repulsion. Molecular orbital theory also predicted the stability of aromatic dications, including the cyclobutadiene dication, the biphenylene dication, and the cyclooctatetrene dication. These dicationic species (**11–13**),

11   **12**   **13**

as well as many other related systems, have since been observed as stable ions in under high-acidity, low nucleophilicity conditions.[19a,b]

Superelectrophilic onium dications have been the subject of extensive studies and their chemistry is discussed in chapters 4–7. Other multiply charged carbocationic species are shown in Table 2. These include: Hogeveen's bridging, nonclassical dication (**14**)[26]; the pagodane dication (**15**)[27]; Schleyer's 1,3-dehydro-5,7-adamantane dication (**16**)[28]; the bis(fluoreenyl) dication (**18**)[29]; dications (**17** and **19**);[19a] trications (**20–21**)[19a,30]; and tetracations (**22–23**).[31] Despite the highly electrophilic character of these carbocations, they have been characterized as persistent ions in superacids.

In solutions of low nucleophilicity, multiply charged electrophiles can frequently exhibit deep-seated rearrangements and fragmentation reactions. These reactions often stem from the electrostatic repulsive effects involving the charge centers, and they have precluded the observation of

Table 2. Some persistent multiply charged carbocations

Dications		Higher Cations

14 **15** **20** **21**

16 **17**

18 **19** **22** **23**

some multi-charged ions. For example, a long-sought goal in carbocation studies was the generation of an aliphatic 1,3-carbodication. Whereas ionization of 2,5-dichloro-2,5-dimethylhexane gives the stable 1,4-dication (**24**, eq 18) ionization of 2,4-dichloro-2,4-dimethylpentane (**25**) leads only to the 2-pentenyl cation (**27**, eq 19).[32] It is thought that the 1,3-dication (**26**) if formed undergoes rapid deprotonation to give ion **27**. Similarly, ionization of the diol (**28**) gives the two monocations instead of the expected 1,3-dication (eq 20).[32] Other reactions of dicationic species are described in chapters 4–7.

$$(18)$$

$$(19)$$

$$(20)$$

The concept of superelectrophilic activation was first proposed 30 years ago.[20] Since these early publications from the Olah group, superelectrophilic activation has been recognized in many organic, inorganic, and biochemical reactions.[22] Due to the unusual reactivities observed of superelectrophiles, they have been exploited in varied synthetic reactions and in mechanistic studies. Superelectrophiles have also been the subject of numerous theoretical investigations and some have been directly observed by physical methods (spectroscopic, gas-phase methods, etc.). The results of kinetic studies also support the role of superelectrophilic activation. Because of the importance of electrophilic chemistry in general and superacidic catalysis in particular, there continues to be substantial interest in the chemistry of these reactive species. It is thus timely to review their chemistry.

REFERENCES

(1) (a) A. Lapworth *Nature*, **1925**, *115*, 625. (b) C. D. Nenitzescu, in *Carbonium Ions*, vol 1; G. A. Olah and P. v. R. Schleyer Eds.; Wiley, New York, 1968; chap 1 and references therein. (c) C. K. Ingold *Recl. Trav. Chim. Pays-Bas.* **1929**, *42*, 797. (d) C. K. Ingold *Chem. Rev.* **1934**, *15*, 225. (e) C. K. Ingold *Structure and Mechanism in Organic Chemistry*, Cornell University Press, 1953.

(2) (a) J. F. Norris *Am. Chem. J.* **1901**, *25*, 117. (b) F. Kehrman; F. Wentzel *Ber. Dtsch. Chem. Ges.* **1901**, *34*, 3815. (c) A. Baeyer; V. Villiger *Ber. Dtsch. Chem. Ges*. **1901**, *35*, 1189; 3013.

(3) (a) G. A. Olah *Angew. Chem. Int. Ed. Engl.* **1973**, *12*, 173. (b) G. A. Olah, "Carbocation and Electrophilic Reactions," VCH-Wiley, Weinheim, New York 1973. (c) G. A. Olah "Nobel Lecture" *Angew. Chem. Int. Ed. Engl.* **1995**, *34*, 1393. (d) G. A. Olah; K. K. Laali; Q. Wang; G. K. S. Prakash *Onium Ions*, Wiley, New York, 1998.

(4) (a) L. P. Hammet *Physical Organic Chemistry*, 2nd Ed., McGraw Hill, New York, 1970. (b) H. H. Jaffé *Chem. Rev.* **1953**, *53*, 191.

(5) C. G. Swain; C. B. Scott *J. Am. Chem. Soc.* **1953**, *75*, 141.

(6) R. Lucius; R. Loos; H. Mayr *Angew. Chem. Int. Ed.* **2002**, *41*, 92, and references cited therein.

(7) (a) N. C. Deno; J. J. Jaruzelski; A. Schriesheim *J. Am. Chem. Soc.* **1955**, *77*, 3044. (b) E. M. Arnett; T. C. Hofelich *J. Am. Chem. Soc.* **1983**, *105*, 2889 and references cited therein.

(8) D. H. Aue; M. T. Bowers in *Gas Phase Ion Chemistry*, M. T. Bowers Ed., Academic Press, New York, 1979; Vol 2, chap 9.

(9) (a) J. E. Hofmann; A. Schriesheim, A. in *Friedel-Crafts and Related Reaction*; G. A. Olah, Ed.; Wiley, New York, NY, 1964; vol 2.; pp 597-640.

(b) J. March, *Advanced Organic Chemistry*, 4th Ed; Wiley, New York, 1992; pp 548–549.

(10) (a) *Chem. Abstr.* **1979**, *90*, 86986u. (b) *U. S. Pat. Appl. 891,872 (1978)*. (c) W. D. Kray; R. W. Rosser *J. Org. Chem.* **1977**, *42*, 1186.

(11) (a) G. A. Olah; G. K. S. Prakash; J. Sommer, in *Superacids*; Wiley, New York, 1985. (b) G. A. Olah; A. Molnar; G. K. S. Prakash; J. Sommer, in *Superacids*, Revised 2nd Ed., Wiley, New York, in preparation.

(12) (a) G. A. Olah *Angew. Chem.*, **1963**, *75*, 800. (b) G. A. Olah in *Carbocation Chemistry* G. A. Olah and G. K. S. Prakash, Eds; Wiley, New York, 2004; chap 2.

(13) (a) G. A. Olah *J. Org. Chem.* **2001**, *66*, 5943. (b) Reference 3b.

(14) (a) J. B. Lambert; S. Zhang; C. L. Stern; J. C. Huffman *Science*, **1993**, *260*, 1917. (b) C. A. Reed *Acc. Chem Res.* **1998**, *31*, 325.

(15) (a) M. R. Rosenthal *J. Chem. Ed.* **1973**, *50*, 33. (b) G. A. Olah; G. Rasul; H. A. Buchholz; X.-Y. Li; G. K. S. Prakash *Bull. Soc. Chim. Fr.* **1995**, *132*, 569. (c) I. Krossing; I. Raabe *Angew. Chem. Int. Ed.* **2004**, *43*, 2066.

(16) S. Strauss *Chem. Rev.* **1993**, *93*, 927.

(17) E. Y.-X. Chen; T. J. Marks *Chem. Rev.* **2000**, *100*, 1391.

(18) (a) T. Kato; C. R. Reed *Angew. Chem. Int. Ed.* **2004**, *43*, 2908. (b) K.-C. Kim; C. A. Reed; D. W. Elliot; L. J. Mueller; F. Tham; L. Lin; J. B. Lambert *Science*, **2002**, *297*, 825. (c) G. K. S. Prakash; C. Bae; G. Rasul; G. A. Olah *J. Org. Chem.* **2002**, *67*, 1297. (d) I. Zharov; B. T. King; Z. Havlas; A. Pardi; J. Michl *J. Am. Chem. Soc.* **2000**, *122*, 10253.

(19) (a) G. K. S. Prakash; T. N. Rawdah; G. A. Olah *Angew. Chem. Int. Ed. Engl.* **1983**, *22*, 390. (c) R. M. Pagni *Tetrahedron* **1984**, *49*, 4161. (c) Reference 3c, chap 10. (d) V. G. Nenajdenko; N. E. Shevchenko; E. S. Balenkova *Chem. Rev.* **2003**, *103*, 229.

(20) G. A. Olah; A. Germain; H. C. Lin; D. Forsyth *J. Am. Chem. Soc.* **1975**, *97*, 2928.

(21) (a) D. M. Brouwer; A. A. Kiffen *Recl. Trav Chim. Pays-Bas* **1973**, *92*, 689. (b) D. M. Brouwer; A. A. Kiffen *Recl. Trav Chim. Pays-Bas* **1973**, *92*, 809. (c) M. Brouwer; A. A. Kiffen *Recl. Trav Chim. Pays-Bas* **1973**, *92*, 906.

(22) (a) G. A. Olah; G. K. S. Prakash; K. Lammertsma *Res. Chem. Intermed.* **1989**, *12*, 141, (b) G. A. Olah *Angew. Chem. Int. Ed. Engl.* **1993**, *32*, 767. (c) G. A. Olah; D. A. Klumpp *Acc. Chem. Res.* **2004**, *37*, 211.

(23) (a) M. Vol'pin; I. Akhrem; A. Orlinkov *New J. Chem.* **1989**, *13*, 771, (b) S. Saito; Y. Sato; T. Ohwada; K. Shudo *J. Am. Chem. Soc.* **1994**, *116*, 2312.

(24) D. A. Klumpp; Y. Zhang; P. J. Kindelin; S. Lau *Tetrahedron* **2006**, *62*, 5915.

(25) (a) L. Pauling *J. Chem Phys.*, **1933**, *1*, 56. (b) J. D. Dunitz; T. K. Ha *J. Chem. Soc. Chem. Commun.* **1972**, 568.

(26) H. Hogeveen; P. W. Kwant *Acc. Chem Res.* **1975**, *8*, 413.

(27) G. K. S. Prakash; V. V. Krishnamurthy; R. Herges; R. Bau; H. Yuan; G. A. Olah; W.-D. Fessner; H. Prinzbach *J. Am. Chem. Soc.* **1986**, *108*, 836.

(28) M. Bremer; P. v. R. Schleyer; K. Schoetz; M. Kausch; M. Schindler *Angew. Chem. Int. Ed. Engl.* **1987**, *26*, 761.

(29) (a) J. L. Malandra; N. S. Mills; D. E. Kadlecek; J. A. Lowery *J. Am. Chem. Soc.* **1994**, *116*, 11622. (b) N. S. Mills; J. L. Malandra; E. E. Burns; A. Green; K. Unruh; D. E. Kadlecek; J. A. Lowery *J. Org. Chem.* **1997**, *62*, 9318.

(30) S. Ito; N. Morita; T. Asao *Bull. Chem. Soc. Jpn.* **2000**, *73*, 1865.

(31) (a) N. J. Head; G. K. S. Prakash; A. Bashir-Hashemi; G. A. Olah *J. Am. Chem. Soc.* **1995**, *117*, 12005. (b) R. Rathore; C. L. Burns; I. A. Green *J. Org. Chem.* **2004**, *69*, 1524.

(32) G. A. Olah; J. L. Grant; R. J. Spear; J. M. Bollinger; A. Serianz; G. Sipos *J. Am. Chem. Soc.* **1976**, *98*, 2501.

2

STUDY OF SUPERELECTROPHILES

2.1 INTRODUCTION

Many of the methods used in the study of electrophiles can also be applied
in studies of superelectrophiles. As noted in the introduction, superelectrophiles often exhibit very high reactivities when compared with conventional electrophiles. The heightened reactivities can consequently be used
to estimate the degree of superelectrophilic activation in the chemistry
of superelectrophiles. In this respect, several reports have characterized
superelectrophiles based on kinetic studies. Reaction rates were shown
to increase with acidity for certain types of acid-catalyzed conversions
indicative of superelectrophilic activation. Although superelectrophiles
affecting reactions are generally present only in low concentrations, some
persistent superelectrophiles have been studied using varied spectroscopic
techniques. Low-temperature NMR and gas-phase techniques have been
used in studies of several types of superelectrophiles. Theoretical calculations have been used to characterize many superelectrophilic systems. In
addition to proving their intermediate nature as energy minima, they also
have provided ground state geometries and energies. Calculations have
also indicated significant energy barriers to dissociation for a variety of
dicationic and tricationic superelectrophiles. This chapter describes the

Superelectrophiles and Their Chemistry, by George A. Olah and Douglas A. Klumpp
Copyright © 2008 John Wiley & Sons, Inc.

various methods of study of characteristic superelectrophilic systems that have been studied. Other techniques, such as isotopic labeling studies, calorimetric measurements, and various physical measurements, are also discussed when relevant.

2.2 REACTIVITY PROFILES

One of the defining features of superelectrophiles is the often-observed high level of reactivity towards nucleophiles of low strength.[1] This experimental observation is frequently used as an indication for the involvement of a superelectrophiles. To illustrate, the following examples show how the electrophile's reactivity can be characterized to indicate superelectrophilic chemistry.

It was the enhanced reactivities of nitronium and acetylium salts in superacidic media that lead Olah to first propose in 1975 the concept of superelectrophilic activation.[2] As mentioned in Chapter 1, nitronium salts exhibit markedly enhanced reactivities in strong acids when compared with reactions in aprotic solvents. Even the slow nitration of methane can be accomplished using nitronium hexafluorophosphate ($NO_2^+PF_6^-$) in superacidic FSO_3H, while nitronium salts in aprotic solvents are unreactive. This increased reactivity in superacid was suggested to involve protosolvation of the nitronium cation, forming the limiting superelectrophile NO_2H^{2+}. In a similar respect, nitration of strongly deactivated arenes is only effected by nitronium salts in superacidic media. For example, nitronium tetrafluoroborate ($NO_2^+ BF_4^-$) does not nitrate m-dinitrobenzene in nitromethane solvent, but in FSO_3H it gives 1,3,5-trinitrotoluene (eq 1).[3]

$$O_2N\text{-}\underset{}{\bigcirc}\text{-}NO_2 + NO_2BF_4 \xrightarrow[150°C]{FSO_3H} O_2N\text{-}\underset{62\%\ \ NO_2}{\bigcirc}\text{-}NO_2 \qquad (1)$$

When nitronium tetrafluoroborate was attempted to be reacted with pentafluorobenzene in the triflic acid, CF_3SO_3H ($H_0 -14.1$), no nitration occured. However, using the much stronger superacid, triflatoboric acid (2 CF_3SO_3H-$B(O_3SCF_3)_3$, $H_0 -20.5$), nitration occurs in high yield (the nitronium ion being *in situ* generated from nitric acid; eq 2).[3]

$$\underset{F}{\overset{F}{\underset{F}{\bigcirc}}}F + HNO_3 \xrightarrow[25°C]{2\ CF_3SO_3H–B(O_3SCF_3)_3} \underset{89\%}{O_2N\overset{F}{\underset{F}{\bigcirc}}F} \qquad (2)$$

Even the trityl cation has been successfully nitrated using superacidic activation of nitronium salts (eq 3).[4]

Due to the delocalization of the cationic charge into the phenyl rings, they are only very weakly nucleophilic. The reaction of nitronium tetrafluoroborate with triphenylcarbenium tetrafluoroborate in excess CF_3SO_3H, however, gives the mononitration product, which can then be reduced using ionic hydrogenation (Et_3SiH). The same reaction in an aprotic medium (CH_2Cl_2) gives no nitration product. These results are all consistent with the protolytic activation of the nitronium cation in the superacids. Methane, m-dinitrobenzene, pentafluorobenzene, and the trityl cation, are all weak nucleophiles, and the nitronium cation (**1**) does not possess sufficient electrophilic reactivity to attack these nucleophiles. Increasing the acidity of the reaction medium, however, leads to an equilibrium with the protosolvated nitronium ion (**2**) (or even the protonitronium ion, **3**),

which is a significantly more reactive electrophile. Thus, nitration of these weak nucleophiles suggest the involvement of the superelectrophilic species (**2** or **3**). A more detailed discussion of the superelectrophilic nitronium ion is found in Chapter 5.

Friedel-Crafts type reactions of strongly deactivated arenes have been the subject of several recent studies indicating involvement of superelectrophilic intermediates. Numerous electrophilic aromatic substitution reactions only work with activated or electron-rich arenes, such as phenols, alkylated arenes, or aryl ethers.[5] Since these reactions involve weak electrophiles, aromatic compounds such as benzene, chlorobenzene, or nitrobenzene, either do not react, or give only low yields of products. For example, electrophilic alkylthioalkylation generally works well only with phenolic substrates.[6] This can be understood by considering the resonance stabilization of the involved thioalkylcarbenium ion and the delocalization of the electrophilic center (eq 4). With the use of excess Lewis acid, however, the electrophilic reactivity of the alkylthiocarbenium ion can be greatly enhanced, indicating superelectrophilic activated species (**4**).[6c] The greater electrophilic reactivity allows for the electrophilic alkylthioalkylation of even benzene and halogenated arenes in good yields (eq 5).

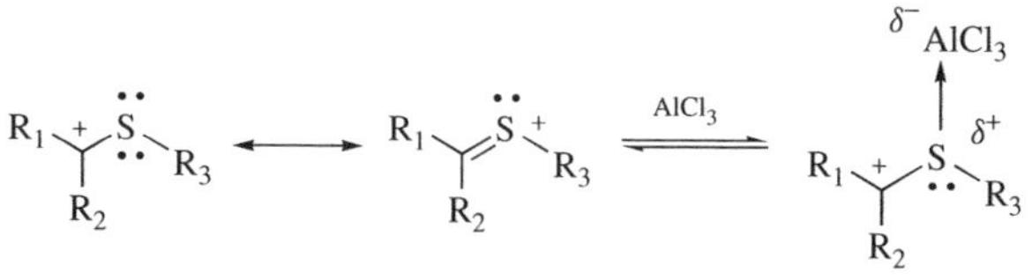

(4)

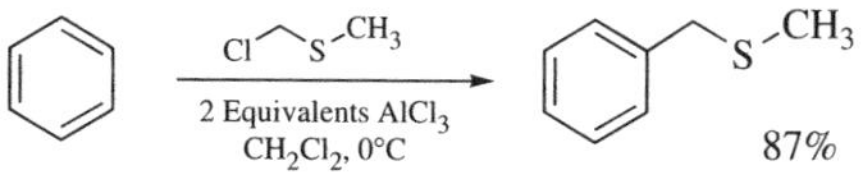

(5)

Similarly, hydroxyalkylation with aldehydes or ketones is best accomplished with activated aromatic compounds, such as phenols.[7] However, reaction even with chlorobenzene has been carried out with either parabanic acid or isatin using triflic acid (eqs 6–7).[8,9]

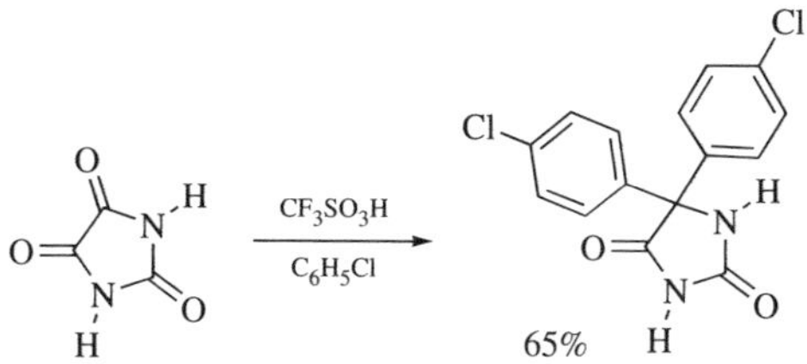

(6)

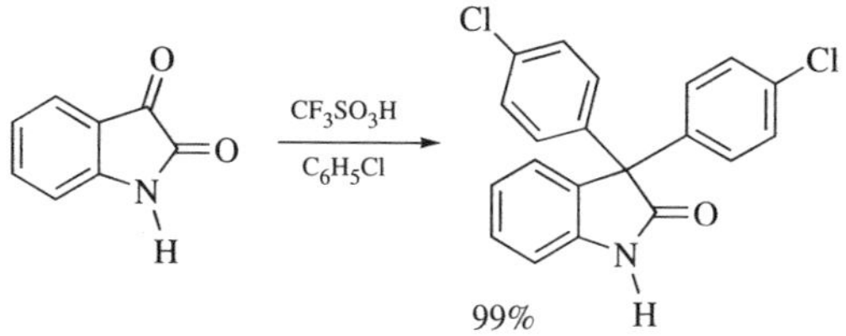

(7)

Diprotonated, superelectrophilic intermediates were suggested to be involved in both conversions. Considering protonated aldehydes, benzaldehyde gives a carboxonium ion that is significantly resonance stabilized and thus unreactive towards aromatic substrates such as o-dichlorobenzene or nitrobenzene. Pyridinecarboxaldehydes, however, show much higher electrophilic reactivities due to their ability to form via N-protonation the superelectrophile (**5**, eq 8).[10] A similar situation is seen in the hydroxyalkylation reactions of acetyl-substituted arenes. Acetophenone is fully protonated in excess triflic acid, but the resulting carboxonium ion (**6**) is not sufficiently electrophilic to react with benzene (eq 9).[11] Yet, acetyl-substituted pyridines (and other N-heterocycles) give the related condensation products in excellent yield under the same conditions (eq 10). The higher reactivity of the pyridine system is attributed to the formation of dication (**7**). In the superacid catalyzed acylation of arenes with methyl benzoate, even nitrobenzene is acylated in high yield, suggesting the formation of a superelectrophile (**8**, eq 11).[12]

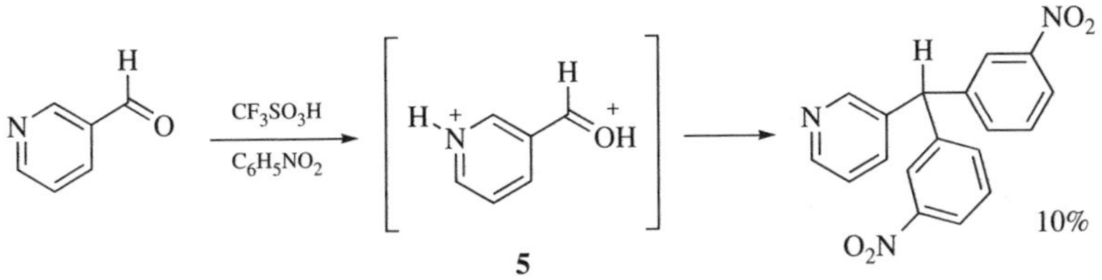

(8)

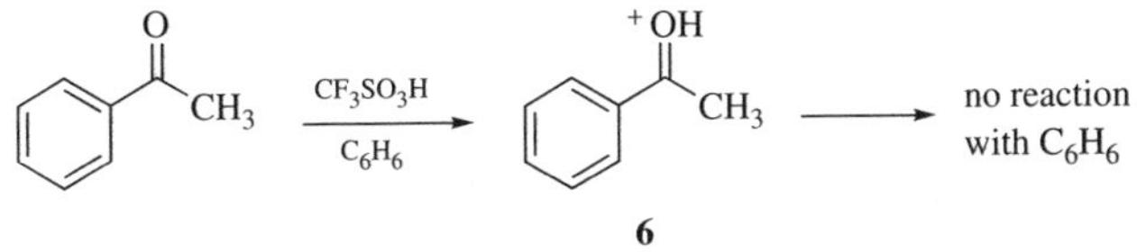

(9)

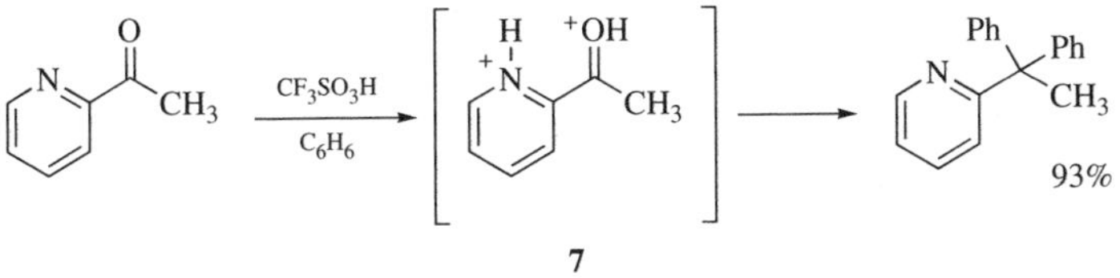

(10)

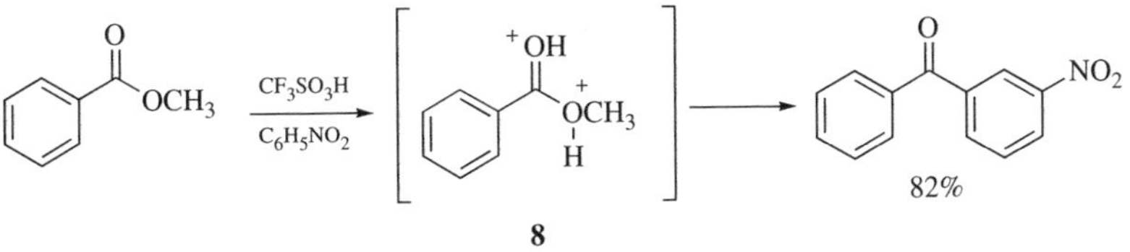

(11)

Although electrophilic reactions involving dications with deactivated arenes may suggest the formation of superelectrophilic intermediates, there are a number of well-known examples of monocationic electrophiles that are capable of reacting with benzene or with deactivated aromatic compounds. For example, 2,2,2-trifluoroacetophenone condenses with benzene in triflic acid (eq 12).[13] A similar activation is likely involved in the H_2SO_4 catalyzed reaction of chloral (or its hydrate) with chlorobenzene giving DDT (eq 13).

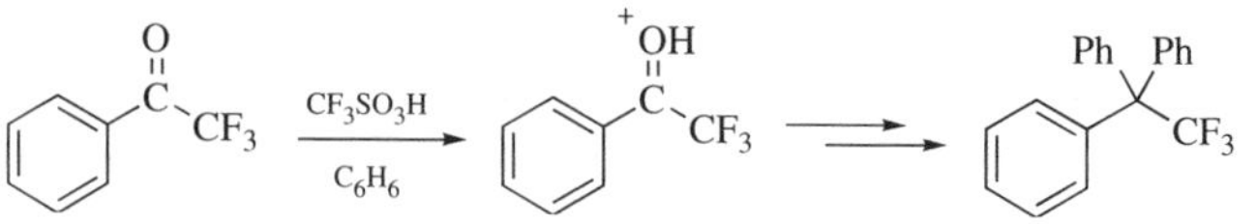

$$(12)$$

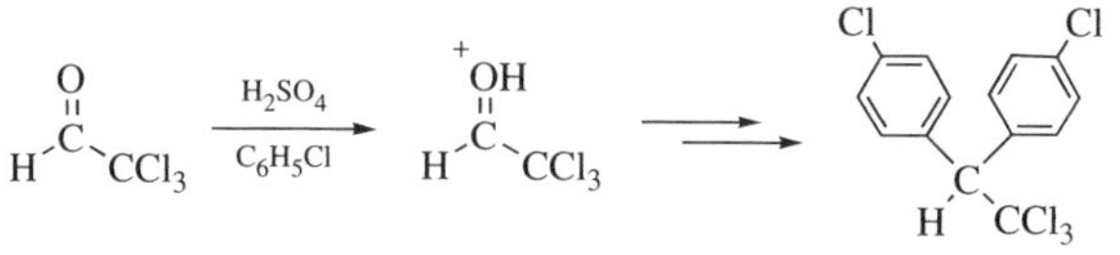

$$(13)$$

The inductive effects of the trifluoromethyl and trichloromethyl groups increase the electrophilic reactivities of the carboxonium ions when compared with those formed from acetophenone or acetaldehyde.

As mentioned in Chapter 1, some superelectrophiles are even capable of reacting with the σ-bonds of alkanes or alkyl groups. σ-Donor alkanes are of course weak nucleophiles. But in reactions with exceedingly strong acids or electrophiles, σ-bonds can act as electron donors. This was demonstrated in the study of the superacid catalyzed reactions of alkanes with the protosolvated acetyl cation (Scheme 1, Chapter 1). Olah et al. showed carbon monooxide can react with isobutane to give methylisopropyl ketone in high yield upon rearrangement of formed pivaldehyde (Scheme 1).[14] Other branched alkanes react similarly. The results are consistent with the formation of the protosolvated formyl cation (9), which reacts directly with the tertiary CH bond via a five coordinate carbocation (10). Subsequent rearrangement yields the methylisopropyl ketone product. Although it is conceivable that the carbonylation might occur by initial formation of a *tert*-butyl cation (via reaction of isobutane with superacid) followed by trapping it with carbon monooxide, this pathway is unlikely because no carboxylic acids are detectable in the product mixtures. An analogous superelectrophilic formylation has also

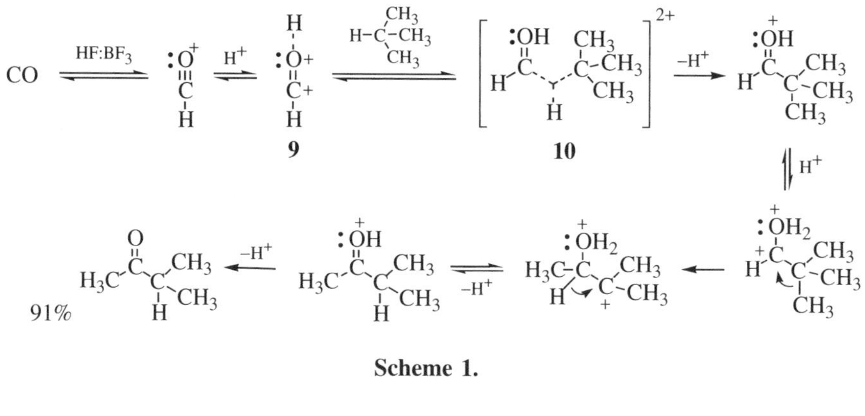

Scheme 1.

been observed in the reactions of adamantane and carbon monooxide to give 1-formyladamantane (eq 14).[15]

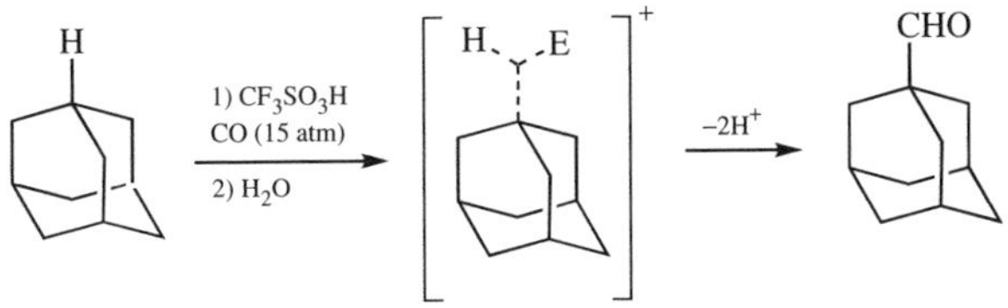

$$E^+ = HCO^+ \text{ or } HCOH^{2+}$$

(14)

The reaction yield increases with acid strength, further suggesting the intermediacy of the protosolvated, superelectrophilic formylating species (**9**).

Superelectrophilic activation has also been proposed to be involved, based upon the reactivity of carbocations with molecular hydrogen (a σ-donor).[16] This chemistry is probably even involved in an enzymatic system that converts CO_2 to methane. It was found that $N^5.N^{10}$–menthyl tetrahydromethanopterin (**11**) undergoes an enzyme-catalyzed reaction with H_2 by hydride transfer to the pro-R position and releases a proton to give the reduced product **12** (eq 15). Despite the low nucleophilicity of H_2, cations like the *tert*-butyl cation (**13**) are sufficiently electrophilic to react with H_2 via 2 electron-3 center bond interaction (eq 16). However, due to stabilization (and thus delocalization) by adjacent nitrogen atoms, cations like the guanidinium ion system (**14**) do not react with H_2 (eq 17).

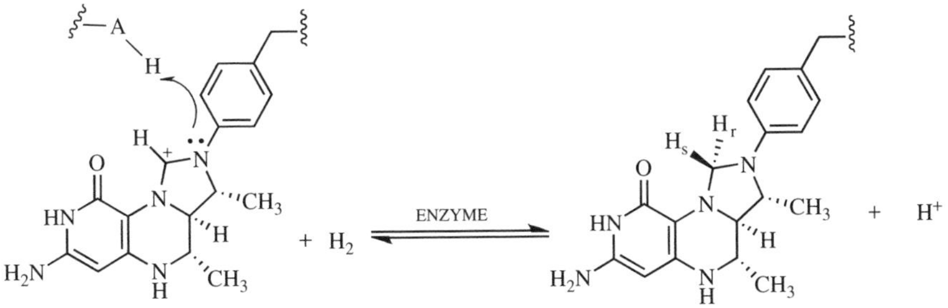

(15)

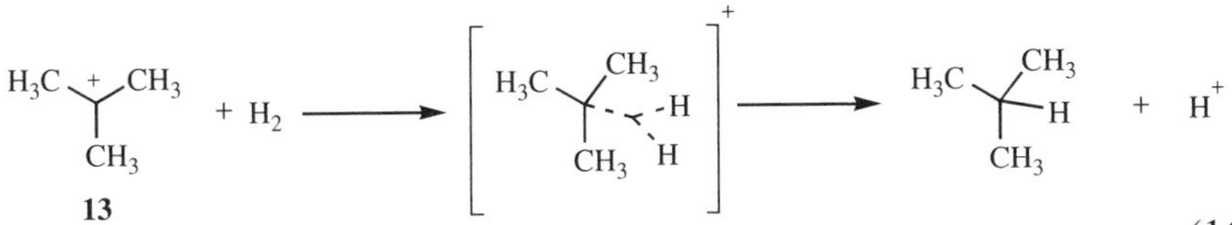

(16)

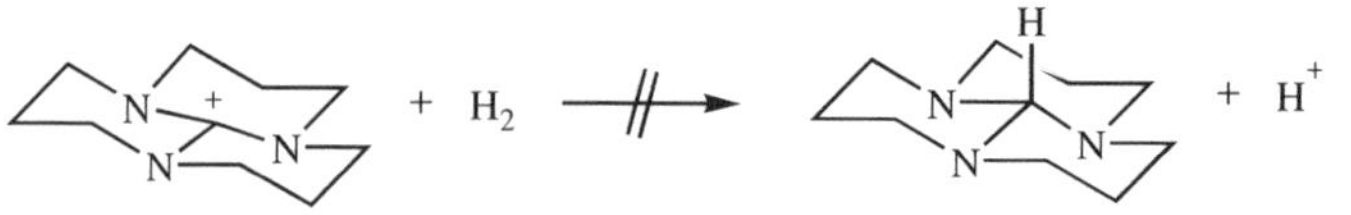

$$\tag{17}$$

14

Since $N^5.N^{10}$–menthyl tetrahydromethanopterin (**11**) should also exhibit significant delocalization of the cationic site, its reactivity with molecular hydrogen is remarkable. It is suggested that protonation (protosolvation) of one or both of the adjacent basic nitrogen sites by the enzymatic system could generate an enhanced electrophilic system (Scheme 2). This superelectrophilic system would be the result of decreased neighboring group participation and thus resulting in increased reactivity. A more detailed discussion of this system is found in Chapter 5.

The above examples are illustrative of how activated electrophilic chemistry involving weak nucleophiles can indicate involvement of superelectrophiles. Chemical reactivity with weak nucleophiles, however, by itself cannot be considered sufficient evidence for superelectrophilic intermediates or reactive dicationic intermediates. Many reactions of alkanes, for example, occur by oxidative, free radical reaction pathways. Strong acid systems may also cause medium effects, such as, for example, increasing the solubility of reagents, which can lead to higher yields or increased reaction rates. However, when these effects can be ruled out and the electrophilic reactions of weak nucleophiles give higher yields and increased reaction rates (e.g., in conversions dependent on acid strength) superelectrophilic activation is indicated.

2.3 KINETIC STUDIES

Kinetic studies have been important in the characterization of varied electrophilic reactions, such as electrophilic aromatic substitutions, addition

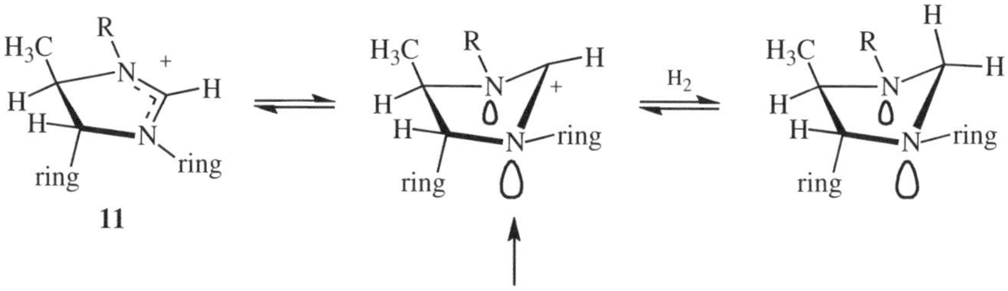

Scheme 2. Proposed mechanism for the activation of the stabilized cation (**11**).

reactions to alkenes, and various carbonyl reactions. Several superelectrophilic reactions have also been studied using kinetics experiments. These reactions typically show dramatic increases in reactions rates (and yields) with increasing acidity. Some representative examples are discussed here. Divinyl ketones are known to cyclize to 2-cyclopentanones in acid-catalyzed electrocylization reactions.[17] The Nazarov and related reactions often require forcing conditions of high acidity and elevated temperatures. O-Protonated cationic intermediates have been proposed in the reactions. With a series of 1-phenyl-2-propen-1-ones in superacid-catalyzed reactions, 1-indanones were obtained in generally good yields (eq 18).[18]

$$(18)$$

The reactions were also carried out in solutions of varying acidity and, in all cases, the cyclization required highly acidic conditions. Cyclization of **15** gives 91% yield of **18** in 100% CF_3SO_3H ($H_0 -14.1$) and 48% yield in 6% w/w CF_3SO_3H : 94% CF_3CO_2H ($H_0 -8.7$). Attempted reactions in CF_3CO_2H (100%, $H_0 -2.7$) gave no cyclization. In kinetic studies, 1-phenyl-2-propen-1-ones were reacted in solutions of varying acidity and in all cases the cyclizations were found to exhibit pseudo first-order kinetics. The rate constants were found to be directly proportional to acid strength over the range of acidities from $H_0 -9$ to -13. These results are consistent with the Zucker-Hammett hypothesis, which states that reaction rates are linearly proportional to acidity if the reactive species (i.e., the present protonated intermediates) are formed in low concentrations and are involved in the rate-determining step. Given, however, that the carbonyl group is essentially fully protonated at $H_0 -9$, an increasing rate constant with acidity indicates formation of the diprotonated intermediate (**17**).

1,3-Diphenyl-1-propanone (**19**) is found to give the cyclization product **22** in good yield when the reaction is carried out in superacid.[19] This acid-catalyzed cyclization also shows a dramatic dependence on the acidity of the reaction medium: a yield of 72% in 100% CF_3SO_3H ($H_0 -14.1$); a yield of 7% in 6% w/w CF_3SO_3H : 94% CF_3CO_2H ($H_0 -8.7$); a yield

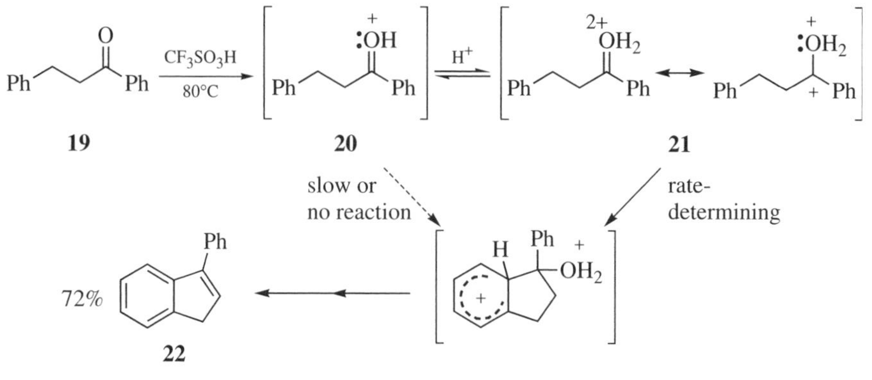

Scheme 3. Proposed mechanism for the cyclization of 1,3-diphenyl-l-propanone (19).

of 0% in 100% CF_3CO_2H. Using NMR and UV titration methods, the pK_{BH+} value of ketone **19** was estimated to be -5.9. Thus, **19** is fully protonated in the solution of acidity $H_0 -9$, and at higher levels of acidity, the monocationic species (**20**) is in equilibrium with the diprotonated species (**21**). The cyclodehydration reaction was also examined in kinetics studies. Between the range of $H_0 -8.2$ and -14.1, the reaction rate constant was found to increase linearly with acidity. These results indicate that the superelectrophilic species **21** is formed and is involved in the rate-determining cyclization step (Scheme 3). The monoprotonated species (**20**) itself is stabilized by inductive and resonance effects and therefore the cyclization does not occur. A second protonation at the carbonyl group is needed to give the superelectrophile (**21**), which leads to the cyclization.

In a study involving an intramolecular Houben-Hoesch reaction (eq. 19),

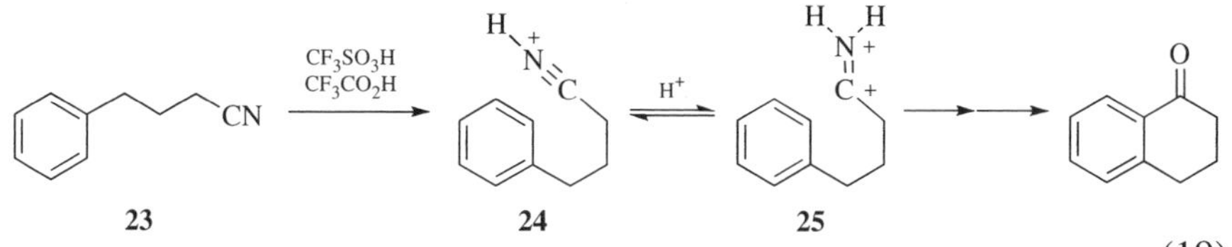

(19)

kinetic evidence suggested the involvement of superelectrophilic diprotonation of nitriles.[20] 4-Phenylbutyronitrile (**23**) was found to cyclize at an appreciable rate only in solutions more acidic than $H_0 = -10$. The first-order rate constant was found to be linearly proportional to acidity over a range of $H_0 = -10.5$ to -13. In fact, the rate increases 100-fold over this range of acidity. The monoprotonated nitrile (**24**) itself is approximately half protonated in solutions of acidity $H_0 = -10$. However, at this acid strength, there is little or no cyclization. This is consistent with the formation of the superelectrophilic intermediate **25** with increasing acid strength. The kinetic data also indicate that the superelectrophilic intermediate **25** is formed (even if in low concentration) and is involved in the rate-limiting reaction step. These experiments also suggest that the rapid increase in reaction rate with acidity is probably due to the degree of protonation, where, in this case, protonation is nearly complete. In the study of the intramolecular cyclization of 1,3-diphenyl-1-propanone (Scheme 3), a similar increase was found in reaction rate with acidity, but the reaction rate increase was in this case not as large. This was interpreted in terms of a more limited degree of protonation or the protosolvation of the activated complex.

When the unsaturated acetal **26** was reacted with superacid at $-60°$C and irradiated, a photostationary state is observed consisting of the two stereoisomers **26a** and **26b** (Scheme 4).[21] Evaluation of the kinetics of stereomutation shows good first order kinetics, and the rate constant increases with acidity. Based on the kinetics results, it was proposed that protonation of the carboxonium group of **26b** leads to the dication (**27**), and this facilitates isomerization through delocalization of the positive charge.

The Pictet-Spengler and Bischler-Napieralski reactions are used to prepare biologically and pharmocologically important isoquinoline ring systems. Versions of these reactions have also been proposed in biosynthetic pathways. Under mildly to strongly acidic conditions, the cyclizations

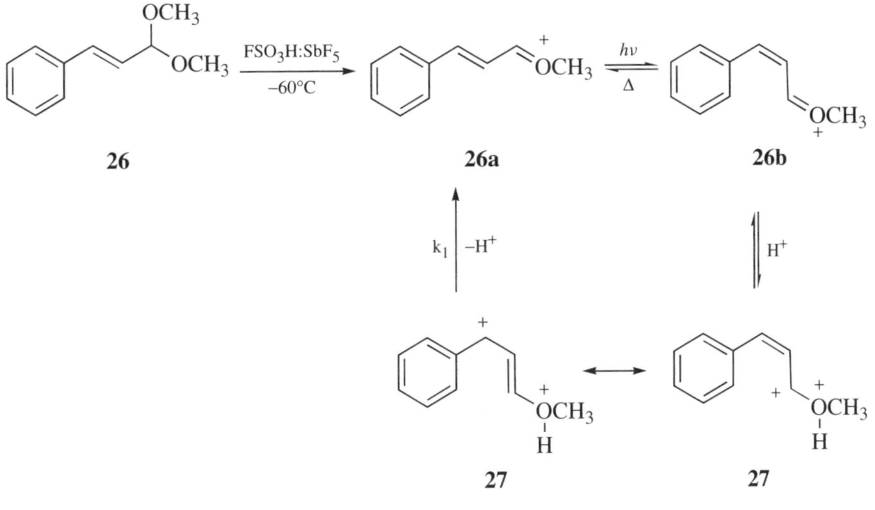

Scheme 4. Role of superlectrophile (**27**) in the stereomutation of the oxonium ion (**26b**).

require activating groups (R) on the aryl-substituent of the imine (eq 20).

$$(20)$$

From studies of the Pictet-Spenger reaction using superacidic conditions, kinetic evidence for superelectrophilic intermediates was reported.[22] While N-methylene-2-phenethylamine (**28a**) reacts slowly or not at all with strong acids, in 90% CF_3SO_3H:10% CF_3CO_2H (w/w; $H_0 -12.5$; 100 equivalents), the cyclization product is obtained in 76% yield (eq 21). When a variety of aldimines were reacted in 100% CF_3SO_3H, the cyclization occurred in good yields. For example, compounds **28b** and **28c** gave the cyclized products (eqs 22 and 23)

$$(21)$$

$$(22)$$

$$(23)$$

in CF_3SO_3H. In 100% CF_3CO_2H, the starting aldimines (**28b-c**) were recovered unreacted. NMR analysis indicated that the imines are completely protonated in CF_3CO_2H. The proton bound to the imine nitrogen appears as a doublet at δ 11.2 for ion **29**. The authors noted that, despite being fully protonated in CF_3CO_2H, **28b** and **28c** were unreactive towards cyclization. In kinetic studies, the imines were dissolved in solutions of varying acidity and the formation of products was monitored by ^{1}H NMR. It was found that cyclizations occur at significant rates when the solutions are more acidic than $H_0 -10.5$. The cyclizations showed good first-order kinetics with reaction rates increasing steadily as the acidity of solution increased (Figure 1). Reaction rates should level off if the monoprotonated imine is the intermediate leading to cyclization, because the imines

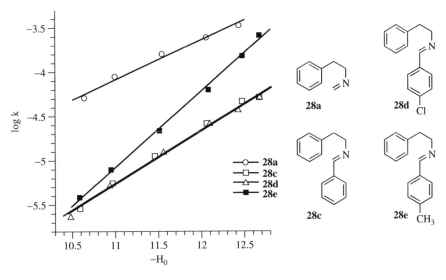

Figure 1. Acidity-rate profiles of the superacid-catalyzed Pictet-Spengler cyclizations.[22]

are fully protonated at acidities as low as H_0 −2.7 (CF_3CO_2H). Electron donating groups (alkyl) on the phenethyl aryl group increased the reaction rate, suggesting that cyclization is involved in the rate-determining step. The linear correlation between acidity and reaction rates indicates that diprotonated, superelectrophilic intermediates such as **30** were involved in the cyclization, leading to the cyclized product **31**.

A hallmark of superelectrophilic reactivity is often the greatly increased yields and reaction rates dependent on increasing acidity. While absolute reaction rate constants can be obtained from kinetic studies and indicate evidence for superelectrophilic activation, qualitative evidence can also be obtained from realitive reaction rates. For example, in a study involving an acetophenone derivative, 2,2,2-triphenylacetophenone (**32**) was converted to 9,10-diphenylphenanthrene (**36**) in quantitative yield in reactions carried out in superacid (Scheme 5).[23] The relative reaction rate is found to increase with acidity and the exclusive product (**36**) is formed only in solutions more acidic than about H_0 −12 (limit of superacidic solution; Table 1). Acetophenone is fully protonated in acidic media stronger than H_0 −12,[24] and 2,2,2-triphenylacetophenone should exhibit a similar basicity. This suggests the involvement of the diprotonated, superelectrophilic species **34**. Formation of the superelectrophile **34** leads to a retro-pinacol rearrangement and subsequent cyclization.

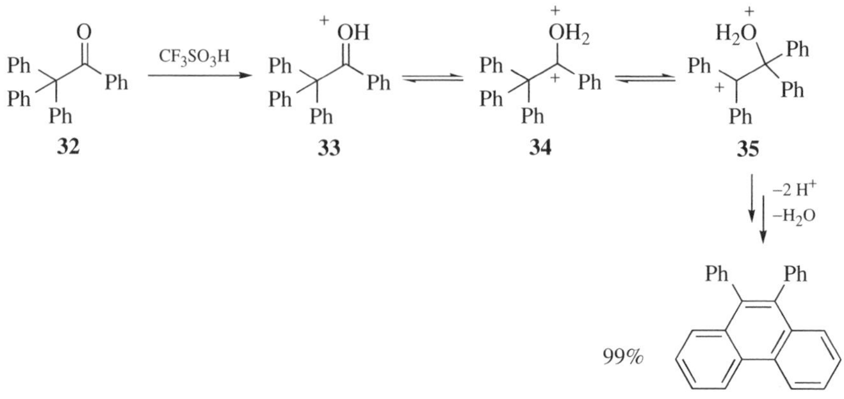

Scheme 5. Proposed mechanism for the condensation of ketone **32**.

Table 1. Relative yields of product 36 with varying acidity

Acidity (H_0)	Acid System $\left(\dfrac{CF_3SO_3H\text{-}CF_3CO_2H}{w/w\%}\right)$	Product Ratio (**32:36**)
-14.1	100 % CF_3SO_3H	0:100
-12.5	72.8	0:100
-11.5	43.5	30:70
-10.6	22.1	100:0
-2.7	100 % CF_3CO_2H	100:0

Reaction conditions: compound **32** (50 mg), benzene (0.5 mL), and acid (0.5 mL), reacted at 25°C for 24 h.

Superelectrophilic intermediates were also proposed in the cyclization reaction of α-(methoxycarbonyl)diphenylmethanol (**37**) with superacid (eq 24).[25] The yield of cyclized product is found to increase dramatically with increasing acidity of the reaction medium. The conversion is believed

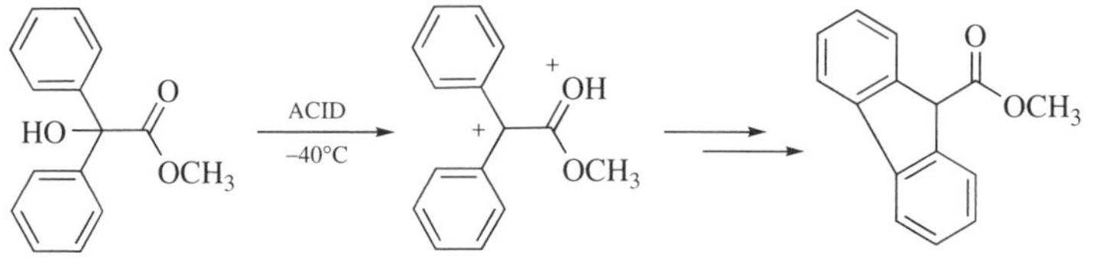

(24)

H_0:	-2.7	-11.8	-13.2
% YIELD:	0%	20%	94%

to involve the dicationic intermediate (**38**) that further reacts by a con-rotatory electrocyclization. The involvement of the superelectrophile was further supported by kinetic experiments and theoretical calculations, as described in more detail in Chapter 5.

Several types of 1,2-dicarbonyl compounds have been studied in super-acid and their chemistry indicates the involvement of superelectrophiles. One of the earliest studies involved the chemistry of 2,3-butanedione (**39**), which condenses rapidly with benzene in CF_3SO_3H (Scheme 6) in good yield.[26] It was proposed that the dicationic intermediates (**40**) and (**41**) are formed under the superacidic conditions and these superelec-trophiles lead to product formation. As expected with superelectrophilic activation, the reactivity of **39** drops off rapidly with decreasing acid strength. While the yield of product **42** is 94% in CF_3SO_3H (H_0 −14.1) and 94% in 50% CF_3SO_3H : 50% CF_3CO_2H (H_0 −12.0), the reaction is slower in the later case. The product yield falls to 62% at H_0 −11 (20% CF_3SO_3H : 80% CF_3CO_2H) and no reaction is observed at H_0 −8 (6% CF_3SO_3H : 94% CF_3CO_2H). Further evidence for superelectrophile **40** was obtained from ^1H and ^{13}C NMR studies in superacid solutions of **39**.

In acid-catalyzed reactions of cyanides with arenes (Gattermann reac-tion), the yield of products and relative reactions rates are found to cor-relate with the acid strength of the media.[20] The reaction of either NaCN or trimethylsilyl cyanide (TMS-CN) with benzene in superacid gave the benzaldimine, which upon hydrolysis yielded benzaldehyde (eq 25). The reactions were found to give reasonable yields (> 40%) of product only

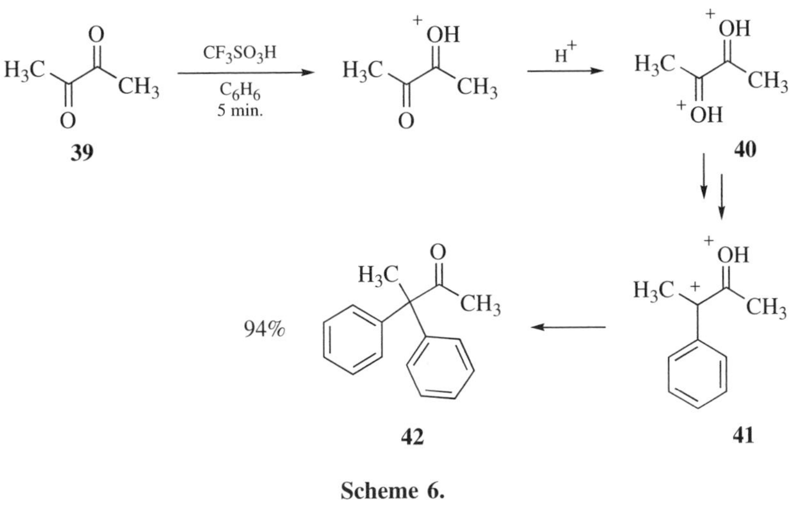

Scheme 6.

when the acidity of the solution was stronger than $H_0 = -14$ and gave excellent yields when acidity was $H_0 < -18$ (Table 2). Since the pK_{BH^+} of hydrogen cyanide (43) is between -10 and -11, hydrogen cyanide is more than half protonated in $CF_3SO_3H:CF_3CO_2H$ (50:50, $H_0 = -12$). This suggests that it is not the monoprotoned hydrogen cyanide (44, eq 26)

$$\text{(25)}$$

$$\text{(26)}$$

that reacts with benzene, but a second protonation generates the superelectrophilic species (45), which is sufficiently reactive to give the product.

As noted previously in Chapter 1, the electrophilic reactivities of acetyl salts increase dramatically as the acidity of the reaction medium increases. This was one of the observations that lead Olah and co-workers to first propose the concept of superelectrophilic activation, or protosolvation of the acetyl cation, in 1975.[2] This seminal paper described the chemistry of acetyl hexafluoroantimonate ($CH_3CO^+SbF_6^-$) and the reaction with alkanes in various solvents. In aprotic solvents such as SO_2, SO_2ClF, AsF_3, and CH_2Cl_2, there was no reaction. However in $HF-BF_3$, acetyl salts react with iso-alkanes and efficient hydride abstraction is observed.[27] This was interpreted by Olah as evidence for protonation of the acetyl

Table 2. Reaction of benzene with cyanides to give benzaldehde

Cyanide	Acid	H_0	Time, min	% Yield
NaCN	23% TfOH : 77% TFA	-10.6	300	0
	55% TfOH : 45% TFA	-11.8	300	3
	80% TfOH : 20% TFA	-12.7	300	11
	100% TfOH	-14.0	30	44
	1% SbF$_5$: 99% TfOH	-16.8	30	65
	5% SbF$_5$: 95% TfOH	< -18	5	92
TMS-CN	20% TfOH : 80% TFA	-10.0	720	0
	55% TfOH : 45% TFA	-11.8	300	2
	80% TfOH : 20% TFA	-12.7	30	17
	100% TfOH	-14.0	30	55
	5% SbF$_5$: 95% TfOH	< -18	5	99

TfOH:CF$_3$SO$_3$H; TFA:CF$_3$CO$_2$H

Table 3. Yields of acetylated product from the reaction of chlorobenzene with $CH_3CO^+SbF_5^-$ in solutions of varying acid strength

Acid	H_0	Yield[a]
100% CF_3CO_2H	−2.7	0%
20% CF_3SO_3H : 80% CF_3CO_2H	−10.0	9%
40% CF_3SO_3H : 60% CF_3CO_2H	−12.0	21%
100% CF_3SO_3H	−14.1	32%
1% SbF_5 : 99% CF_3SO_3H	−16.8	48%
5% SbF_5 : 95% CF_3SO_3H	< −18	82%

salts. Similarly, acetyl salts show heightened reactivities with aromatic substrates in solutions of increasing acidity. For example, it was found by Shudo and Ohwada that acetyl hexafluoroantimonate ($CH_3CO^+SbF_6^-$) and benzoyl hexafluoroantimonate ($C_6H_5CO^+SbF_6^-$) salts acylate aromatic compounds most effectively in highly acidic solution.[20] In the case of chlorobenzene, the acetylated product is formed in 0 to 32 to 82% yield, for progressively stronger acid systems (Table 3). NMR studies ruled out solubility of the salts as an explanation for the increased reactivity. Similar results were obtained with the benzoyl cation. Together with the results obtained from other work, these data suggest the formation of the protioacyl dications **46** and **47** in the highly acidic solutions.

These examples illustrate how electrophilic systems can exhibit enhanced reaction rates and yields with increasing strength of the acidic reaction media. Both qualitative and quantitative kinetic studies strongly suggest the involvement of superelectrophilic species in reactions.

2.4 SPECTROSCOPIC STUDIES

Similarly, as spectroscopic techniques have been used to study many long-lived, electrophilic species such as carbocations, acyl and carboxonium ions, and various onium ions, they have also been used in a number of reports directed to characterization of superelectrophiles. Both condensed and gas phase techniques have been used to study superelectrophilic systems. In the condensed phase, however, superelectrophiles are generally formed in equilibria containing only low concentrations, or with very short lifetimes. Consequently, their direct observation has frequently not yet been possible. In some cases, however, they are persistent enough for spectroscopic methods, even NMR studies.

Nuclear magnetic resonance (NMR) spectroscopy has been used to directly observe varied persistent superelectrophilic species. Although ^{1}H and ^{13}C NMR have been the most often used techniques, there have also been applications of ^{15}N, ^{17}O, and ^{19}F NMR in their structural characterization. Coupled with theoretical computational methods capable of estimating NMR chemical shifts, these studies have been very useful in the study of superelectrophiles.

It was reported that arylpinacols (48a) can undergo a superacid-catalyzed dehydrative cyclization to give the aryl-substituted phenanthrenes (eq 27).[23] Superelectrophilic intermediates were proposed in the conversion. Tetraarylethylene dications have been studied by several methods and were observed directly by NMR as well as by UV-vis spectroscopy and X-ray crystallography.[28] The low temperature oxidation of tetraarylethylenes gives the dicationic species (50, eq 28).

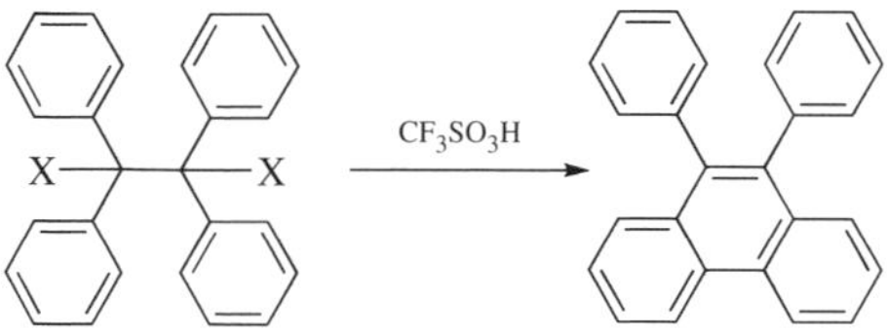

(27)

48a, X = OH   49

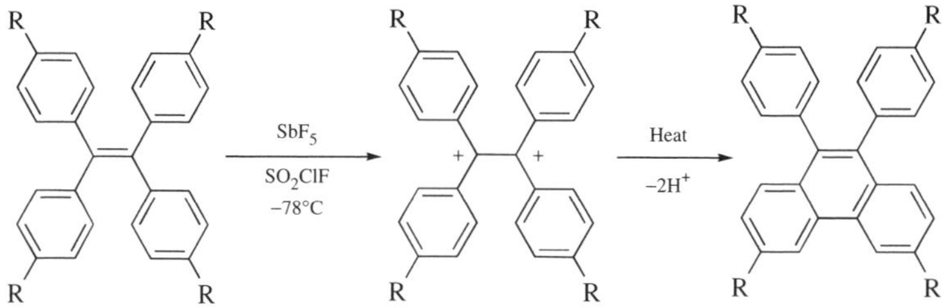

(28)

50

R = H, NMe$_2$, OCH$_3$

^{1}H and ^{13}C NMR studies of the dichloride (48b, X = Cl) with Lewis acid (SbF$_5$ or SbCl$_5$) showed only the cyclized phenanthrene product. The incipient dication (50) in this case was not observed.

In several recent studies, nitro-substituted olefins have been shown to exhibit high electrophilic reactivities in superacid-promoted reactions.[29] NMR studies have been used to identify some of the superelectrophilic intermediates in these reactions. For example, it was found that nitroethylene reacts with benzene in the presence of 10 equivalents of CF_3SO_3H to give deoxybenzoin oxime in 96% yield (eq 29). Since the reaction does not occur with only one equivalent of TfOH, it was proposed that the N,N-dihydroxy-iminium-methylium dication (**51**) is generated. In spectroscopic studies, 1-nitro-2-methyl-1-propene (**52**) was dissolved in CF_3SO_3H, and at $-5°C$ the stable dication (**53**) could be directly observed by ^1H and ^{13}C NMR spectroscopy (eq 30).

(29)

(30)

Based on chemical shift data, it is clear that a considerable amount of positive charge resides at the C-2 position of the nitroolefin. The two methyl groups on dication **53** are, however, nonequivalent, indicating slow rotation around the C1–C2 bond (partial double bond character). Similarly, diprotonated intermediates were suggested to be involved in the reactions of β-nitrostyrene and nitro-substituted naphthalenes (Scheme 7 and 8). Superelectrophilic intermediates **54** and **55** were proposed in the reactions with superacidic CF_3SO_3H. The involvement of these intermediates was supported by their direct observation using ^1H and ^{13}C NMR in superacidic solvents. When (E)-β-nitrostyrene is dissolved in CF_3SO_3H at $-20°C$, spectral data are consistent with the diprotonated species **54**. Double protonation at the nitro group generates positive charge at the C-2 position, which is reflected in the deshielding of this carbon. When the same NMR experiment was done with (Z)-β-nitrostyrene, the N,N-dihydroxyiminium benzyl dication (**54**) is likewise observed,

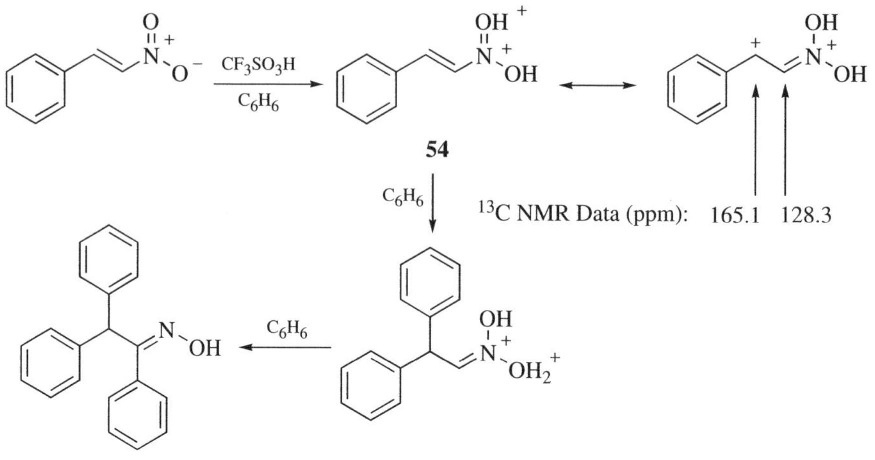

Scheme 7.

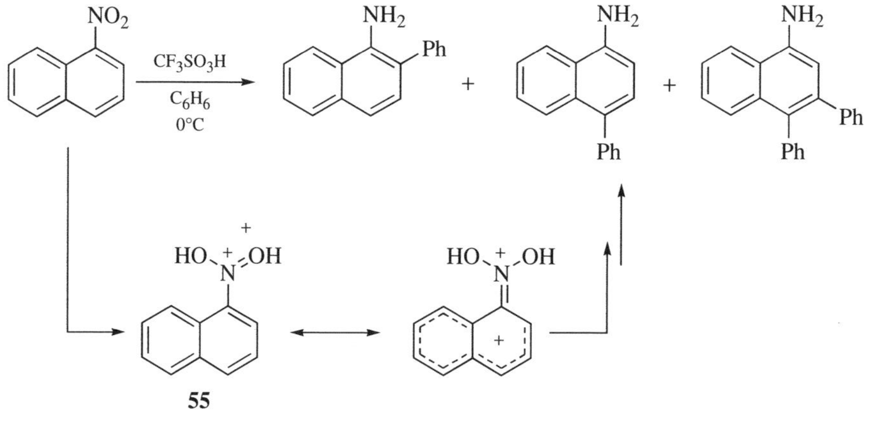

¹³C NMR Data (ppm): 165.1 128.3

Scheme 8.

indicating free rotation about the C1-C2 bond. In the reaction of 1-nitronaphthalene with CF_3SO_3H (30 equiv.) and benzene, phenylated products were obtained (Scheme 8).[30] These Freidel-Crafts-type reactions are consistent with the formation of dicationic electrophilic intermediates (**55**). The dication **55** was characterized by ^{1}H, ^{13}C, and ^{15}N NMR spectra. When compared with the spectrum from the weaker acid CF_3CO_2H, the naphthalene ring carbons are more deshielded in the superacid (CF_3SO_3H), indicating delocalization of positive charge into the ring system. Similar results were obtained for the 2-nitronaphthalene system.

In a study involving the superacid-catalyzed reaction of amino-alcohols, a chiral, dicationic electrophile was observed by low temperature ^{13}C NMR.[31] Ionization of benzylic alcohols in superacids can generate stable carbocations, such as the trityl cation. Because of the resonance stabilization of the carbocationic centers, they are fairly weak electrophiles, incapable of reacting with benzene (eq 31). However, it was shown that adjacent ammonium groups can increase the electrophilic reactivities of the diphenylethyl cations (eq 32).

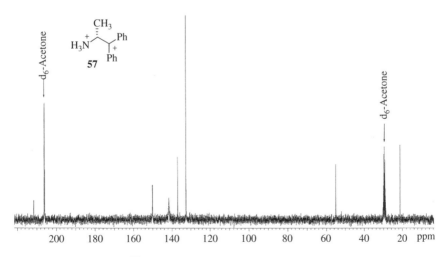

When the amino alcohol (**56**) is ionized in $FSO_3H-SbF_5-SO_2ClF$ at $-40°C$, a clean NMR spectrum is observed for the reactive dicationic electrophile (**57**, Figure 2), showing the carbocationic resonance at $\delta^{13}C$ 211.5.

Carboxonium ions are indicated to be involved in a number of super-electrophilic reactions. In several cases, the direct observation of the superelectrophiles and reactive dications has been possible using low

Figure 2. ^{13}C NMR spectrum of dicationic electrophile **57**.

temperature, stable ion conditions. For example, the condensations of the ketones and aldehydes occur in high yields in CF_3SO_3H (eqs 33–37).[32–34]

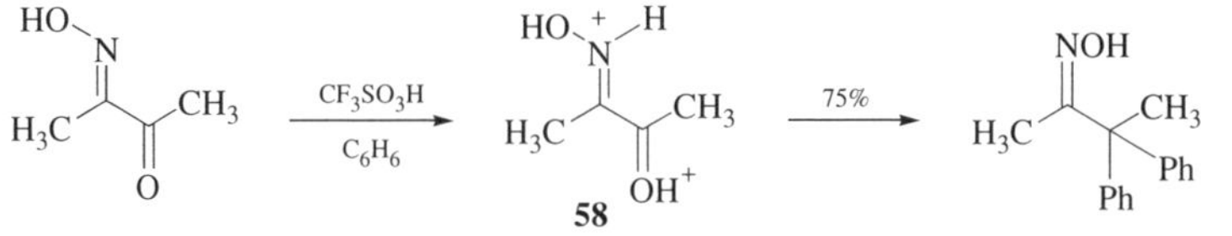

(33)

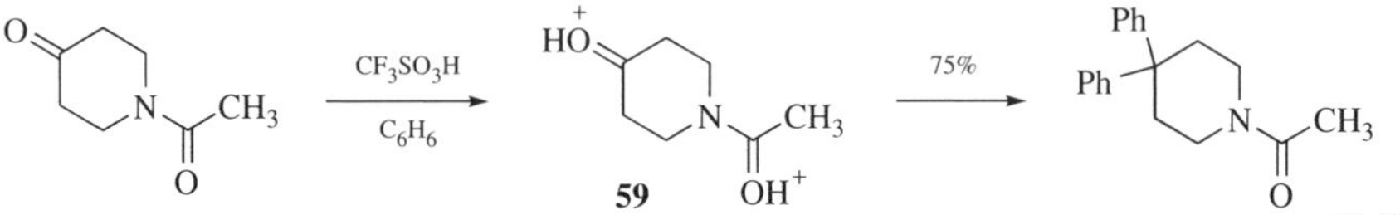

(34)

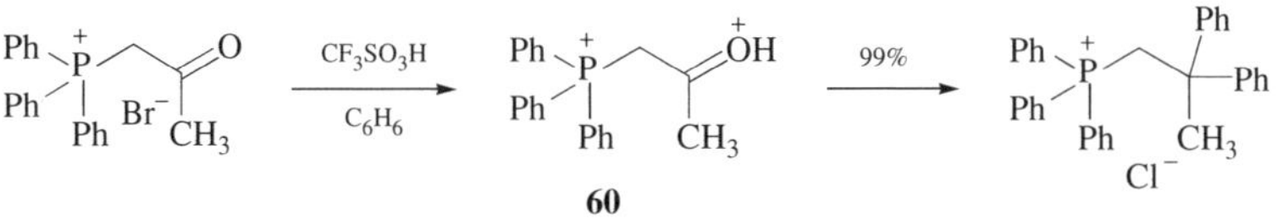

(35)

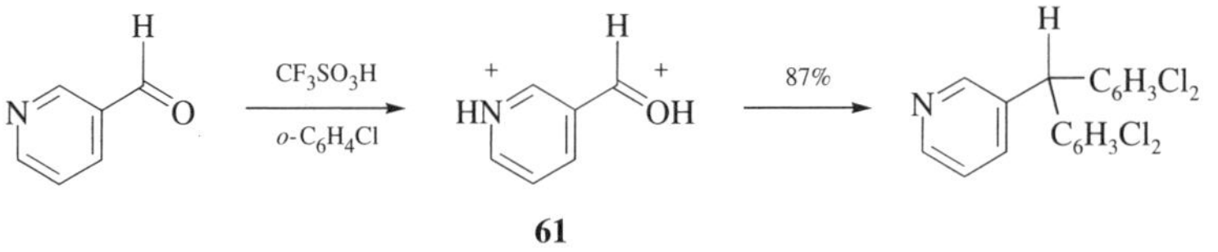

(36)

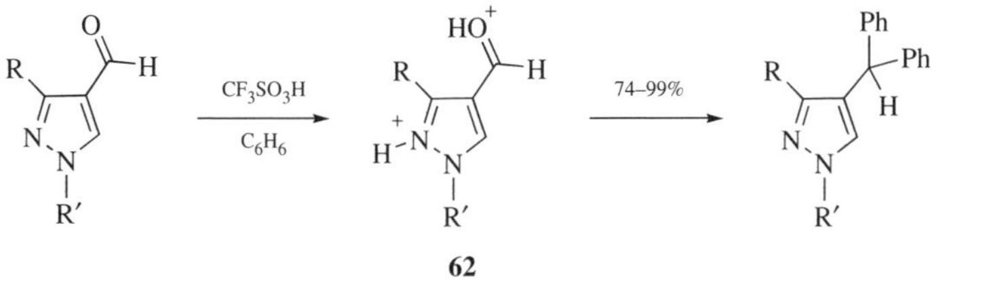

(37)

The proposed dicationic, carboxonium ion intermediates (**58–62**) have been directly observed by ^{13}C NMR. A more detailed description of this superelectrophilic chemistry is found in chapters 5-7.

A comprehensive series of ionic hydrogenation reactions have been studied by Koltunov, Repinskaya, and co-workers, and superelectrophilic intermediates have been proposed.[34] Some of these intermediates have been characterized by ^{1}H and ^{13}C NMR (Table 4). Many of these dicationic intermediates have been shown to react with cyclohexane by hydride abstraction, indicating superelectrophilic character.

As mentioned previously, superelectrophilic intermediates were proposed in the reaction of α-(methoxycarbonyl)diphenylmethanol (**37**)

Table 4. Superelectrophilic intermediates observed by ^{1}H and ^{13}C NMR spectroscopy and capable of hydride abstraction from cyclohexane[34]

Entry	Superelectrophile

(1)

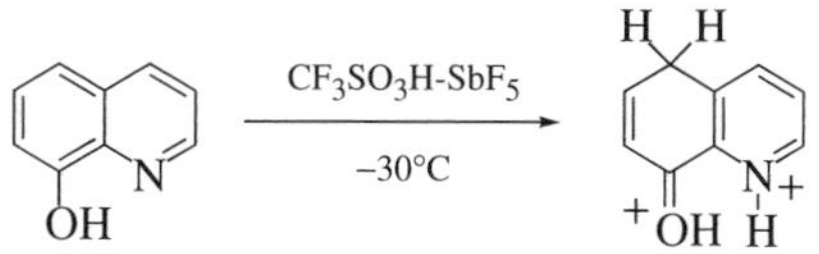

(2)

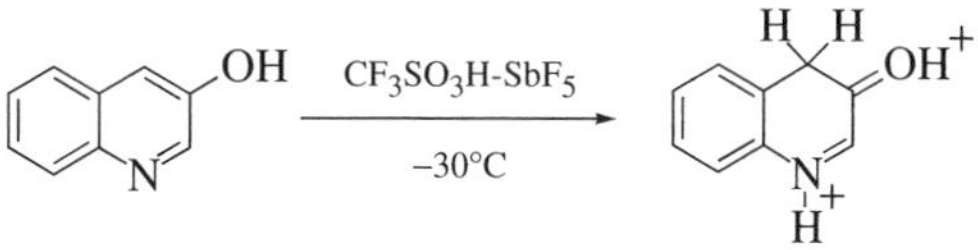

(3)

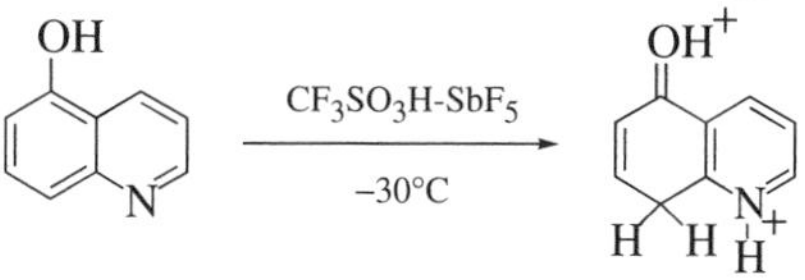

(4)

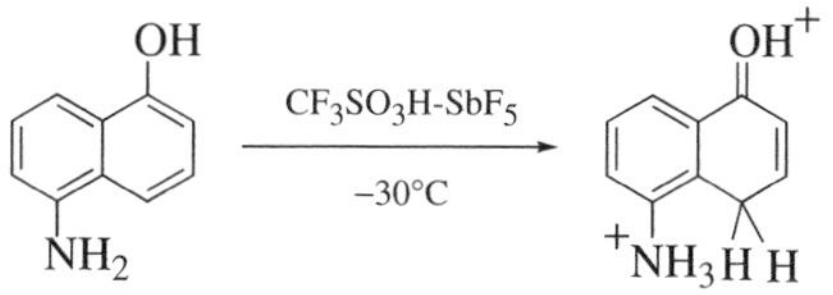

(5)

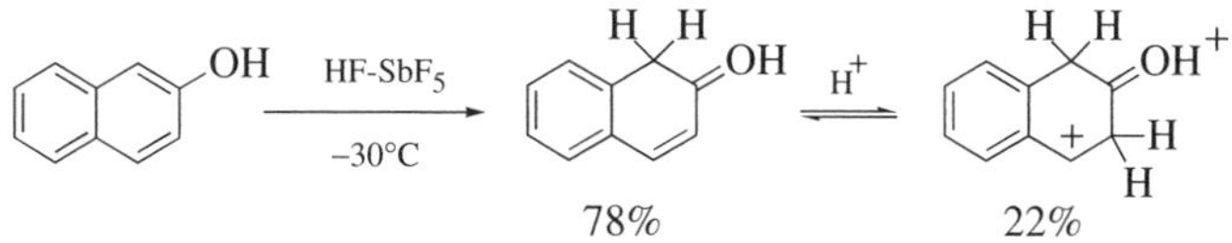

78%   22%

(6)

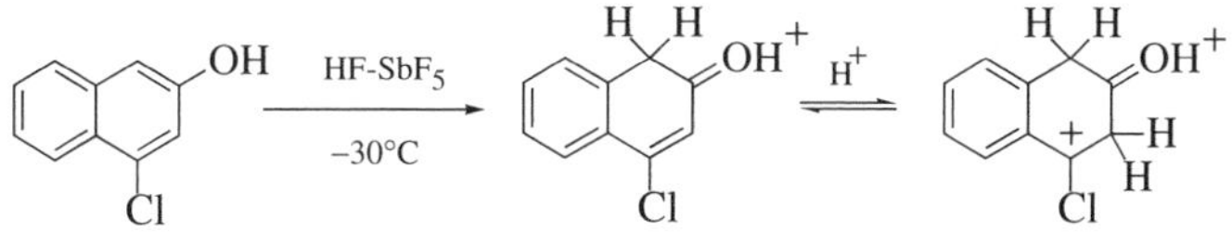

33%   67%

(7)

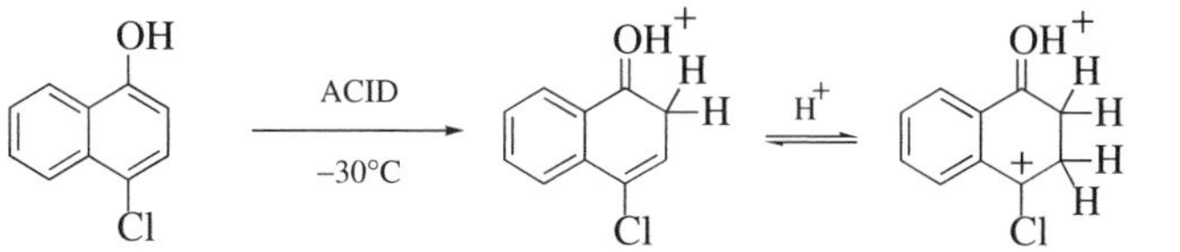

80%   20%

FSO$_3$H-SbF$_5$ 80%   20%

HF-SbF$_5$ 0%   100%

and related compounds with superacid (eq 38, R=H).[25]

$$\underset{\substack{37,\ R = H \\ 64,\ R = -OCH_3}}{} \xrightarrow[-20°C]{CF_3SO_3H} \underset{65,\ R = -OCH_3}{} \longrightarrow \underset{63,\ R = H}{}$$

(38)

Both kinetic and theoretical studies support the involvement of dicationic intermediates. Spectroscopic studies were also used to confirm the involvement of superelectrophiles. However, compound 37 does not produce an NMR observable superelectrophilic intermediate in FSO_3H-SbF_5 at $-50°C$. Because the cyclization reaction occurs so rapidly, only the cyclized product 63 (in its protonated form) is seen in the low temperature NMR experiments. When the system was further stabilized by *para*-methoxy groups, the dicationic intermediate (65), however, was observed as a cleanly formed species in CF_3SO_3H. In the ^{13}C NMR spectrum, the carbenium ion resonance is observed at $\delta^{13}C^+$ 176.0, while the *para*-carbons are at $\delta^{13}C$ 174.7. Based on the chemical shifts, the authors concluded that the carbocation is largely delocalized in the *para*-methoxy phenyl rings. The delocalization of charge is thought to be important in the activation of the electrocyclization reaction step leading to product formation.

In addition to the direct observations of some dicationic superelectrophilic intermediates, NMR spectroscopy has been also used in some of the previously discussed kinetic studies. For example, in the case of the fluorene cyclization (eq 38), the alcohol (37) was dissolved in mixtures of CF_3SO_3H and CF_3CO_2H (using large excess of acid) at low temperature.[25] The cyclization reaction rates were then determined by ^1H NMR by following the disappearance of the starting material. The cyclization was found to exhibit good pseudo-first order kinetic behavior. By repeating the experiments with differing ratios of CF_3SO_3H and CF_3CO_2H, the acidity of the media was systematically varied and the reactions rate constants were shown to increase linearly with acidity. This observation strongly supports the proposed role of superelectrophiles in the conversions.

In superacids, many superelectrophiles capable of π-delocalization generate vividly colored solutions. Despite the strong absorptions in the visible range of the electromagnetic spectrum, there have been so far no systematic studies of the UV-visible spectra of superelectrophilic species. Neither has UV-visible spectroscopy by itself been used to study super-electrophiles. UV-visible spectroscopy has been, however, used in some studies to estimate the $pK_{BH}+$ of monoprotonated species, such as pro-tonated ketones.[36] Based on $pK_{BH}+$ values, it then becomes possible to estimate the acidity range at which a substrate is completely protonated. If increasing acidity of the medium leads to new chemistry or increasing reaction rates, then it can be used to probe for diprotonated, or superelec-trophilic intermediates. Superelectrophiles, as mentioned, generally are formed in low concentration and may be in equilibrium with strongly absorbing monocationic species. Therefore, UV-visible spectroscopy is expected to have only limited value in the studies of π-delocalized super-electrophiles.

Infrared and Raman are also rapid spectroscopic techniques that have been useful in the characterization of electrophiles in the condensed phase. Many superelectrophiles are expected to possess characteristic or new vibrational modes. The harmonic vibrational frequencies and infrared intensities for the nitronium ion (NO_2^+) and protonitronium ion (HNO_2^{2+}) have been estimated using *ab initio* molecular orbital calculations (Table 5).[37] Although the vibrational modes for the superelectrophile (HNO_2^{2+}) clearly differ from that of the monocation, data were so far not reported for the superelectrophile using infrared and Raman spectroscopy. When nitronium salts were dissolved in excess HF-SbF5, no apparent

Table 5. Calculated harmonic vibrational frequencies (cm_1) and infrared intensities (KM mol_1)[37]

	HF/6-31G*// HF/6-31G*		MP2/6-31G**// MP2/6-31G**	
	ω	I	ω	I
$O = \overset{+}{N} = O$	2760	190	2526	1717
	2660	1773	2415	19
	1426	3	1246	2
	1006	540	884	408
	682	0	553	21
	630	50	541	51
$O = \overset{+}{N} = OH$	2654	919	2566	62
	1682	0	1302	0
	750	5	601	3

changes were observed in the Raman frequencies and no new absorptions were observed in the infrared spectrum. The failure to observe the super-electrophile (HNO_2^{2+}) by vibrational spectroscopy may be attributed to its formation in only low equilibium concentration.

An impressive application of infrared and Raman spectroscopy was demonstrated in studies of superelectrophilic diprotonated thiourea, $[H_3NC(SH)NH_2]2^+ \cdot 2AsF_6^-$. The Raman spectrum (taken at $-110°C$) corresponded reasonably well with calculated vibrational bands predicted by density functional theory.[38] Coupled with computational methods for predicting vibrational frequencies, it is expected that vibrational spectro-scopic techniques will be useful for the observations of these and other superelectrophiles.

2.5 GAS PHASE STUDIES OF SUPERELECTROPHILES AND RELATED IONS

There has been continuing interest in the study of multiply charged species in the gas phase using mass spectroscopic techniques.[39] The small dica-tionic and tricationic structures in these gas phase studies are clearly related to superelectrophiles generated in the superacidic condensed phase. Among the techniques used in these studies, electron ionization (EI) is well known as a source of persistent, multiply charged ions of small size. In the case of dicationic species, these ions can be generated by direct ion-ization of neutral molecules with the loss of two electrons, or by stepwise electron ejection processes involving first a monocation and then a dica-tion. Multiply charged ions have also been generated by photoionization. However, this technique can be problematic because the multiply charged cations are sometimes generated in an excited state, which can lead to rapid dissociation due to Franck-Condon effects. Another useful approach has been through the use of charge stripping (CS) techniques.[39] Good thermochemical data have also been obtained using CS experiments. In a typical CS experiment, a mass-selected monocation M^+ is collided with a suitable target gas (often O_2) to provide the dication M^{2+} (eq 39). Anal-ysis of the kinetic energies of the cation M^+ and dication M^{2+} can then provide a good approximation of the vertical ionization energy of mono-cation M^+. A similar approach is used in double-charge transfer (DCT) spectroscopy (eq 40).

$$M^+ + T \longrightarrow M^{2+} + T + e^- \tag{39}$$

$$P^+ + M \longrightarrow P^- + M^{2+} \tag{40}$$

DCT experiments involve colliding a monocationic projectile (P^+) with a target molecule of interest (M). The dicationic product (M^{2+}) is formed from the collision and kinetic energy analysis of the anionic product (P^-) can give information about the electronic states of M^{2+}. Electrospray ionization (ESI) is another method of generating multiply charged ions in mass spectral studies. This technique involves generating ions from a solvation sphere. However, there is the limitation that the highly charged species can either react with the solvent, or resist decomplexation from the solvent. An interesting approach suggested by Schröder and Schwarz involves the direct use of liquid superacids as solvents for electrospray ionization techniques with the goal of generating superelectrophiles from desolvation.[39a] Present ESI technology, however, limits the use of highly corrosive superacidic solvents.

Unlike superelectrophiles generated in the condensed phase, the gas phase species are not stabilized by solvation or association to counter ions. The stabilities of multiply charged small ions are dependent on their tendencies to dissociate into fragments in the gas phase.[39] For diatomic systems, there are two possible routes of dissociation to atomic fragments: the cleavage to a pair of monocations (eq 41) or cleavage to a dication and neutral species (eq 42).

$$XY^{2+} \longrightarrow X^+ + Y^+ \tag{41}$$

$$XY^{2+} \longrightarrow X^{2+} + Y \tag{42}$$

Cleavage of the dication (XY^{2+}) to a pair of monocations is sometimes referred to as "coulombic explosion." Cleavage of the dication (XY^{2+}) to a dication and neutral species is simply the reverse reaction of an electrostatic intereaction between the dicationic X^{2+} and a neutral atom. Small dicationic (or tricationic) species may be observable in gas phase studies when the dication rests in a global energy minimum or rests in a local minimum protected from dissociation by an energy barrier.

Several types of densely charged cationic molecules have been generated and observed using mass spectrometry techniques (Table 6). Rare gas clusters have been of interest to experimentalists and theoreticians since as mentioned Linius Pauling first predicted the stability of He_2^{2+} in the 1930s. A number of these clusters have been experimentally observed.[40] Other types of dicationic and trication species studied in the gas phase include metal halides, oxides, sulfides, selenides, diatomic species, and halogenated carbons. Among the oxides, the protonitronium dication ($HONO^{2+}$) was observed by Schwarz and associates in the gas phase by dissociative electron ionization.[41] Attempts to observe $HONO^{2+}$ by

Table 6. Experimentally observed gas phase dicationic and tricationic species[a]

Rare Gas Clusters	$GeNe^{2+}$, $GeAr^{2+}$, $GeKr^{2+}$, $PtHe^{2+}$, He_2^{2+}, $XeNe^{2+}$, $KrHe^{2+}$, $XeHe^{2+}$, $ArNe^{2+}$, VHe^{2+}
Metal Halides	MgF^{2+}, AlF^{2+}, SiF^{2+}, TiF^{2+}, BaF^{2+}, UF^{2+}, $MgCl^{2+}$, $FeCl^{2+}$, $SrCl^{2+}$, $BaCl^{2+}$, $CaBr^{2+}$, $BaBr^{2+}$, BaI^{2+}, SF^{3+}, TlF^{3+}, VF^{3+}, UF^{3+}, UF_2^{3+}
Oxides, Sulfides; and Selenides	UO^{2+}, $HONO^{2+}$, VO^{3+}, COS^{3+}, SO_2^{3+}, CS^{3+}, CS_2^{3+}, CSe_2^{3+}
Diatomics	B_2^{3+}, S_2^{3+}, Cl_2^{3+}, Br_2^{3+}, I_2^{3+}
Halogenated Carbon	CCl_2^{2+}, CF_2^{2+}, CF_3^{2+}

[a] Adapted from reference 39a; references cited therewithin.

charge stripping methods were so far not successful, as this ionization method produces a species in a vibrationally excited state leading to rapid dissociation to monocationic fragments. The protonitronium dication was generated in the gas phase from electron impact (100-eV electrons) of HNO_3. The resulting ion at m/z 23.5 is assigned to the protonitronium dication ($HONO^{2+}$). The observation of the protonitronium dication is consistent with theoretical calculations that predict the structure to reside in a potential energy well, slowing the deprotonation reaction path to H^+ and NO_2^+ or the dissociation reaction path leading to OH^+ and NO^+. In another study, the $C_2H_4O^{2+}$ dications were studied by theoretical and experimental methods.[42] These gas phase experiments found evidence for the protioacetyl dication (CH_3COH^{2+}), which could be produced by charge stripping experiments from $CH_3COH^{+\bullet}$. The mass spectroscopy studies confirmed theoretical predictions that the protioacetyl dication was at a potential energy minimum protected by a significant barrier to dissociation. As mentioned, the protonitronium dication and the protioacetyl dication were the first superelectrophiles to be proposed in superacid-catalyzed chemistry. Subsequent gas phase techniques have proven to be useful in the direct study of these superelectrophiles.

Mass spectrometric techniques also enable experimentalists to react dications and trications with neutral substrates in the gas phase to explore the chemical reactions of these multiply charged species. Although only a few superelectrophiles have thus far been examined experimentally for their gas phase chemical reactivities, other electrophilic gas phase reactions have demonstrated the potential of these methods. For example, the nobium dication can be generated in the gas phase and it reacts with methane by a dehydrogenation reaction to produce a stable $NbCH_2^{2+}$ species (eq 43).[43]

$$Nb^{2+} + CH_4 \longrightarrow NbCH_2^{2+} + H_2 \xrightarrow[-H_2]{CH_4} \xrightarrow[-H_2]{CH_4} NbC_5H_6^{2+} + NbC_6H_6^{2+}$$

(43)

The dicationic product ($NbCH_2^{2+}$) is itself highly electrophilic and reacts further with up to six molecules of methane to yield dicationic products like $NbC_6H_6^{2+}$. There have also been a number of reports of stable dicationic systems in which an electrophilic species coordinates with substrates (**66–68**).[44]

$$[CF_2{\bullet}H_2]^{2+} \quad [CCl_2{\bullet}Cl_2]^{2+} \quad [Cr{\bullet}O_2]^{2+}$$
$$\textbf{66} \qquad\qquad \textbf{67} \qquad\qquad \textbf{68}$$

For example, when CF_2^{2+} was generated and reacted with H_2, a long-lived intermediate (**66**) was observed. It is proposed to have a structure in which H_2 is coordinated with the CF_2^{2+}, although new covalent bonds are not believed to be formed. The complex itself dissociates to the monocations, CHF_2^+ and H^+. When dicationic carbon tetrachloride is generated, the ion rearranges to a C_{2v} structure thought to involve coordination of Cl_2 and CCl_2^{2+} (**67**). Similarly, side-on coordination of O_2 was proposed in the chromium complex **68**. A gas phase reaction has also been described for $HCBr^{2+}$ and molecular hydrogen.[44e]

Roithova and Schröder have demonstrated that a dicationic species can participate in carbon-carbon bond forming reactions in the gas-phase and that this is a potential route to polycyclic aromatic hydrocarbons.[45] When $C_7H_6^{2+}$ is generated (from double ionization of toluene) and allowed to react with acetylene, new ions, $C_9H_7^{2+}$ and $C_9H_6^{2+}$, are detected (eq 44).

$$C_6H_5CH_3 \xrightarrow{-2e^-} C_7H_6^{2+} \xrightarrow{HC\equiv CH} C_9H_7^{2+} + C_9H_6^{2+}$$

(44)

Experiments with deuterated species show extensive H/D scrambling, which implies longer-lived intermediates in the reaction. Calculations at the B3LYP/6-311+G(2d,p) level suggest that the reaction pathway involves bond formation between the cycloheptatrienylidene dication and acetylene, followed by isomerization and hydrogen loss (Scheme 9). The authors of this study note the possibility of similar chemistry generating polycyclic aromatic compounds in interstellar space.

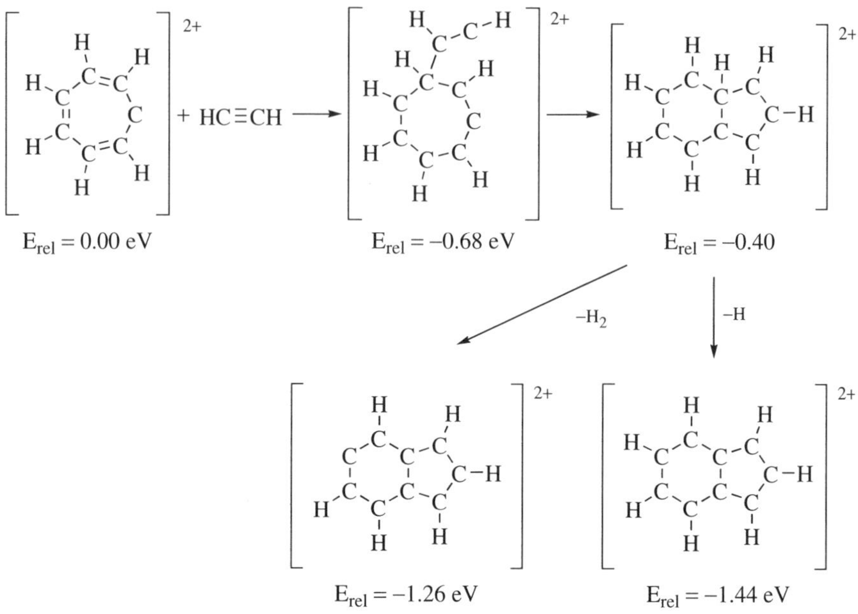

Scheme 9.

Although there are many experimental challenges to observing super-electrophiles by mass spectrometry, gas phase techniques have some apparent advantages over condensed phase studies. Mass spectroscopic studies are extremely sensitive, and, consequently, dicationic or tricationic superelectrophiles can be detected and studied despite being short-lived. As noted previously, many superelectrophiles are formed in very low concentrations in equilibria with conventional, monocationic electrophiles (in the condensed phase). Depending on the methods by which the super-electrophiles are generated, gas phase studies may be able to exclude the effects of these equilibrium reactions. Finally, gas phase studies have already been demonstrated to provide good thermochemical data and these data can be compared with those from theoretical calculations.

2.6 CALCULATIONAL METHODS

Quantum mechanical calculations are an essential part of chemistry, and these methods have been extremely useful in studies of superelectrophilic chemistry. For example, computations have been used in some studies to show that the formation of dicationic superelectrophiles lowers the activation energy for chemical conversions, when compared with the similar reaction involving singly charged cationic electrophiles. In other studies, high-level calculations were used to estimate the amount of positive charge at each of the relevant atoms of a superelectrophile. The centers of highest positive charge were then shown to also be the locations of nucleophilic attack. Studies have also been done to relate superelectrophilic character to global electrophilicity indexes and local electrophilicity (related to its electrophilic Fukui function, f_k^+). In conjunction with spectroscopic studies, theoretical methods are often used to predict NMR chemical shifts or spectral absorption frequencies of superelectrophiles. Many superelectrophiles have also been studied using high level *ab initio* calculations to find optimized geometries and the energies of the optimized structures.

An example of the application of theoretical methods comes from a study of diphenylmethyl cations having electron-withdrawing substituents. Over the years, somewhat conflicting reports were published related to the chemistry involving diphenylmethyl cations. Depending on the conditions of the experiment, three types of major products have been reported for the reaction of the diphenylmethyl cation **69** (eq 45).[25] Products **70** and **71** are thought to form by conrotatory electrocyclization reactions involving cationic intermediates. For monocation **69**, the transition state energy for the fluorene cyclization is estimated to be 8.05 kcal/mol above the transition state for the benzofuran cyclization (B3LYP/6-31*+ZPE). These results are consistent with experimental studies in that less acidic conditions tend to favor monocationic intermediates and benzofuran (**70**) products. When the O-protonated, dicationic species (**72**) is studied, the energy barrier to fluorene cyclization dramatically decreases.

(45)

Dication **72** is estimated to have a fluorene cyclization barrier of 8.52 kcal/mol, compared with a value of 25.04 kcal/mol for the cyclization of monocation **69**. The theoretically predicted barriers to cyclization are also in accord with the thermodynamic values obtained from kinetic experiments. Moreover, these results are consistent with the high yield formation of the fluorene product (**71**) in superacid.

In a study of the Nazarov cyclization and related reactions, theoretical calculations were used to show that formation of dicationic superelectrophiles significantly lowers the energy barriers for concerted electrocyclization reactions.[18] In *ab initio* calculations at the B3LYP/6-31G* theory level, 1,4-pentadien-3-one (**73**) and 1-phenyl-2-propen-1-one (**78**) were used as models for the acid-catalyzed cyclization (Scheme 10). When comparing monocationic (**74** and **79**) and dicationic intermediates (**75** and **80**), the energy barriers for the conrotatory, 4π-electrocyclizations are considerably lower for the dications. Cyclization of the dication **75** is estimated to have an energy barrier about 12 kcal/mol less than the cyclization involving the monocation **74**.

Computational methods have also been used frequently to estimate the thermodynamic stabilities of superelectrophiles. These calculations have often involved the estimation of barriers to gas phase dissociation or deprotonation, and the proton affinities of conventional electrophilic intermediates. Other useful studies have calculated the heats of reactions for isodesmic processes. An interesting example of these calculations comes from a study of the protoacetyl dication (CH_3COH^{2+}).[42] In calculations at the 6-31G*//4-31G level of theory, the protoacetyl dication (**83**) is estimated to react with methane by hydride abstraction with a very favorable

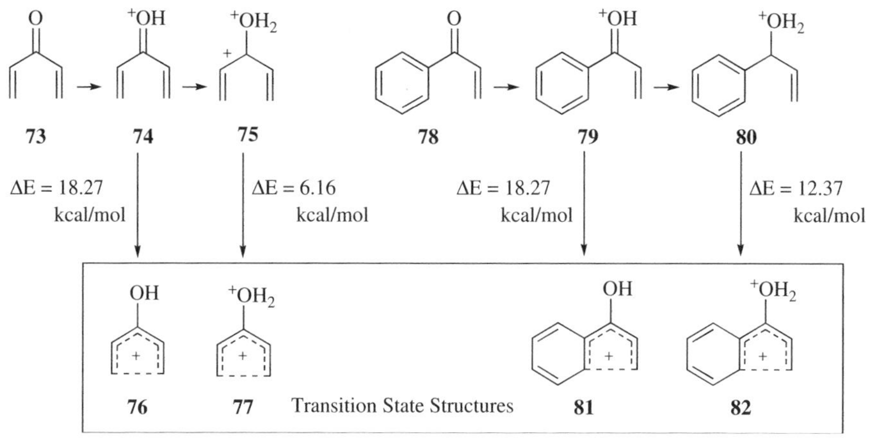

Scheme 10.

enthalpy of reaction (eq 46).

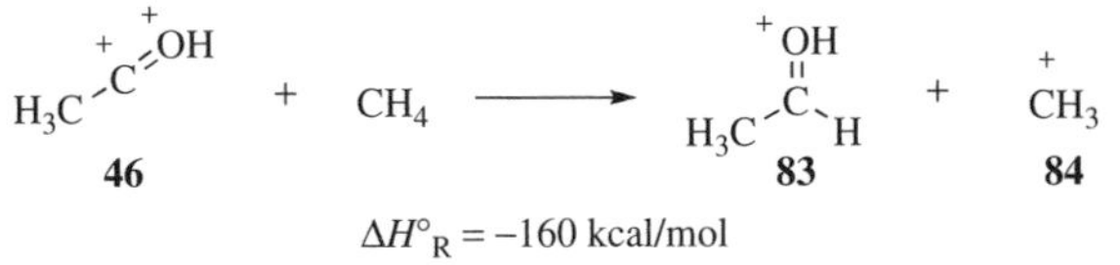

$$\Delta H^{\circ}_{R} = -160 \text{ kcal/mol}$$

It was concluded that separation of charge is the dominant driving force for the reaction.

In superacid catalyzed reactions of hydroxyquinolines and isoquinolines, dicationic superelectrophiles were proposed as intermediates in their reactions (see Table 4).[35d] In order to explain differences in relative reactivities between the isomeric superelectrophiles, the energies of the lowest unoccupied molecular orbitals (ε_{LUMO}), the square of the coefficients (c^2) at the reactive carbon atoms, and the NBO charges (q) on CH groups were determined by MNDO and DFT computational methods. For example, 8-hydroxyquinoline (**85**) is found to be more reactive than 6-hydroxyquinoline (**87**) in the superacid catalyzed reactions with benzene and cyclohexane (eqs 47–48).

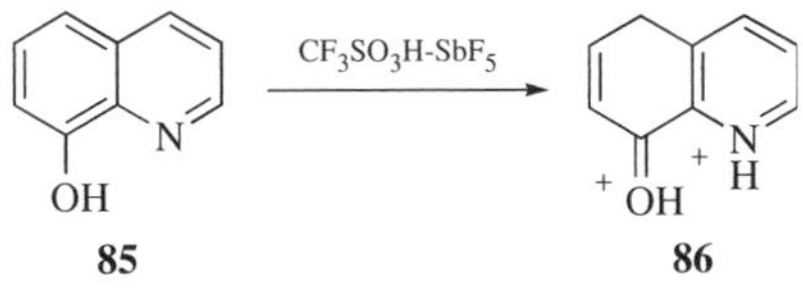

(47)

(48)

When the ε_{LUMO} for dication **86** is compared with the ε_{LUMO} for dication **88**, dication **86** has a LUMO of much lower energy. Thus, lower ε_{LUMO} values correspond to increased electrophilic reactivities. These studies also suggested the importance of atomic charge, q, at a reaction center in determining the reactivity from a kinetic point of view. A more detailed discussion of this chemistry is found in Chapter 7.

Calculations by Pérez have examined superelectrophilic species within the context of both electron deficiency and polarizability of the active sites of superelectrophiles.[46] Using the model equation proposed by Parr and associates,

$$\omega = \mu^2/2\eta$$

where μ is the electronic chemical potential and η is chemical hardness (these values being estimated by the one-electron energies derived from the frontier molecular orbitals HOMO and LUMO), the global electrophilicity index (ω) was calculated for several types of superelectrophilic systems (Table 7). In all cases, the superelectrophilic species (H_4O^{2+}, $H_2C(OH_2)^{2+}$, and $HC(OH)(OH_2)^{2+}$) are found to have significantly greater electrophilicities than the corresponding neutral or monocation species. It is noted that the calculations indicate the "enhanced electrophilicity mainly results from their remarkably high electronegativity." Further calculations were done to estimate the local electrophilicities at potential reaction centers and this was done by taking into account the electrophilic Fukui function (f_k^+). The results from these calculations show good agreement between the local electrophilicity index (ω_c) and calculated NMR chemical shifts.

Over the years, computational methods have become increasingly accurate in their predictions of NMR chemical shifts for given structures.[47] The combination of *ab initio*-optimized geometries, theoretically calculated chemical shifts, and experimental NMR data affords a powerful tool for structural determination. *Noncorrelated* chemical shift methods such as IGLO (individual gauge for localized orbitals) and LORG (localized orbital/local origin gauge) allow the prediction of NMR chemical shifts and have been applied extensively to carbocationic structures. More recently, *correlated* chemical shift calculations have been used (GIAO-MP2 and SOPPA, gauge-independent atomic orbital method and second-order polarization propagator approximation, respectively) and a

Table 7. Estimated chemical hardness (η), electronic chemical potential (μ), and electrophilicity indexes (ω)[46]

	HOMO	LUMO	η (eV)	μ (eV)	ω (eV)
H_4O^{2+}	−1.2954	−0.6021	18.86	−25.82	17.67
H_4O^+	−0.7489	−0.2700	13.03	−13.86	7.37
H_2O	−0.2911	−0.0627	9.63	−3.11	0.50
$H_2C(OH_2)^{2+}$	−1.0243	−0.6891	9.12	−23.31	29.97
H_2COH^+	−0.6727	−0.3773	8.04	−14.29	12.70
H_2CO	−0.2685	−0.0421	6.16	−4.23	1.45
$HC(OH)(OH_2)^{2+}$	−0.9377	−0.5899	9.46	−20.78	22.82
$HC(OH)_2^+$	−0.6400	−0.3085	9.02	−12.90	9.22
HCO_2H	−0.2917	−0.0008	7.92	−3.98	1.00

[a]Calculations done at the B3LY/6-31G(d) level.

multiconfigurational SCF version of IGLO has been applied to problems requiring an MCSCF wave function. There has been generally good agreement between calculated NMR chemical shifts and experimentally determined values for a variety of NMR active nuclei and electrophiles. For example, acetic acid, mono and diprotonated acetic acid (**89**), the acetyl cation, and the protioacetyl dication (**46**) have all been studied using computational methods (Table 8; MP2/6-31G*//GIAO-MP2/tzp/dz level of theory).[48] In the case of protonated acetic acid, fast proton exchange leads only to one averaged ^{17}O NMR signal, and the experimentally determined ^{13}C and ^{17}O NMR chemical shifts agree reasonably well with those calculated by theory. Although the superelectrophilic species (**89**, **46**) have not been observable using NMR experiments, theory predicts the ^{17}O NMR signals to be significantly deshielded by protonation. Computational studies were also done on nitric acid, the nitronium cation, and the protionitronium dication (**3**), with respect to ^{15}N NMR spectra.[49] Comparsion to the experimentally determined ^{15}N NMR chemical shifts of HNO_3 and NO_2^+ shows good agreement between experiment and theory. Although ^{15}N NMR has not yet provided evidence for the protionitronium dication (**3**), ^{17}O NMR suggests protosolvation in superacid and formation of **3**.[49b] The geminal superelectrophile H_4O^{2+} (**90**) has likewise not been observed by NMR experiments, but theory has been used to estimate the chemical shift values of $\delta^{17}O$ 50.7 and δ^1H 13.6 for ^1H NMR. Although with theoretical methods it has become possible to include solvation spheres in calculations of molecular structures and properties, little or no effort has been made to apply these types of calculations to the solvation of electrophiles in superacidic solvents. If such calculations were to be done, then the predicted NMR chemical shifts for electrophiles in superacids would become increasingly accurate.

Many of the superelectrophiles that have been proposed in chemical conversions have been studied by *ab initio*-computational methods. These calculations generally determine the optimized geometry of the superelectrophile and calculate the potential energy of the species. These results are then used to further understand the superelectrophilic chemistry. Many of the structures calculated to date are shown in Table 9. The table generally lists only the highest-level calculations, and references to earlier calculations or those at lower-levels may be found in the cited manuscripts. Moreover, a vast number of doubly and multiply charged species have been proposed from mass spectroscopy studies and related plasma studies. These charged species have often been studied in both ground and excited states by high-level computational methods. A comprehensive description of these species is outside the scope of this text, but the interested reader may consult one of the recent reviews.[39]

Table 8. Calculated and experimental NMR chemical shifts[a]

Compound	Calculated NMR Chemical Shifts (δ, ppm)[b]		Experimental NMR Chemical Shifts (δ, ppm)[b]	
CH_3CO_2H	$^{13}C_1 : 171.9$	$^{13}C_2 : 20.7$	$^{13}C_1 : 176.9$	$^{13}C_2 : 20.8$
	$^{17}O_1 : 194.3$	$^{17}O_2 : 413.4$	$^{17}O_{1,2} : 251.0$	
$CH_3C(OH)_2+$	$^{13}C_1 : 202.0$	$^{13}C_2 : 21.7$	$^{13}C_1 : 203.7$	
	$^{17}O_1 : 228.3$	$^{17}O_2 : 231.2$	$^{17}O : 193.0$	
$CH_3C(OH)(OH_2)^{2+}$ **89**	$^{13}C_1 : 203.4$	$^{13}C_2 : 24.7$	Not Observed	
	$^{17}O_1 : 148.2$	$^{17}O_2 : 294.2$		
CH_3CO^+	$^{13}C_1 : 155.7$	$^{13}C_2 : 6.3$	$^{13}C_1 : 150.3$	
	$^{17}O : 343.8$		$^{17}O : 299.5$	
$CH_3C(OH)^{2+}$ **46**	$^{13}C_1 : 157.6$	$^{13}C_2 : 4.8$	Not Observed	
	$^{17}O : 194.6$			
HNO_3	$^{15}N : 366.8$		$^{15}N : 377.0$	
NO_2^+ **1**	$^{15}N : 268.3$		$^{15}N : 251.0$	
NO_2H^{2+} **3**	$^{15}N : 272.8$		Not Observed	
H_3O^+	$^{17}O : 30.0$		$^{17}O : 10.2$	
	$^1H : 7.2$		$^1H : 9.5-10.6$	
H_4O^{2+} **90**	$^{17}O : 50.8$		Not Observed	
	$^1H : 13.6$			

[a] Acetic acid, protonated acetic, diprotonated acetic acid (**89**), acetyl cation, and the protioacteyl cation (**46**) were calculated at the MP2/6-31G*//GIAO-MP2/tzp/dz level of theory; nitric acid, nitronium cation, and the protionitronium cation (**3**) were calculated at the HF/6-31G*// II//6-31G* level of theory; hydronium ion and the tetrahydridooxonium ion (**90**) were calculated at the MP2/6−31G*//GIAO-MP2/tzp/dz level of theory.
[b] Chemical shifts referenced to the following: ^{13}C, $(CH_3)_4Si$; ^{17}O, H_2O; ^{15}N, NH_3.

Table 9. Superelectrophiles studied by theoretical methods

Structure[a]	Level of Theory	Calculated Properties	References
Carboxonium-Based Systems			
$\overset{+}{O}H_2$ on C with H, $\overset{+}{H}$	MP2/6-31G**// MP2/6-31G**	Geometry, energy, frequency, proton affinity of H_2COH^+, isodesmic reaction, deprotonation barrier, IGLO-II and GIAO-MP2 NMR chemical shifts	48,50,51,52
$\overset{+}{O}H_2$ on C with H_3C, $\overset{+}{H}$	MP2/6-31G**// MP2/6-31G**	Geometry, energy, isodesmic reaction, frequency, proton affinity of CH_3CHOH^+, IGLO-II and GIAO-MP2 NMR chemical shifts	48,50
$\overset{+}{O}H_2$ on C with H_3C, $\overset{+}{}CH_3$	MP2/6-31G**// MP2/6-31G**	Geometry, energy, isodesmic reaction, frequency, proton affinity of $(CH_3)_2COH^+$, IGLO-II and GIAO-MP2 NMR chemical shifts	48,50
$\overset{+}{O}H$ (=O) on C with H, $\overset{+}{O}H_2$	MP2/6-31G**// MP2/6-31G**	Geometry, energy, isodesmic reaction, frequency, proton affinity of $HC(OH)_2^+$, IGLO-II and GIAO-MP2 NMR chemical shifts	48,50
$\overset{+}{O}H$ (=O) on C with H_3C, $\overset{+}{O}H_2$	MP2/6-31G**// MP2/6-31G**	Geometry, energy, isodesmic reaction, frequency, proton affinity of $CH_3C(OH)_2^+$, IGLO-II and GIAO-MP2 NMR chemical shifts	48,50
$\overset{+}{O}H$ (=O) on C with HO, $\overset{+}{O}H_2$	MP2/6-31G**// MP2/6-31G**	Geometry, energy, isodesmic reaction, frequency, proton affinity of $C(OH)_3^+$, IGLO-II and GIAO-MP2 NMR chemical shifts	48,50,53
$H\overset{+}{O}{=}C{=}\overset{+}{O}H$	MP2/6-31G**// MP2/6-31G**	Geometry, energy, isodesmic reaction, frequency, proton affinity of $OC(OH)^+$, deprotonation barrier, IGLO-II and GIAO-MP2 NMR chemical shifts	48,50
$CO_2 \bullet 2AlCl_3$	B3LYP/6-311+G*// MP2/6-311+G*	Geometry, energy, vibrational frequencies	54

Table 9. (*continued*)

Structure[a]	Level of Theory	Calculated Properties	References
Carboxonium-Based Systems			
	MP4(SDTQ)/6-311G*// MP2/6-311G*	Geometry, energy, IGLO and GIAOMP2 NMR chemical shifts	55
	MP2/6-31G*// MP2(fu)/6-31G*	Geometry, energy, enthalpy of $CF_3C(OH)_2{}^+$ protonation	56
	MP2(SDTQ)/6-31G**// MP2/6-31G**	Geometry, energy, transition state structure leading to proton loss and barrier to deprotonation, IGLO and GIAO-MP2 NMR chemical shifts	57
	MP4(SDTQ)/6-31G*// MP2/6-31G*	Geometry, energy	58
	MP4(SDTQ)/6-31G*// MP2/6-31G*	Geometry, energy	58
	MP4(SDTQ)/6-31G*// MP2/6-31G*	Geometry, energy	58
	B3LYP/6-31G*// B3LYP/6-31G*	Geometry, charge densities, IGLO NMR chemical shifts	59

Table 9. (*continued*)

Structure[a]	Level of Theory	Calculated Properties	References
Carboxonium-Based Systems			
	B3LYP/6-31G*// B3LYP/6-31G*	Geometry, energy, NBO charges, LUMO energy (MNDO)	35c
	B3LYP/6-31G*// B3LYP/6-31G*	Geometry, energy, NBO charges, LUMO energy (MNDO)	35c
	B3LYP/6-31G*// B3LYP/6-31G*	Geometry, energy, NBO charges, LUMO energy (MNDO)	35c
	B3LYP/6-31G*// B3LYP/6-31G*	Geometry, energy, NBO charges, LUMO energy (MNDO)	35c
	B3LYP/6-31G*// B3LYP/6-31G*	Geometry, energy	60
	B3LYP/6-31G*// B3LYP/6-31G*	Geometry, energy, rotational energy barriers, IGLO and GIAO-MP2 NMR chemical shifts	61
	B3LYP/6-31G*// B3LYP/6-31G*	Geometry, energy, rotational energy barriers, IGLO and GIAO-MP2 NMR chemical shifts	61

Table 9. (*continued*)

Structure[a]	Level of Theory	Calculated Properties	References
Carboxonium-Based Systems			
H_2CCHO^+ structure	HF/6-31G*//HF 4-31G	Geometry, energy, proton affinity of H_2CCHO^+, charge densities	62
structure	HF/6-31G*//HF 4-31G	Geometry, energy, proton affinity, charge densities	62
structure	MP2/6-31G*// MP2/6-31G*	Geometry, energy, IGLO NMR chemical shifts	63
structure	MP2/6-31G*// MP2/6-31G*	Geometry, energy, IGLO NMR chemical shifts	63
structure	HF/6-31G*//HF 6-31G*	Geometry, energy, IGLO NMR chemical shifts	63
structure	HF/6-31G*//HF 6-31G*	Geometry, energy, IGLO NMR chemical shifts	63
structure	MP4(SDTQ)/6-311G*// MP2/6-311G*	Geometry, energy, IGLO NMR chemical shifts	64
structure	MP4(SDTQ)/6-311G*// MP2/6-311G*	Geometry, energy, IGLO NMR chemical shifts	64

Table 9. (*continued*)

Structure[a]	Level of Theory	Calculated Properties	References
Carboxonium-Based Systems			
HO—(ring with N–H, two $+$), OH	B3LYP/6-31G*// B3LYP/6-31G*	Geometry, energy, NBO charges, LUMO energy	65
HO—(bicyclic ring with N–H, two $+$), OH	B3LYP/6-31G*// B3LYP/6-31G*	Geometry, energy, NBO charges, LUMO energy	65
HO—(ring with N–Cl, two $+$), OH	B3LYP/6-31G*// B3LYP/6-31G*	Geometry, energy, energetics for Cl^+ dissociation	66
HO—(ring with N, H Cl, three $+$), OH	B3LYP/6-31G*// B3LYP/6-31G*	Geometry, energy, energetics for Cl^+ dissociation	66
$\overset{+}{O}H_2$ (diene cation)	B3LYP/6-31G*// B3LYP/6-31G*	Geometry, energy, transition state structure leading to electrocyclization and energy barrier for cyclization	18
$\overset{+}{O}H_2$ (phenyl with allyl cation)	B3LYP/6-31G*// B3LYP/6-31G*	Geometry, energy, transition state structure leading to electrocyclization and energy barrier for cyclization	18
$\overset{+}{O}H_2$ (phenyl with CH_3 allyl cation)	B3LYP/6-31G*// B3LYP/6-31G*	Geometry, energy, transition state structure leading to electrocyclization and energy barrier for cyclization	18
H_3C, CH_3 H, C–C–$\overset{+}{O}$–H, CH_3	B3LYP/6-31G*// B3LYP/6-31G*	Geometry, energy	14b
H H, H–C–$\overset{+}{C}$–C–$\overset{+}{O}$–H, H H	HF/6-31G*// HF/4-31G	Geometry, energy, proton affinity of $CH_2CHCHOH^+$	24

Table 9. (*continued*)

Structure[a]	Level of Theory	Calculated Properties	References
Carboxonium-Based Systems			
	MP2/6-31G*// HF/6-31G*	Geometry, energy, energetics of bond rotation	67
Acylium and Related Systems			
$H_3C-\overset{+}{C}=\overset{+}{O}-H$	MP2/6-31G*// MP2/6-31G* B3LYP/6-311G**// B3LYP/6-311G**	Geometry, energy, IGLO and GIAO-MP2 NMR chemical shifts	42,52,70
	MP2/6-311+G**// MP2/6-311+G**	Geometry, energy, transition state structure and energy barrier leading to proton loss, NBO charges, GIAO-CCSD(T) and GIAO-MP2 NMR chemical shifts	69
	MP4(SDTQ)/ 6-311+G**// MP2/6-311+G**	Geometry, energy, transition state structure and energy barrier leading to proton loss, NBO charges, GIAO-CCSD(T) and GIAO-MP2 NMR chemical shifts	69
	B3LYP/6-31G**// B3LYP/6-31G**	Geometry, charge densities, IGLO NMR chemical shifts	59
$H-\overset{+}{C}=\overset{+}{O}-H$	MP2/6-31G*// HF/6-31G*	Geometry, energy, energy barrier leading to proton loss, proton affinity of HCO+, isodesmic reaction	50
$F-\overset{+}{C}=\overset{+}{O}-H$	MP4(SDTQ)/ 6-31G**// MP2/6-31G**	Geometry, energy, IGLO and GIAO-MP2 NMR chemical shifts	57

Table 9. (*continued*)

Structure[a]	Level of Theory	Calculated Properties	References
Oxonium-Based Systems			
H–O(2+)H with H above and H below (H₃O²⁺)	MP4(STDQ)/ 6-311++G**// MP2/6-311G**	Geometry, energy, proton affinity of H_3O^+, transition state structures for dissociation, IGLO-II and GIAO-MP2 NMR chemical shifts	48,50,70,71
F–O(2+)H with H above and H below (H₂FO²⁺)	QCISD(T)/6-311G**// QCISD(T)/6-311G**	Geometry, energy, proton affinity of H_2FO^+, transition state structure and energy barrier to proton loss, GIAO-MP2 NMR chemical shifts	72
F–O(2+)H with H above and F below (HF₂O²⁺)	QCISD(T)/6-311G**// QCISD(T)/6-311G**	Geometry, energy, proton affinity of HF_2O^+, transition state structure and energy barrier to proton loss, GIAO-MP2 NMR chemical shifts	72
H_3C–O(2+)H with H above and CH_3 below	MP2/6-31G*/ MP2/6-31G*	Geometry, energy, proton affinity of $(CH_3)_2OH^+$, IGLO-II and GIAO-MP2 NMR chemical shifts	48,50
H_3C–O(2+)CH_3 with H above and CH_3 below	MP2/6-31G*// HF6-31G* MP4(STDQ)/ 6-311++G**// MP2/6-311G**	Geometry, energy, isodesmic reaction, frequency, proton affinity of $(CH_3)_3O^+$, IGLO-II and GIAO-MP2 NMR chemical shifts	48,50
H_3C–C(+)–C structure with CH_2, OH_2, CH_3, H	B3LYP/6-31G*// B3LYP/6-31G*	Geometry, energy	14b
H_3C–O(+)–O(+)–CH_3 structure with CH_3 groups	MP2/6-31G*/ MP2/6-31G*	Geometry, energy, IGLO and GIAO-MP2 NMR chemical shifts	73

Table 9. (*continued*)

Structure[a]	Level of Theory	Calculated Properties	References
Hypercoordinate Systems			
	B3LYP/6-31G**// B3LYP/6-31G**	Geometry, energy	68
	B3LYP/6-31G**// B3LYP/6-31G**	Geometry, energy	68
	B3LYP/6-31G**// B3LYP/6-31G**	Geometry, energy	68
	B3LYP/6-31G**// B3LYP/6-31G**	Geometry, energy	68
	B3LYP/6-31G**// B3LYP/6-31G**	Geometry, energy	14b
	HF/6-31G*//HF/6-31G*	Geometry, energy	74
	HF/6-31G*//HF/6-31G*	Geometry, energy	68
	MP4(SDTQ)/6-311+G**// MP2/6-311+G**	Geometry, energy	75

Table 9. (*continued*)

Structure[a]	Level of Theory	Calculated Properties	References
Hypercoordinate Systems			
(structure: protonated system with N, C, H)	MP4(SDTQ)/ 6-311+G**// MP2/6-311+G**	Geometry, energy	75
(structure with N, C, CH₃)	MP4(SDTQ)/ 6-311+G**// MP2/6-311+G**	Geometry, energy	75
(structure H₃C–N=C–H)	MP4(SDTQ)/ 6-311+G**// MP2/6-311+G**	Geometry, energy	75
(structure H–C–F–H)	B3LYP/6-31G**// B3LYP/6-31G**	Geometry, energy, NBO charges, GIAO-MP2 NMR chemical shifts	76,77
(structure H–C–Cl–H)	B3LYP/6-31G**// B3LYP/6-31G**	Geometry, energy, NBO charges, GIAO-MP2 NMR chemical shifts	76,77
(structure H–C–Br–H)	B3LYP/6-31G**// B3LYP/6-31G**	Geometry, energy, NBO charges, GIAO-MP2 NMR chemical shifts	76,77
(structure H–C–I–H)	B3LYP/6-31G**// B3LYP/6-31G**	Geometry, energy, NBO charges, GIAO-MP2 NMR chemical shifts	76,77
(structure H–Si–C)	QCISD(T)/ 6-311G**// MP2/6-311G**	Geometry, energy, energetics of dissociation	78
(structure H–Si–Si)	QCISD(T)/ 6-311G**// MP2/6-311G**	Geometry, energy, energetics of dissociation	78

Table 9. (*continued*)

Structure[a]	Level of Theory	Calculated Properties	References
Hypercoordinate Systems			
CH_5^{3+}	MP2/6-31G**// MP2/6-31G**	Geometry, energy, NBO charges	79
CH_6^{2+}	MP4(SDTQ)/6-311G**// MP2/6-311G**	Geometry, energy, transition state structure and energy barrier to proton loss	80
CH_7^{3+}	MP4(SDTQ)/6-311G**// MP2/6-311G**	Geometry, energy, energetics of cleavage processes	81
H H—H \\\|+/ H–C–C—H /+\ H—H H	MP4(SDTQ)/6-311G**// MP2/6-311G**	Geometry, energy, energetics of cleavage processes	80
H H H H \\ |/ |+C C+| H C H H H	MP4(SDTQ)/6-311G**// MP2/6-311G**	Geometry, energy, energetics of cleavage processes	80
H—H H H \\+/ |/ H H–C C C–H H H H H—H /+\	MP4(SDTQ)/6-311G**// MP2/6-311G**	Geometry, energy, energetics of cleavage processes	80
H H H H \\ |/ |+C C+| H C H H CH₃	MP4(SDTQ)/6-311G**// MP2/6-311G**	Geometry, energy, energetics of cleavage processes	80
H H \ +/ H–C–C\ /+\ H H—H	QCISD(T)/6-311G**// MP2/6-311G**	Geometry, energy, energetics of dissociation	78
H CH₃ \ +/ H–C–C\ /+\ H H—H	MP4/6-31G**// MP2/6-31G**	Geometry, energy	82
H H H \ |+C\ H H–C–C\ H /+\ H H—H	MP4/6-31G**// MP2/6-31G**	Geometry, energy	82

Table 9. (*continued*)

Structure[a]	Level of Theory	Calculated Properties	References
Hypercoordinate Systems			
H H H–C–O+ H⁻H H	MP2/6-31G*/MP2/ 6-31G*	Geometry, energy, dissociation pathways, GIAO-MP2 NMR chemical shifts; proton affinity of $CH_3OH_2^+$, isodesmic enthalpies	48,50,74,77
H H H–C–S + H⁻H H	MP2/6-31G*// HF/6-31G*	Geometry, energy, GIAO-MP2 NMR chemical shifts; proton affinity of $CH_3SH_2^+$, isodesmic enthalpies	50,77
H CH_3 H–C–O + H⁻H CH_3	MP4(STDQ)/ 6-311++G**// MP2/6-311G**	Geometry, energy	75
SH_5^{3+}	CCST(T)/cc-pVTZ// QCISD(T)/6-311G*	Geometry, energy, energetics of dissociation pathways	83
BH_5^{2+} and BH_6^{3+}	CCST(T)/6-311G**// QCISD(T)/6-311G**	Geometry, energy, NBO charges	79a
NH_5^{2+} and NH_6^{3+}	CCST(T)/6-311G**// QCISD(T)/6-311G**	Geometry, energy, energetics of dissociation pathways	71,84,85
PH_5^{2+} and PH_6^{2+}	CCST(T)/6-311G**// QCISD(T)/6-311G**	Geometry, energy, energetics of dissociation pathways	71,85
AsH_6^{3+} and AsH_6^{3+}	CCST(T)/6-311G**// QCISD(T)/6-311G**	Geometry, energy, energetics of dissociation pathways	71,85
H–H CH_3 ⁺C–N⁺CH₃ H H CH_3	QCISD/6-311G**// QCISD/6-311G**	Geometry, energy	84

Table 9. (*continued*)

Structure[a]	Level of Theory	Calculated Properties	References
Hypercoordinate Systems			
	MP4(SDTQ)/ 6-31G**// MP2(fu)/6-31G**	Geometry, energy	86
	B3LYP/6-31G**// B3LYP/6-31G**	Geometry, energy, GIAO NMR chemical shifts	87
Carbocationic Systems			
	MP3/6-31G**// HF/6-31G*	Geometry, energy, energetics of cleavage processes, transition state structure and energy barrier to proton loss.	88
	MP2/6-31G*// HF/6-31G*	Geometry, energy, energetics of bond rotation	67
	MP4(SDTQ)/ 6-31G*// MP2(fu)/6-31G*	Geometry, energy, IGLO NMR chemical shifts	89
	B3LYP/6-31G**// B3LYP/6-31G**	Geometry, IGLO NMR chemical shifts	90
	B3LYP/6-31G**// B3LYP/6-31G**	Geometry, energy, transition state structure and energy barrier leading to electrocyclization, NBO charges	25

Table 9. (*continued*)

Structure[a]	Level of Theory	Calculated Properties	References
Carbocationic Systems			
	B3LYP/6-31G**// B3LYP/6-31G**	Geometry, energy, transition state structure and energy barrier leading to electrocyclization, NBO charges	25
	B3LYP/6-31G**// B3LYP/6-31G**	Geometry, energy, transition state structure and energy barrier leading to electrocyclization, NBO charges	25
	B3LYP/6-31G**// B3LYP/6-31G**	Geometry, energy	87
	B3LYP/6-31G**// B3LYP/6-31G**	Geometry, energy, GIAO NMR chemical shifts	91
	B3LYP/6-31G**// B3LYP/6-31G**	Geometry, energy, GIAO NMR chemical shifts	91
	MP2/3-21G//HF/3-21G	Geometry, energy, IGLO NMR chemical shifts	89
	B3LYP/6-311G**// B3LYP/6-311G**	Geometry, energy, IGLO NMR chemical shifts, NBO charges	92

Table 9. (*continued*)

Structure[a]	Level of Theory	Calculated Properties	References
Carbocationic Systems			
	B3LYP/6-31G**// B3LYP/6-31G**	Geometry, energy, NBO charges, energy of LUMO (MNDO)	35c
	B3LYP/6-31G**// B3LYP/6-31G**	Geometry, energy, NBO charges, energy of LUMO (MNDO)	35c
	B3LYP/6-311G**// B3LYP/6-311G**	Geometry, energy	93
	B3LYP/6-311G**// B3LYP/6-311G**	Geometry, energy	93
Nitrogen-Based Systems			
	B3LYP/6-31G**// B3LYP/6-31G**	Geometry, energy, NBO charges, transition state structure and energy barrier to proton loss, vibrational frequencies and IR intensities	94
	B3LYP/6-31G**// B3LYP/6-31G**	Geometry, energy, NBO charges, transition state structure and energy barrier to proton loss, vibrational frequencies and IR intensities	94
	MP4(SDTQ)/6-31G*// MP2/6-31G*	Geometry, energy, energy barrier to proton loss, IGLO and GIAOMP2 NMR chemical shifts	94

Table 9. (*continued*)

Structure[a]	Level of Theory	Calculated Properties	References
Nitrogen-Based Systems			
$H_3N-C(NH_2^+)=NH_2^+$ (see structure)	MP4(SDTQ)/6-31G*// MP2/6-31G*	Geometry, energy, energy barrier to proton loss, IGLO and GIAO-MP2 NMR chemical shifts	95
(see structure)	B3LYP/ 6-311+G**// B3LYP/6-311+G**	Geometry, energy, Wiberg bond indices, NBO charges, GIAO-MP2 NMR chemical shifts	96
(see structure)	B3LYP/6-311+G**// B3LYP/ 6-311+G**	Geometry, energy, Wiberg bond indices, NBO charges, GIAO-MP2 NMR chemical shifts, energetic of dissociation processes	96
(see structure)	B3LYP/6-311+G**// B3LYP/6-311+G**	Geometry, energy, Wiberg bond indices, NBO charges, GIAO-MP2 NMR chemical shifts, energetic of dissociation processes	96
$O=N^+=O^+H$	MP2/6-31G**// MP2/6-31G** CASSCF/6-31G*	Geometry, energy, atomic charges, transition state structure and energy barrier to proton loss, vibrational frequencies and IR intensities, energetics of dissociation pathways.	37,41
$O=N^+=O^+-He^+$	B3LYP/6-31G**// B3LYP/6-31G**	Geometry, energy, NBO charges, transition state structure and energy barrier to proton loss, vibrational frequencies and IR intensities	97
$[O=N-He]^{3+}$	B3LYP/6-31G**// B3LYP/6-31G**	Geometry, energy, NBO charges, transition state structure and energy barrier to proton loss, vibrational frequencies and IR intensities	97

Table 9. (*continued*)

Structure[a]	Level of Theory	Calculated Properties	References
Nitrogen-Based Systems			
$H-\overset{+}{N}\equiv\overset{+}{N}-H$	MP4(SDTQ)/6-31G*// MP2/6-31G*	Geometry, energy, NBO charges, transition state structure and energy barrier to proton loss, IGLO NMR chemical shifts	98
$H-\overset{+}{N}\equiv\overset{+}{N}-OH$	MP4(SDTQ)/6-31G*// MP2/6-31G*	Geometry, energy, NBO charges, IGLO NMR chemical shifts	98
$N\equiv\overset{+}{N}-\overset{+}{N}H_3$	MP4(SDTQ)/6-31G*// MP2/6-31G*	Geometry, energy, NBO charges, IGLO NMR chemical shifts	98
$\underset{H}{\overset{H}{>}}C=\overset{+}{N}=\overset{+}{N}\underset{H}{\overset{H}{<}}$	MP4(SDTQ)/6-31G*// MP2/6-31G*	Geometry, energy, NBO charges, transition state structure and energy barrier to proton loss, GIAO-MP2 NMR chemical shifts	98
$\underset{H}{\overset{H}{>}}\overset{+}{C}-C\equiv\overset{+}{N}-H$	HF/6-31G*// HF/4-31G	Geometry, energy, charges densities, proton affinity of $+CH_2CN$	62
$H_3C-\overset{+}{C}\underset{H}{\overset{\overset{+}{N}H_3}{<}}$	MP4(SDTQ)/6-31G*// MP2/6-31G*	Geometry, energy	99
$\underset{H}{\overset{H}{>}}\overset{+}{C}-C\underset{H}{\overset{NH_3}{<}}H$	MP4(SDTQ)/6-31G*// MP2/6-31G*	Geometry, energy	99
$H_3C-\overset{+}{C}\underset{CH_3}{\overset{\overset{+}{N}H_3}{<}}$	MP4(SDTQ)/6-31G*// MP2/6-31G*	Geometry, energy, NBO charges, transition state structure and energy barrier to proton loss, IGLO NMR chemical shifts	99
$\underset{H}{\overset{H}{>}}\overset{+}{C}-\overset{+}{N}\underset{H}{\overset{H}{<}}H$	MP2/6-31G**// MP2/6-31G**	Geometry, energy, transition state structure and energy barrier to proton loss, isodesmic reaction with hydride donor, proton affinity of CH_2NH_{2+}	51

Table 9. (*continued*)

Structure[a]	Level of Theory	Calculated Properties	References
Nitrogen-Based Systems			
	MP2/6-31G**// MP2/6-31G**	Geometry, energy, energetics of bond rotation	67
	MP2/6-31G**// MP2/6-31G**	Geometry, energy, energetics of bond rotation	67
	MP2/ 6-31G**// MP2/6-31G**	Geometry, energy, energetics of bond rotation	67
	HF/6-31G*// HF/6-31G*	Geometry, charge, Mulliken charges	100
Phosphonium Systems			
	MP2/6-31G**// MP2/6-31G**	Geometry, energy, transition state structure and energy barrier to proton loss, isodesmic reaction with hydride donor, proton affinity of CH_2PH_{2+}, GIAO-MP2 NMR chemical shifts	51
	MP2/6-31G**// MP2/6-31G**	Geometry, energy, transition state structure and energy barrier to proton loss, isodesmic reaction with hydride donor, proton affinity of CH_2PH_{2+}, GIAO-MP2 NMR chemical shifts	51
Sulfur-Based Systems			
S_8^{2+} and S_n^{2+} (n = 3–7)	B3PW91/ 6-311+G(3 d,f)// B3PW91/6-311+G*	Geometry, energy, spectral absorption bands	101

Table 9. (*continued*)

Structure[a]	Level of Theory	Calculated Properties	References
Sulfur-Based Systems			
$H_2S(H)_2{}^{2+}$ (H–S with four H, 2+)	MP4(SDTQ)/6-31G**// HF/6-31G*	Geometry, energy, transition state structure and energy barrier for proton loss	50,102
$H_3C-S(H)_2{}^{2+}$ with CH_3 (H, H, CH₃ on S, 2+)	MP2/6-31G**// HF/6-31G*	Geometry, energy, energetics for dissociation	50
$H_3C-S(H)(CH_3)_2{}^{2+}$ (H, CH₃, CH₃ on S, 2+)	MP2/6-31G**// HF/6-31G*	Geometry, energy, energetics for dissociation, proton affininty of $+S(CH_3)_3$	50
$H_3C-S(CH_3)_2(CH_3){}^{2+}$ (CH₃, CH₃, CH₃ on S, 2+)	MP2/6-31G**// HF/6-31G*	Geometry, energy, energetics for dissociation	50
$H_3C-S(CH_3)(OH){}^{2+}$ (CH₃, OH, CH₃ on S, 2+)	B3LYP/6-311+G**// B3LYP/6-311+G**	Geometry, energy, energetics for dissociation, GIAO NMR chemical shifts	103
$[(H_3C)_2S-O(CH_3)(H)]^{2+}$	B3LYP/6-311+G**// B3LYP/6-311+G**	Geometry, energy, energetics for dissociation, GIAO NMR chemical shifts	103
$[(H_3C)_2S-O(H)(H)]^{2+}$	B3LYP/6-311+G**// B3LYP/6-311+G**	Geometry, energy, energetics for dissociation, GIAO NMR chemical shifts	103
$[(H_3C)_2S-O(CH_3)(CH_3)]^{2+}$	B3LYP/6-311+G**// B3LYP/6-311+G**	Geometry, energy, energetics for dissociation, GIAO NMR chemical shifts	103
$H_2N-C(^+SH)-^+NH_3$	B3LYP/6-31G*// B3LYP/6-31G*	Geometry, energy, energetics for proton loss, IGLO and GIAO-MP2 NMR chemical shifts, vibrational frequencies and IR intensities	38

Table 9. (*continued*)

Structure[a]	Level of Theory	Calculated Properties	References
Sulfur-Based Systems			
H_3N $\overset{+}{C}$ $\overset{+}{NH_3}$ with $\overset{+}{S}H$	B3LYP/6-31G*// B3LYP/6-31G*	Geometry, energy, energetics for proton loss, IGLO and GIAO-MP2 NMR chemical shifts, vibrational frequencies and IR intensities	38
H_2N $\overset{+}{C}$ $\overset{+}{NH_3}$ with $\overset{+}{S}H_2$	B3LYP/6-31G*// B3LYP/6-31G*	Geometry, energy, energetics for proton loss, IGLO and GIAO-MP2 NMR chemical shifts, vibrational frequencies and IR intensities	38
$\left[HS \overset{SH}{\underset{}{C}} SH \right]^{•,\,2+}$	(P)MP4(SDTQ)// HF/6-31G*	Geometry, energy, hydride affinity, rotational barriers	104
$H_3C - \overset{\overset{H}{\mid}}{\underset{\underset{CH_3}{\mid}}{S}} - H \quad 2+$	MP2/6-31G*// HF/6-31G*	Geometry, energy, energetics for dissociation	50
$HS \overset{+}{=} \overset{\overset{SH}{\mid}}{C} \underset{\underset{SH}{\mid}}{\overset{}{C}} \overset{+}{=} SH$	MP2/6-31G**// HF/6-31G*	Geometry, energy, energetics of bond rotation	67
Boron-Based Systems			
$\overset{H}{\underset{H}{\overset{+}{N}}} \overset{\overset{+}{B} - H}{\underset{H}{}}$	MP4/6-311G(d,p)// MP4/6-311G(d,p)	Geometry, energy, transition state structure and energy barrier for proton loss	105

Table 9. (*continued*)

Structure[a]	Level of Theory	Calculated Properties	References
Halonium Systems			
$\begin{array}{c} H \\ | \\ H-F^{2+} \\ | \\ H \end{array}$	B3LYP/LANL2DZ// B3LYP/LANL2DZ	Geometry, energy, transition state structure and energy barrier leading to proton loss, NBO charges, proton affinity of H_2F_+	76,106
$\begin{array}{c} H \\ | \\ H-Cl^{2+} \\ | \\ H \end{array}$	B3LYP/LANL2DZ// B3LYP/LANL2DZ	Geometry, energy, transition state structure and energy barrier leading to proton loss, NBO charges, proton affinity of H_2Cl_+	76,106
$\begin{array}{c} H \\ | \\ H-Br^{2+} \\ | \\ H \end{array}$	B3LYP/LANL2DZ// B3LYP/LANL2DZ	Geometry, energy, transition state structure and energy barrier leading to proton loss, NBO charges, proton affinity of H_2Br_+	76,106
$\begin{array}{c} H \\ | \\ H-I^{2+} \\ | \\ H \end{array}$	B3LYP/LANL2DZ// B3LYP/LANL2DZ	Geometry, energy, transition state structure and energy barrier leading to proton loss, proton affinity of H_2I_+, NBO charges	76,106
$\begin{array}{c} CH_3 \\ | \\ H_3C-F^{2+} \\ | \\ CH_3 \end{array}$	B3LYP/LANL2DZ// B3LYP/LANL2DZ	Geometry, energy, NBO charges	76
$\begin{array}{c} CH_3 \\ | \\ H_3C-Cl^{2+} \\ | \\ CH_3 \end{array}$	B3LYP/LANL2DZ// B3LYP/LANL2DZ	Geometry, energy, NBO charges	76
$\begin{array}{c} CH_3 \\ | \\ H_3C-Br^{2+} \\ | \\ CH_3 \end{array}$	B3LYP/LANL2DZ// B3LYP/LANL2DZ	Geometry, energy, NBO charges	76
$\begin{array}{c} CH_3 \\ | \\ H_3C-I^{2+} \\ | \\ CH_3 \end{array}$	B3LYP/LANL2DZ// B3LYP/LANL2DZ	Geometry, energy, NBO charges	76

Table 9. (*continued*)

Structure[a]	Level of Theory	Calculated Properties	References
Halonium Systems			
H, +C+, F, H (structure)	MP2/6-31G**// MP2/6-31G**	Geometry, energy, transition state structure and energy barriers of dissociation processes, proton affinity of CH_2F_+, GIAOMP2/tzp/dz NMR chemical shifts, isodesmic reactions	51
H, +C+, Cl, H (structure)	MP2/6-31G**// MP2/6-31G**	Geometry, energy, transition state structure and energy barriers of dissociation processes, proton affinity of CH_2F+, GIAOMP2/tzp/dz NMR chemical shifts, isodesmic reactions	51,107
F, +C, F, F, H (structure)	MP4(SDTQ)/6-31G**// MP2/6-31G**	Geometry, energy, transition state structure and energy barrier for proton loss	107
Cl, +C, Cl, Cl, H (structure)	MP4(SDTQ)/6-31G**// MP2/6-31G**	Geometry, energy, transition state structure and energy barrier for proton loss, GIAO-MP2 NMR chemical shifts, isodesmic reaction	107
Br, +C, Br, Br, H (structure)	MP4(SDTQ)/LANL2DZ// MP2/LANL2DZ	Geometry, energy, transition state structure and energy barrier for proton loss	107
I, +C, I, I, H (structure)	MP4(SDTQ)/LANL2DZ// MP2/LANL2DZ	Geometry, energy, transition state structure and energy barrier for proton loss	107
H, Cl, +C, Cl, Cl, H (structure)	MP4(SDTQ)/6-31G**// MP2/6-31G**	Geometry, energy, transition state structure and energy barrier for proton loss, GIAO-MP2 NMR chemical shifts, isodesmic reaction	107
H, Br, +C, Br, Br, H (structure)	MP4(SDTQ)/LANL2DZ// MP2/LANL2DZ	Geometry, energy, transition state structure and energy barrier for proton loss	107

Table 9. (*continued*)

Structure[a]	Level of Theory	Calculated Properties	References
Halonium Systems			
	MP4(SDTQ)/LANL2DZ// MP2/LANL2DZ	Geometry, energy, transition state structure and energy barrier for proton loss	107
	MP4(SDTQ)/6-31G**// MP2/6-31G**	Geometry, energy, transition state structure and energy barrier for proton loss, GIAO-MP2 NMR chemical shifts, isodesmic reaction	107
	MP4(SDTQ)/LANL2DZ// MP2/LANL2DZ	Geometry, energy, transition state structure and energy barrier for proton loss	107
	MP4(SDTQ)/LANL2DZ// MP2/LANL2DZ	Geometry, energy, transition state structure and energy barrier for proton loss	107
	MP2/6-31G**// MP2/6-31G**	Geometry, energy, transition state structure and energy barriers of dissociation processes, proton affinity of CH_2F+, GIAOMP2/tzp/dz NMR chemical shifts, isodesmic reactions	107
	MP2/6-31G**// MP2/6-31G**	Geometry, energy, transition state structure and energy barriers of dissociation processes, proton affinity of CH_2F+, GIAOMP2/tzp/dz NMR chemical shifts, isodesmic reactions	107

[a]Calculated structures found to be at potential energy minima (zero imaginary frequencies).

REFERENCES

(1) (a) G. A. Olah *Angew. Chem. Int. Ed. Engl*. **1993**, *32*, 767. (b) G. A. Olah; D. A. Klumpp *Acc. Chem. Res*. **2004**, *37*, 211. (c) G. A. Olah; G. K. S. Prakash; K. Lammertsma *Res. Chem Intermed*. **1989**, *12*, 141.

(2) G. A. Olah; A. Germain; H. C. Lin; D. Forsyth *J. Am. Chem. Soc*. **1975**, *97*, 2928.

(3) (a) G. A. Olah; H. C. Lin *J. Am. Chem. Soc*. **1974**, *96*, 549. (b) G. A. Olah; H. C. Lin *Synthesis* **1974**, 444. (c) G. A. Olah; A. Orlinkov; A. B. Oxyzoglou; G. K. S. Prakash *J. Org. Chem*. **1995**, *60*, 7348.

(4) G. A. Olah; Q. Wang; A. Orlinkov; P. Ramaiah *J. Org. Chem*. **1993**, *58*, 5017.

(5) (a) *Friedel-Crafts and Related Reaction*, G. A. Olah, Ed.; Wiley: New York, 1964. (b) J. March, *Advanced Organic Chemistry*, 4th Ed; Wiley: New York, 1992, chap 12.

(6) (a) R. Taylor, *Electrophilic Aromatic Substitution*, Wiley: Chichester, 1990, p. 218. (b) D. J. R. Massy *Synthesis*, **1987**, 589. (c) G. A. Olah; Q. Wang; G. Neyer *Synthesis*, **1994**, 276.

(7) (a) J. E. Hofmann; A. Schriesheim, A. in *Friedel-Crafts and Related Reaction*; G. A. Olah, Ed.; Wiley: New York, 1964; vol 2.; pp 597–640. (b) J. March, *Advanced Organic Chemistry*, 4th Ed; Wiley: New York, 1992; pp 548–549.

(8) D. A. Klumpp; K. Y. Yeung; G. K. S. Prakash; G. A. Olah *Synlett* **1998**, 918.

(9) D. A. Klumpp; K. Y. Yeung; G. K. S. Prakash; G. A. Olah *J. Org. Chem*. **1998**, *63*, 4481.

(10) D. A. Klumpp; S. Lau *J. Org. Chem*. **1999**, *64*, 7309.

(11) D. A. Klumpp; M. Garza; G. V. Sanchez; S. Lau; S. DeLeon *J. Org. Chem*. **2000**, *65*, 8997.

(12) J. P. Hwang; G. K. S. Prakash; G. A. Olah *Tetrahedron* **2000**, *56*, 7199.

(13) (a) *Chem. Abstr*. **1979**, *90*, 86986u. (b) *U. S. Pat. Appl. 891,872 (1978)*. (c) W. D. Kray; R. W. Rosser *J. Org. Chem*. **1977**, *42*, 1186.

(14) (a) G. A. Olah; G. K. S. Prakash; T. Mathew; E. R. Marinez *Angew. Chem., Int. Ed*. **2000**, *39*, 2547. (b) G. A. Olah; T. Mathew; E. R. Marinez; P. M. Esteves; M. Etzkorn; G. Rasul; G. K. S. Prakash; *J. Am. Chem. Soc*. **2001**, *123*, 11556.

(15) O. Farooq; M. Marcelli; G. K. S. Prakash; G. A. Olah *J. Am. Chem. Soc*. **1988**, *110*, 864.

(16) A. Berkessel; R. K. Thauer *Angew. Chem. Int. Ed. Engl*. **1995**, *34*, 2247.

(17) H. Pellissier *Tetrahedron* **2005**, *61*, 6479.

(18) T. Suzuki; T. Ohwada; K. Shudo *J. Am. Chem. Soc*. **1997**, *119*, 6774.

(19) S. Saito; Y. Sato; T. Ohwada; K. Shudo *J. Am. Chem. Soc*. **1994**, *116*, 2312.

(20) Y. Sato; M. Yato; T. Ohwada; S. Saito; K. Shudo *J. Am. Chem. Soc.* **1995**, *117*, 3037.

(21) C. Blackburn; R. F. Childs *J. Chem. Soc., Chem. Commun.* **1984**, 812.

(22) A. Yokoyama; T. Ohwada; K. Shudo *J. Org. Chem.* **1999**, *64*, 611.

(23) D. A. Klumpp; D. N. Baek; G. K. S. Prakash; G. A. Olah *J. Org. Chem.* **1997**, *62*, 6666.

(24) T. Ohwada; N. Yamagata; K. Shudo *J. Am. Chem. Soc.* **1991**, *113*, 1364.

(25) T. Ohwada; T. Suzuki; K. Shudo *J. Am. Chem. Soc.* **1998**, *120*, 4629.

(26) T. Yamazaki; S. Saito; T. Ohwada *Tetrahedron Lett.* **1995**, *36*, 5749.

(27) D. M. Brouwer; A. A. Kiffen *Recl. Trav Chim. Pays-Bas* **1973**, *92*, 689.

(28) G. A. Olah; J. L. Grant; R. J. Spear; J. M. Bollinger; A. Serianz; G. Sipos *J. Am. Chem. Soc.* **1976**, *98*, 2501.

(29) (a) T. Ohwada; A. Itai; T. Ohta; K. Shudo *J. Am. Chem. Soc.* **1987**, *109*, 7036. (b) T. Ohwada; K. Okabe; T. Ohta; K. Shudo *Tetrahedron* **1990**, *46*, 7539.

(30) T. Ohta; K. Shudo; T. Okamoto *Tetrahedron Lett.* **1984**, *25*, 325.

(31) D. A. Klumpp; S. L. Aguirre; G. V. Sanchez, Jr.; S. de Leon *Org. Lett.* **2001**, *3*, 2781.

(32) T. Ohwada; T. Yamazaki; T. Suzuki; S. Saito; K. Shudo *J. Am. Chem. Soc.* **1996**, *118*, 6220.

(33) (a) D. A. Klumpp; R. Rendy; Y. Zhang; A. Gomez; A. McElrea *Org. Lett.* **2004**, *6*, 1789. (b) Y. Zhang; S. A. Aguirre; D. A. Klumpp *Tetrahedron Lett.* **2002**, *43*, 6837.

(34) (a) D. A. Klumpp; Y. Zhang; P. J. Kindelin; S. Lau *Tetrahedron* **2006**, *62*, 5915. (b) D. A. Klumpp; P. J. Kindelin; A. Li *Tetrahedron Lett.* **2005**, *46*, 2931.

(35) (a) K. Y. Koltunov; L. A. Ostashevskaya; I. B. Repinskaya, *Russ. J. Org. Chem. (Engl. Transl.)* **1998**, *34*, 1796. (b) K. Y. Koltunov; G. K. S. Prakash; G. Rasul; G. A. Olah *J. Org. Chem.* **2002**, *67*, 8943. (c) K. Y. Koltunov; G. K. S. Prakash; G. Rasul; G. A. Olah *J. Org. Chem.* **2002**, *67*, 4330. (d) K. Y. Koltunov; G. K. S. Prakash; G. Rasul; G. A. Olah *Heterocycles* **2004**, *62*, 757. (e) K. Y. Koltunov; I. B. Repinskaya *Zhur. Org. Khim.* **1995**, *31*, 1579. (f) K. Y. Koltunov; I. B. Repinskaya *Russ. J. Org. Chem. (Engl. Transl.)* **2000**, *36*, 446. (g) K. Y. Koltunov; I. B. Repinskaya *Russ. J. Org. Chem. (Engl. Transl.)* **2002**, *38*, 437. (h) L. A. Ostashevskaya; K. Y. Koltunov; I. B. Repinskaya *Russ. J. Org. Chem. (Engl. Transl.)* **2000**, *36*, 1474.

(36) E. M. Arnett *Prog. Phys. Org. Chem.* **1963**, *1*, 223.

(37) G. A. Olah; G. Rasul; R. Aniszfeld; G. K. S. Prakash *J. Am. Chem. Soc.* **1992**, *114*, 5608.

(38) G. A. Olah, A. Burrichter, G. Rasul, K. O. Christe, G. K. S. Prakash *J. Am. Chem. Soc.* **1997**, *117*, 4345.

(39) (a) K. Lammertsma; P. v. R. Schleyer; H. Schwarz *Angew. Chem. Int. Ed. Engl*. **1989**, *28*, 1321. (b) W. Koch; F. Maquin; D. Stahl; H. Schwarz *Chimia* **1985**, *39*, 376. (c) D. Schröder; H. Schwarz *J. Phys. Chem. A*. **1999**, *103*, 7385.

(40) (a) L. Pauling *J. Chem Phys*., **1933**, *1*, 56. (b) J. D. Dunitz; T. K. Ha *J. Chem. Soc. Chem. Commun*. **1972**, 568. (c) See also reference 39c.

(41) T. Weiske; W. Koch; H. Schwarz *J. Am. Chem. Soc*. **1993**, *115*, 6312.

(42) W. Koch; G. Frenking; H. Schwarz; F. Maquin; D. Stahl *Int. J. Mass Spec. Ion Proc*. **1985**, *63*, 59.

(43) H. Schwarz *Angew. Chem. Int. Ed. Engl*. **1991**, *30*, 820.

(44) (a) K. A. Newson; S. D. Price *Chem. Phys. Lett*. **1998**, *294*, 223. (b) Z. Dolejsek; M. Farnik; Z. Herman *Chem. Phys. Lett*. **1995**, *235*, 99. (c) C. Guenat; F. Maquin; D. Stahl; W. Koch; H. Schwaz *Int. J. Mass Spec. Ion Proc*. **1985**, *63*, 265. (d) A. Fiedler; I. Kretzschmar; D. Schröder; H. Schwarz *J. Am. Chem. Soc*. **1996**, *118*, 9941. (e) J. Roithova; J. Zabka; Z. Herman; R. Thissen; D. Schröder; H. Schwarz *J. Phys. Chem. A*. **2006**, *110*, 6447.

(45) J. Roithova; D. Schröder *J. Am. Chem Soc*. **2006**, *128*, 4208.

(46) P. Perez *J. Org. Chem*. **2004**, *69*, 5048.

(47) H.-U. Siehl; V. Vrcek in *Calculation of NMR and EPR Parameters (2004)*, M. Kaupp; M. Buehl; V. G. Malkin, Eds.; Wiley-VHC, Weinheim, 2004; pp 371–394.

(48) G. A. Olah; A. Burrichter; G. Rasul; R. Gnann; K. O. Christe; G. K. S. Prakash *J. Am. Chem. Soc*. **1997**, *119*, 8035.

(49) (a) G. A. Olah; G. Rasul; R. Aniszfeld; G. K. S. Prakash *J. Am. Chem. Soc*. **1992**, *114*, 8035. (b) G. K. S. Prakash; G. Rasul; A. Burrichter; G. A. Olah in *Nitration-Recent Laboratory and Industrial Developments*, L. F. Albright; R. V. C. Carr; R. J. Schmitt (Eds.), ACS Symposium Series 623, American Chemical Society, Washington D.C., 1996, p. 10.

(50) N. Hartz; G. Rasul; G. A. Olah *J. Am. Chem. Soc*. **1993**, *115*, 1277.

(51) G. Rasul; G. K. S. Prakash; G. A. Olah *Theochem*. **1999**, *466*, 245.

(52) B. F. Yates; W. J. Bouma; L. Radom *J. Am. Chem. Soc*. **1986**, *108*, 6545.

(53) G. Rasul; V. Prakash Reddy; L. Z. Zdunek; G. K. S. Prakash; G. A. Olah *J. Am. Chem. Soc*. **1993**, *115*, 2236.

(54) G. A. Olah; B. Torok; J. P. Joschek; I. Bucsi; P. M. Esteves; G. Rasul; G. K. S. Prakash *J. Am. Chem. Soc*. **2002**, *124*, 11379.

(55) G. A. Olah; T. Heiner; G. Rasul; G. K. S. Prakash *J. Am. Chem. Soc*. **1998**, *120*, 7993.

(56) G. K. S. Prakash; G. Rasul; A. Burrichter; K. K. Laali; G. A. Olah *J. Org. Chem*. **1996**, *61*, 9253.

(57) G. A. Olah; A. Burrichter; T. Mathew; Y. D. Vankar; G. Rasul; G. K. S. Prakash *Angew. Chem. Int. Ed. Engl*. **1997**, *36*, 1875.

(58) G. A. Olah; N. Hartz; G. Rasul; A. Burrichter; G. K. S. Prakash *J. Am. Chem. Soc*. **1995**, *117*, 6421.

(59) N. J. Head; G. Rasul; A. Mitra; A. Bashir-Heshemi; G. K. S. Prakash; G. A. Olah *J. Am. Chem. Soc*. **1995**, *117*, 6421.

(60) D. A. Klumpp; Y. Zhang; P. J. Kindelin; S. Lau *Tetrahedron* **2006**, *62*, 5915.

(61) V. Prakash Reddy; G. Rasul; G. K. S. Prakash; G. A. Olah *J. Org. Chem*. **2003**, *68*, 3507.

(62) T. Ohwada; K. Shudo *J. Am. Chem. Soc*. **1989**, *111*, 34.

(63) G. A. Olah; J. Bausch; G. Rasul; H. George; G. K. S. Prakash *J. Am. Chem. Soc*. **1993**, *115*, 8060.

(64) G. Rasul; G. K. S. Prakash; G. A. Olah *J. Org. Chem*. **1994**, *59*, 2552.

(65) K. Y. Koltunov; G. K. S. Prakash; G. Rasul; G. A. Olah *Eur. J. Org. Chem*. **2006**, *21*, 4861.

(66) G. K. S. Prakash; T. Mathew; D. Hoole; P. M. Esteves; Q. Wang; G. Rasul; G. A. Olah *J. Am. Chem. Soc*. **2004**, *126*, 15770.

(67) G. Frenking *J. Am. Chem. Soc*. **1991**, *113*, 2476.

(68) G. A. Olah; A. Burrichter; G. Rasul; G. K. S. Prakash; M. Hachoumy; J. Sommer *J. Am. Chem. Soc*. **1996**, *118*, 10423.

(69) G. Rasul; G. K. S. Prakash; G. A. Olah *J. Phys. Chem. A*. **2006**, *110*, 1041.

(70) (a) A. I. Boldyrev; J. Simons *J. Chem. Phys*. **1992**, *97*, 4272. (b) W. Koch; N. Heinrich; H. Schwarz; F. Maquin; D. Stahl *Int. J. Mass Spec. Ion Proc*. **1985**, *67*, 305.

(71) G. A. Olah; G. K. S. Prakash; M. Barzaghi; K. Lammertsma; P. v. R. Schleyer; J. A. Pople *J. Am. Chem. Soc*. **1986**, *108*, 1032.

(72) V. Prakash Reddy; E. Sinn; G. A. Olah; G. K. S. Prakash; G. Rasul *J. Phys. Chem. A*. **2004**, *108*, 4036.

(73) G. A. Olah; G. Rasul; A. Burrichter; M. Hachoumy; G. K. S. Prakash; R. I. Wagner; K. O. Christe *J. Am. Chem. Soc*. **1997**, *119*, 9572.

(74) K. Lammertsma *J. Am. Chem. Soc*. **1984**, *106*, 4619.

(75) G. A. Olah; G. Rasul; A. Burrichter; G. K. S. Prakash *Proc. Nat. Acad. Sci. USA* **1998**, *95*, 4099.

(76) G. A. Olah; G. Rasul; M. Hachoumy; A. Burrichter; G. K. S. Prakash *J. Am. Chem. Soc*. **2000**, *122*, 2737.

(77) G. Rasul; G. K. S. Prakash; G. A. Olah *Proc. Nat. Acad. Sci. USA* **2002**, *99*, 9635.

(78) G. Rasul; G. K. S. Prakash; G. A. Olah *J. Phys. Chem. A*. **2005**, *109*, 798.

(79) (a) G. A. Olah; G. Rasul *J. Am. Chem. Soc*. **1996**, *118*, 12922. (b) G. A. Olah; G. Rasul *Acc. Chem. Res*. **1997**, *30*, 245.

(80) G. A. Olah; G. K. S. Prakash; G. Rasul *J. Org. Chem*. **2001**, *61*, 2907.

(81) G. A. Olah; G. Rasul *J. Am. Chem. Soc*. **1996**, *118*, 8503.

(82) G. A. Olah; N. Hartz; G. Rasul; G. K. S. Prakash; M. Burkhart; K. Lammertsma *J. Am. Chem. Soc.* **1994**, *116*, 3187.

(83) G. A. Olah; G. Rasul; G. K. S. Prakash *Chem Eur. J*. **1997**, *3*, 1039.

(84) G. A. Olah; A. Burrichter; G. Rasul; G. K. S. Prakash *J. Am. Chem. Soc.* **1997**, *119*, 4594.

(85) (a) G. Rasul; G. K. S. Prakash; G. A. Olah *J. Am. Chem. Soc.* **1997**, *119*, 12984. (b) G. Rasul; G. K. S. Prakash; G. A. Olah *J. Phys. Chem. A.* **1998**, *102*, 8457.

(86) G. A. Olah; N. Hartz; G. Rasul; G. K. S. Prakash *J. Am. Chem. Soc.* **1993**, *115*, 6985.

(87) G. Rasul; G. A. Olah; G. K. S. Prakash *Proc. Nat. Acad. Sci. USA* **2004**, *101*, 10868.

(88) K. Lammertsma; M. Barzaghi; G. A. Olah; J. A. Pople; A. J. Kos; P. v. R. Schleyer *J. Am. Chem. Soc.* **1983**, *105*, 5252.

(89) R. Herges; P. v. R. Schleyer; M. Schindler; W.-D. Fessner *J. Am. Chem. Soc.* **1991**, *113*, 3649.

(90) G. A. Olah; V. Prakash Reddy; G. Rasul; G. K. S. Prakash *J. Am. Chem. Soc.* **1999**, *121*, 9994.

(91) C. Taeschler; T. S. Sorensen *Tetrahedron Lett.* **2001**, *42*, 5339.

(92) (a) G. A. Olah; T. Shamma; A. Burrichter; G. Rasul; G. K. S. Prakash *J. Am. Chem. Soc.* **1997**, *119*, 3407. (b) G. A. Olah; T. Shamma; A. Burrichter; G. Rasul; G. K. S. Prakash *J. Am. Chem. Soc.* **1997**, *119*, 12923.

(93) A. Li; P. J. Kindelin; D. A. Klumpp *Org. Lett.* **2006**, *8*, 1233.

(94) G. Rasul; G. K. S. Prakash; G. A. Olah *Inorg. Chem.* **2002**, *41*, 5589

(95) G. A. Olah; A. Burrichter; G. Rasul; M. Hachoumy; G. K. S. Prakash *J. Am. Chem. Soc.* **1997**, *119*, 12929.

(96) G. Rasul; G. A. Olah; G. K. S. Prakash *Inorg. Chem.* **2003**, *24*, 8059.

(97) G. A. Olah; G. K. S. Prakash; G. Rasul *Proc. Nat. Acad. Sci. USA* **1999**, *96*, 3494.

(98) G. Rasul; G. K. S. Prakash; G. A. Olah *J. Am. Chem. Soc.* **1994**, *116*, 8985.

(99) G. A. Olah; G. K. S. Prakash; G. Rasul *J. Org. Chem.* **2002**, *67*, 8547.

(100) R. A. Cox; D. Y. K. Fung; I. G. Csizmadia; E. Buncel *Can. J. Chem.* **2003**, *81*, 535.

(101) I. Krossing; J. Passmore *Inorg. Chem.* **2004**, *43*, 1000.

(102) G. A. Olah; G. K. S. Prakash; M. Marcelli; K. Lammertsma *J. Phys. Chem.* **1988**, *92*, 878.

(103) G. Rasul; G. K. S. Prakash; G. A. Olah *J. Org. Chem.* **2000**, *65*, 8776.

(104) R. Glaser; G. S.-C. Choy; G. S. Chen; H. Grützmacher *J. Am. Chem. Soc.* **1996**, *118*, 11617.

(105) T. Drewello; W. Koch; C. B. Lebrilla; D. Stahl; H. Schwarz *J. Am. Chem. Soc.* **1987**, *109*, 2922.

(106) A. I. Boldyrev; J. Simons *J. Chem. Phys.* **1993**, *99*, 769.

(107) G. A. Olah; G. Rasul; A. K. Yudin; A. Burrichter; G. K. S. Prakash; A. L. Chistyakov; I. V. Stankevich; I. S. Akhrem; N. P. Gambaryan; M. E. Vol'pin *J. Am. Chem. Soc.* **1996**, *118*, 1446.

3

GENERATING
SUPERELECTROPHILES

As discussed in Chapter 1, many electrophiles are stable in solvents and the presence of substances of moderate Lewis basicity. Even relatively strong electrophiles can be generated and handled under appropriate conditions.[1] A wide variety of electrophilic carbocationic and onium ion salts have been prepared and characterized by spectroscopic methods, X-ray crystallography, kinetic studies, and other techniques. Several types of these electrophilic salts are even available commercially (such as varied onium, iminium, and carbocationic salts, etc.). Electrophiles are generally generated by the reactions of suitable precursors with Brønsted or Lewis acids. They can be sometimes isolated, but for synthetic conversions they are more often generated *in situ* and reacted directly with nucleophilic reagents.

When superelectrophilic reagents are involved in condensed phase reactions, they are usually generated *in situ* in highly acidic systems.[2] Superelectrophiles are generally not isolated as persistent stable salts, although some highly stabilized onium dicationic systems have been isolated (*vide infra*). Both Brønsted and Lewis acid containing systems, including solid and liquid acids, have been shown suitable to form superelectrophiles. Because superelectrophilic activation often involves interaction of a substrate with two or more equivalents of acid, superelectrophilic conversions are often carried out in the presence of excess acid. Among the Brønsted

Superelectrophiles and Their Chemistry, by George A. Olah and Douglas A. Klumpp acids, sulfonic acids and their Lewis acid conjugate systems have been most frequently used for the reactions involving superelectrophiles. Direct spectroscopic observation of superelectrophiles in these acids have also been reported. Although sulfuric acid and oleum are known to catalyze conversions involving reactive dications (*vide infra*), H_2SO_4 ($H_0 - 12$) and H_2SO_4-SO_3 ($H_0 \geq -14.5$) have seen relatively limited use in superelectrophilic reactions, as they themselves can react with many of the involved systems (sulfonation, oxidation, etc.). HF-based superacid systems have been important in the study of many superelectrophilic systems. Their Lewis acid conjugates such as HF-BF_3 and HF-SbF_5 were used in extensive, fundamentally important studies of many systems. Acid systems based on $AlCl_3$ and $AlBr_3$ have also been useful in various superelectrophilic reactions.

Trifluoromethanesulfonic acid (CF_3SO_3H, triflic acid) is an effective and most widely used catalyst and activating agent in superelectrophilic chemistry. Triflic acid is a stable, nonoxidizing (or weakly oxidizing) superacid with a useful liquid range (mp $-40°C$; bp $161-162°C$) and with an acidity of $H_0 - 14.1$.[3] It has been used to protolytically generate superelectrophiles from varied electrophiles (Table 1). Triflic acid has been shown *inter alia* to protolytically activate nitronium salts, as well as nitrosubstituted olefins and arenes (entries 1–3).[4−6] In a study of the cyclization of imines, triflic acid was found to be an effective catalyst and superelectrophilic activating agent (entry 4).[7] A number of carboxonium-type superelectrophiles have been generated using triflic acid, including those from 1,2-dicarbonyl compounds (entries 5–7),[8−10] enones (entries 8–10),[11−13] ketones (entry 11–13),[14−16] unsaturated carboxylic acids (entries 14–15),[17] amides (entries 16–18),[12,18,19] and esters (entry 19).[20] Because the acidity of hygroscopic triflic acid is significantly decreased by water, the importance was noted to freshly distill it prior to use. Water can also be produced in some triflic acid catalyzed reactions, leading to decreased acidity. This water-induced decrease of acidity must be taken into account for reactions requiring high levels of acidity, as well for quantitative studies, such as kinetic experiments. Using the triflic acid–triflic anhydride system can remedy this problem.

The acidity of triflic acid solutions can be greatly increased with added Lewis acids.[2] For example, mixtures of triflic acid with either SbF_5 or $B(O_3SCF_3)_3$ have been estimated to have $H_0 - 16.8$ (with SbF_5) and $H_0 - 18.5$ (with $B(O_3SCF_3)_3$). These conjugate Brønsted-Lewis superacids have been used to generate superelectrophiles from varied substrates. In a study related to the Gattermann and Houben-Hoesch reactions, superelectrophiles **1** and **3** were suggested as the intermediates in reactions with CF_3SO_3H:SbF_5 (eqs 1–2).[21] When the hydroxyester (**4**) is ionized

Table 1. Superelectrophiles generated in solutions of CF_3SO_3H.

Entry	Substrate	Superelectrophile	Entry	Substrate	Superelectrophile
(1)			(10)		
(2)			(11)		
(3)			(12)		
(4)			(13)		
(5)			(14)		
(6)			(15)		
(7)			(16)		
(8)			(17)		
(9)			(18)		
			(19)		

in either $CF_3SO_3H:SbF_5$ or CF_3SO_3H at low temperature, the dication (**5**) can be directly observed by NMR spectroscopy (eq 3).[22]

(1)

$$(2)$$

$$(3)$$

Triflatoboric acid ($2 \, CF_3SO_3H - B(O_3SCF_3)_3$) can be prepared by the reaction of five equivalents of triflic acid with one equivalent of BCl_3 (with HCl gas evolution).[23] This very strong conjugate Lewis acid has been shown to produce superelectrophilic protonitronium ion (**6**), or its equivalent protosolvated species, in equilibria (eq 4).[23] In a study of the C-alkylation with alkyloxonium ions, triflatoboric acid was found to enhance the electrophilic reactivity of a trimethyloxonium salt (**7**). Formation of a the superelectrophilic species (**8**) was indicated (eq 5).[24] Protosolvated formyl cation (**9**) has also been proposed as the active electrophile in reactions of carbon monoxide in CF_3SO_3H, $CF_3SO_3H:SbF_5$, or $2 \, CF_3SO_3H - B(O_3SCF_3)_3$ (eq 6).[25] The methyloxonium ion was similarly suggested to be protosolvated in triflatoboric acid, producing a superelectrophilic species (eq 7).[26]

$$(4)$$

$$(5)$$

$$(6)$$

$$CH_3OH \xrightleftharpoons{H^+} \overset{+}{CH_3OH_2} \xrightleftharpoons{HA} \underset{\underset{HA}{|}}{\overset{+}{CH_3OH_2}} \rightleftharpoons \overset{2+}{CH_3OH_3} \qquad (7)$$

Shudo and Ohwada have developed and used acid systems composed of varying ratios of CF_3SO_3H and CF_3CO_2H in order to obtain solutions having acidities between $H_0 - 7.7$ and $H_0 - 13.7$.[27] These acid systems have been used in kinetic studies related to superelectrophiles $10-13$ (eqs $8-11$).[7,13,14,21]

$$(8)$$

$$(9)$$

$$(10)$$

$$(11)$$

Kinetic evidence suggests the formation of increasing superelectrophilic activity ($10-13$), as the acidity of the reaction media increases from $H_0 - 7.7$ to $H_0 - 13.7$ (*vide supra*).

Superacidic FSO_3H (fluorosulfonic acid, $H_0 - 15$) has also been used in some studies involving superelectrophilic activation. However, due to its tendency for sulfonation and oxidation, this acid has found only limited use in synthetic conversions involving superelectrophiles. Fluorosulfonic acid has been shown effective to activate nitronium salts in their reactions with weak nucleophiles, and again it was suggested that the protosolvated species (**6**) is involved in the reactions.[28] Both fluorosulfonic acid and triflic acid have been reported to give the diprotonated species (**14**) from 3-arylindenones (eq 12).[29]

(12)

The FSO_3H:SbF_5 conjugate superacid ("Magic Acid"; $H_0 - 15$ to -25) has been useful in extensive studies of onium cations and dications.[1,2] Its very high acidity was the basis for the "stable ion conditions" developed by the Olah group to directly observe many types of reactive electrophiles. The acidity of the magic acid system varies according to the proportion of SbF_5 in the mixture (FSO_3H:SbF_5 1:1, $H_0 - 25$).[2] Magic Acid solutions have also been used in several NMR studies of reactive dications and superelectrophiles, such as ions **15** and **16**.[17,18] Among the superelectrophiles generated in FSO_3H:SbF_5 solutions are protosolvated oxonium ions (Table 2), sulfonium ions, and related species (**17**–**18**),[30,31] carboxonium ions (**19**–**20**),[32,33] and others (**21**–**22**).[34,35] In the cases of **17**, **18**, and **21**, data indicate low-equilibrium concentrations of the superelectrophiles, while **19**, **20**, and **22** are formed in sufficiently high-equilibrium concentrations to allow their direct observation by spectroscopic methods.

HF-based conjugate Lewis acid systems have been particularly important in the study of superelectrophiles. Commercial anhydrous HF is a relatively modest Brønsted acid ($H_0 - 11$), but as shown by Gillespie, if carefully dried and handled, anhydrous HF has $H_0 -15$. The acidity of HF is greatly increased with the addition of Lewis acids, such as BF_3 (HF-BF_3, $H_0 - 11$ to -16) and SbF_5 (HF-SbF_5, $H_0 - 11$ to -28).[2] Like Magic Acid, these Lewis acid conjugates vary in strength according to the ratio of HF and the Lewis acid.

As discussed in Chapter 1, Brouwer and Kiffen reported the observation that HF-BF_3 promoted hydride transfer from isoalkanes to acyl cations. These results were later shown by Olah and co-workers to be due to superelectrophilic activation of the acyl cation (**24**, eq 13).[37] Diprotonated acetone and aldehydes were also shown to abstract hydride from isoalkanes in HF-BF_3 solutions.[38] Carboxonium ions (**25**) are generally

Table 2. Superelectrophilic species generated in FSO$_3$H:SbF$_5$ solution.

Entry	Superelectrophile

Entry		Superelectrophile
(1)	H$_3$C—X$^+$—CH$_3$ (CH$_3$) X = O, S, Se, Te $\xrightarrow{\text{FSO}_3\text{H:SbF}_5}$	[H$_3$C—X(H)—CH$_3$ (CH$_3$)]$^{2+}$ **17**
(2)	H—S$^+$—D (H) $\xrightarrow{\text{FSO}_3\text{H:SbF}_5}$	[H—S(D)—D (H)]$^{2+}$ **18**
(3)	$\xrightarrow{\text{FSO}_3\text{H:SbF}_5}$	**19**
(4)	$\xrightarrow{\text{FSO}_3\text{H:SbF}_5}$	**20**
(5)	H$_3$C—C(=O)—O—CH$_3$ $\xrightarrow{\text{FSO}_3\text{H:SbF}_5}$	**21**
(6)	H$_2$N—C(=O)—NH$_2$ $\xrightarrow{\text{FSO}_3\text{H:SbF}_5}$	**22**

not reactive towards weak nucleophiles such as alkanes and deactivated arenes. However, a superelectrophilic species (**26**) is formed in HF-BF$_3$ solution (eq 14) and hydride abstraction occurs. HF-BF$_3$ has also been shown to promote synthetically useful reactions (see Chapter 5) such as involving the protosolvated formyl cation (**9**),[39] diprotonated pivaldehyde (**27**),[39] and the superelectrophile derived from dimethyl ether (**28**).[40]

$$
H_3C-\underset{\underset{OH}{||}}{\overset{O}{C}} \xrightarrow{HF:BF_3}
\underset{\underset{CH_3}{|}}{\overset{:O:}{\overset{|||}{\overset{+}{C}}}}
\longleftrightarrow
\underset{\underset{CH_3}{|}}{\overset{:O:}{\overset{||}{C+}}}
\underset{}{\overset{H}{\rightleftharpoons}}
\underset{\underset{CH_3}{|}}{\overset{:OH}{\overset{||}{\overset{+}{C+}}}}
\tag{13}
$$

23   24

$$
R-\underset{\underset{CH_3}{||}}{\overset{O}{C}} \xrightarrow[\text{or HF:SbF}]{HF:BF_3}
R-\underset{\underset{CH_3}{|}}{\overset{:OH}{\overset{||}{C}}}
\underset{}{\overset{H^+}{\rightleftharpoons}}
R-\underset{\underset{CH_3}{||}}{\overset{\overset{2+}{OH_2}}{C}}
\longleftrightarrow
R-\underset{+}{\underset{\underset{CH_3}{|}}{\overset{\overset{+}{:OH_2}}{C}}}
\tag{14}
$$

R = H, CH$_3$ 25   26

$$
H-\underset{\underset{CH_3}{\underset{CH_3}{|}}}{\overset{O}{\overset{||}{C}}}CH_3 \xrightarrow{HF:BF_3}
H-\underset{\underset{CH_3}{\underset{CH_3}{|}}}{\overset{\overset{2+}{OH_2}}{\overset{||}{C}}}CH_3
\longleftrightarrow
H-\underset{\underset{CH_3}{\underset{CH_3}{|}}}{\overset{\overset{+}{OH_2}}{\overset{+}{C}}}CH_3
\tag{15}
$$

27

$$
CH_3OCH_3 \xrightarrow{HF\text{-}BF_3}
\underset{BF_4^-}{\overset{+}{CH_3}OCH_3\ H}
\rightleftharpoons
\underset{H}{\overset{H}{CH_3}}\overset{2+}{OCH_3}\ 2\,BF_4^-
\tag{16}
$$

28

Besides the synthetic utility of the HF-BF$_3$ system, both HF and BF$_3$ are gases at room temperature and thus they can be readily recovered from reaction mixtures.

The strongest liquid superacid known so far, HF–SbF$_5$ has been used to generate a variety of superelectrophiles. In their study of the protosolvation of oxonium ions, Olah and co-workers found evidence even for diprotonated water.[41] Hydrogen/deuterium exchange of the hydronium ion (H$_3$O$^+$) was observed in DF-SbF$_5$ solution ($H_0 - 25$) where there is no deprotonation equilibrium, and the diprotonated superelectrophile (**29**) is indicated to be the key intermediate (eq 17). Hydrogen/deuterium exchange was not observed in somewhat weaker FSO$_3$H-SbF$_5$ solution. Evidence was reported to suggest that carboxylic acids form diprotonated species in HF-SbF$_5$, and these intermediates can subsequently generate superelectrophilic acyl dications (eq 18).[42]

$$\overset{+}{\underset{D}{:O}}\!\!-\!\!\overset{H}{D} \xrightarrow{\text{HF-SbF}_5} \left[\underset{D}{\overset{2+}{H-O}}\!\!-\!\!\overset{H}{D}\right] \xrightarrow{-D^+} \overset{+}{\underset{D}{:O}}\!\!-\!\!H \qquad (17)$$

29

$$\underset{F_3C}{\overset{O}{\underset{\|}{C}}}\!\!-\!\!OH \xrightarrow{\text{HF:SbF}_5} \underset{F_3C}{\overset{OH^+}{\underset{\|}{C}}}\!\!-\!\!OH \rightleftharpoons \underset{F_3C}{\overset{OH^+}{\underset{\|}{C}}}\!\!-\!\!\overset{+}{OH_2} \xrightarrow{-H_2O} \underset{F_3C}{\overset{OH^+}{\underset{\|}{C}}}\!\!-\!\!\overset{+}{C} \qquad (18)$$

A number of substituted naphthols were studied in HF-SbF$_5$ solution by NMR and their superelectrophilic diprotonated carboxonium ions were directly observed (eqs 19–20).[43] These superelectrophiles are capable of reacting with weak nucleophiles, such as cyclohexane. Even alkyl cations such as the *tert*-butyl cation (**30**) have been shown to undergo protolysis in DF-SbF$_5$.[44] When DF–SbF$_5$ is added to the previously formed *tert*-butyl cation (obtained from the alkyl halide in SbF$_5$), it results in hydrogen/deuterium exchange (eq 21). Under these conditions no deprotonation equilibrium with isobutylene can exist to account for this isotopic exchange. Deuteration at the C-H σ-bond must occur, forming the superelectrophile (**31**). In studies related to trihalomethyl cations, Sommer, Jacquesy, and others have demonstrated that HF–SbF$_5$ strongly activates the trichloromethyl cation in its reactions with C–H σ-bonds.[45] The formation of the superelectrophilic species (**32**) is suggested (eq 22).

$$(19)$$

$$(20)$$

30   **31**

$$(21)$$

$$CCl_4 \xrightarrow{\text{HF-SbF}_5} \underset{Cl}{\overset{Cl}{\underset{|}{\overset{+}{C}}}}\!\!Cl \underset{}{\overset{H^+}{\rightleftharpoons}} \underset{Cl}{\overset{Cl}{\underset{|}{\overset{+}{C}}}}\!\!\overset{Cl}{\underset{}{H}} \qquad (22)$$

32

A similar type of activated trichloromethyl cation has been generated from CCl_4 in excess Lewis acids (SbF_5 or $AlCl_3$, *vide infra*).

In the above examples, the superelectrophiles are formed via protonation or protosolvation of the corresponding electrophiles via their electron donating ligands by superacids. There are also a number of examples of superelectrophiles being formed by coordination with strong Lewis acids, such as SbF_5, $AlCl_3$, and $AlBr_3$. In general, Lewis acids can participate in the formation of superelectrophiles by two routes:

1. The electrophile initially formed is further complexed by a Lewis acid, thus decreasing neighboring group participation into the electrophilic center, generating the superelectrophile.
2. An electron donor Lewis base complexes simultaneously with two Lewis acid sites, forming a doubly electron-deficient species, or superelectrophile.

In most of the examples of superelectrophilic reactions involving Lewis acids, they are conducted using an excess of the Lewis acid. This is in accord with electrophilic solvation by the Lewis acid, i.e. activation of the electrophile requires interaction with two or more equivalents of Lewis acid. As an example, superelectrophilic nitration can be accomplished with NO_2Cl and at least three equivalents of $AlCl_3$ (eq 23).[46] This powerful nitrating reagent involves a superelectrophilic complexed nitronium ion (**33**).

$$NO_2Cl \ + \ 3 \ AlCl_3 \ \longrightarrow \ \overset{..}{O}=\overset{+}{N}=\overset{..}{\underset{..}{O}}\diagup^{AlCl_3}_{\quad Al_2Cl_7^-} \tag{23}$$

$$\mathbf{33}$$

Complexation between the oxygen lone pair electrons and the $AlCl_3$ generates an increasing positive charged bent nitronium ion, creating a low-lying LUMO at the nitrogen. Using excess aluminum halide, Vol'pin, Akhrem, and co-workers have developed several types of "aprotic superacids." These superelectrophilic reagents are capable of cracking hydrocarbons by their reactions with C-C and C-H σ-bonds, *vide infra*. For example, a superelectrophilic acyl cation (**34**) is formed by the reaction of acetyl chloride with excess $AlCl_3$ (eq 24).[47]

$$CH_3COCl \cdot 2AlCl_3 \ \longrightarrow \ \underset{\delta^{++}}{H_3C}\overset{\overset{\textstyle :O \longrightarrow AlCl_3^\delta}{\|}}{\underset{..}{C}}\overset{..}{\underset{..}{Cl}} \longrightarrow AlCl_3^\delta \ \overset{AlCl_3}{\rightleftharpoons} \ H_3C\overset{\overset{\textstyle :O^{\delta+} \longrightarrow AlCl_3^\delta}{\|}}{\underset{+}{C}} \ Al_2Cl_7^- \tag{24}$$

$$\mathbf{34}$$

Highly electrophilic species are also generated in the reactions of halo-genated methanes and excess aluminum halides (eq 25), probably involv-ing the superelectrophilic complex (**35**).[48] A reactive thio-methylating agent has also been developed based on superelectrophilic activation by $AlCl_3$ (eq 26).[49] Similarly, electrophilic solvation was proposed in the carboxylation of arenes with $CO_2/Al/AlCl_3$ (eq 27).[50]

$$(25)$$

$$(26)$$

$$(27)$$

These are representative examples (eqs 23–27) of superelectrophilic species being generated by interactions with Lewis acids. The presence of trace water impurities in Lewis acids, such as $AlCl_3$, creates strong protic Brønsted acids to produce superelectrophiles, as in reactions with enones (eq 28)[51] and α,β-unsaturated amides (eq 29).[12,52] Superelec-trophilic species have been similarly formed from a variety of naphthols with excess aluminum halide (eqs 30–31).[53] These multiply charged inter-mediates are capable of reacting with benzene or cyclohexane (hydride abstraction). Several studies have also described the use of H_2O-BF_3 as a strong acid catalyst (H_0 −12), which is believed to act primarily as a Brønsted acid.[54] It has been shown to produce superelectrophilic species arising from N-halosuccinimides (eq 32).

$$(28)$$

$$(29)$$

$$X = AlCl_3^- \text{ or } Al_nCl_{3n}^-$$

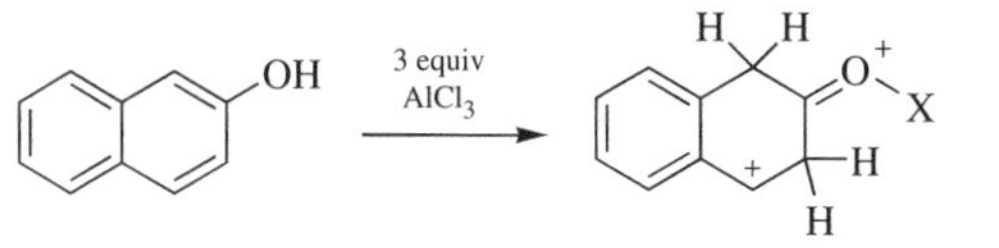

$$X = H, AlCl_3^- \text{ or } Al_nCl_{3n}^- \tag{30}$$

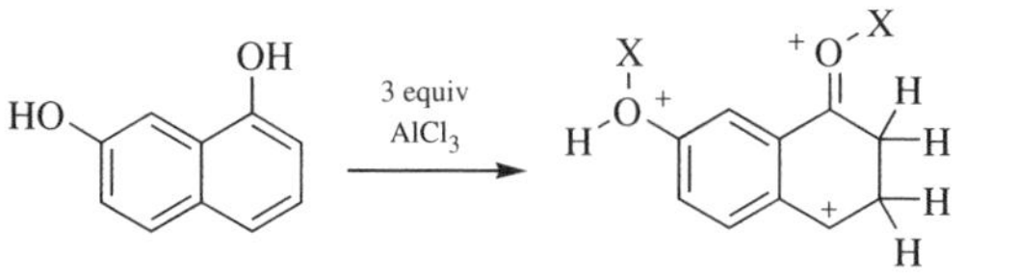

$$X = H, AlBr_3^- \text{ or } Al_nBr_{3n}^- \tag{31}$$

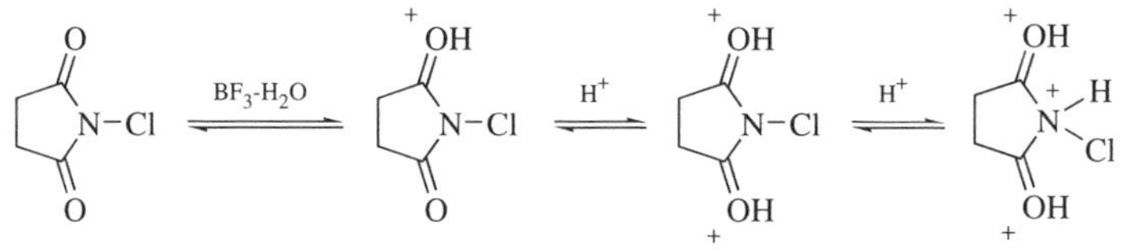

(32)

In addition to the discussed Brønsted or Lewis superacidic activation in solution chemistry, there have been reports to suggest that superelectrophilic species can be formed with solid acids, and even in biochemical systems. For example, Sommer and co-workers have found several examples in which HUSY zeolite has exhibited catalytic activity similar to liquid superacids (eqs 33–34).[12] In the same study, the perfluorinated resinsulfonic acid Nafion-H (SAC-13) was found to give products consistent with the formation of the superelectrophile (**36**, eq 35).

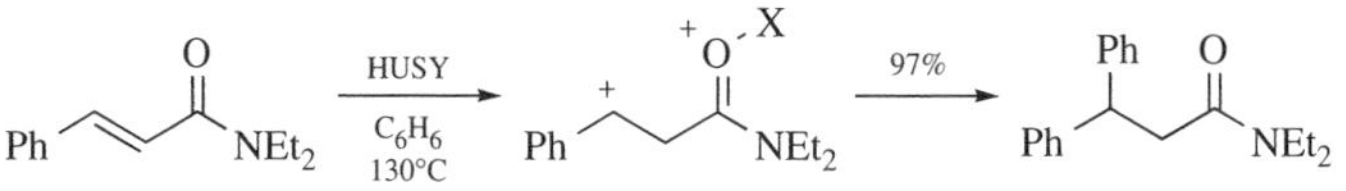

(33)

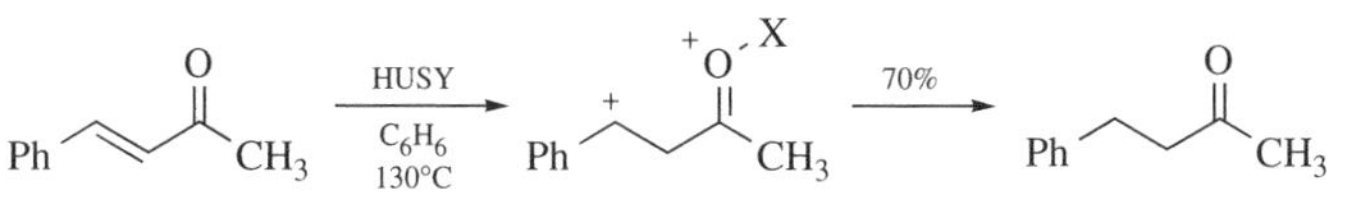

(34)

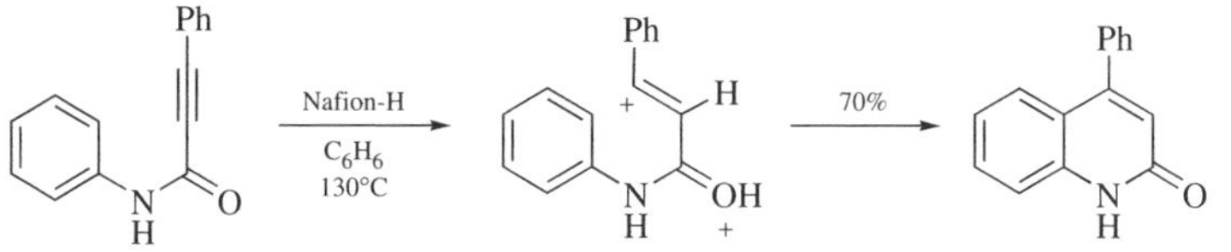

36

(35)

Although it is still difficult to correctly determine or estimate the acidities of solid acids, both Nafion-H and HUSY are considered weaker acids than typical liquid superacids.[2] Favorable geometry in rigid systems (i.e., Nafion-H and HUSY), however, can cause bi-dentate or multi-dentate interaction between the substrate molecule and the solid acid (or acidic enzyme site), which can explain the observed electrophilic activation. Accordingly, superelectrophiles are generated by multiple interactions with closely oriented acid sites (both Brønsted and Lewis) on the surface of the solid acid. Consistent with this, the structure of Nafion-H contains clusters of sulfonic acid groups. This type of multi-dentate complexation is similar to the discussed catalytic activation of electrophiles having suitable electron donor ligands in varied liquid acids systems (*vide supra*). A closely related bi-dentate electrophilic activation has been demonstrated with carbonyl compounds involving double hydrogen bonding. Several similar catalytic systems have been developed, for example, Gong and Wu's enantioselective aldol reaction catalyst (**37**).[55] Theoretical calculations have shown that the catalyst activates the carbonyl group of benzaldehyde by double hydrogen bonding (**38**, Scheme 1), which is shown to be more effective than the proline-catalyzed reaction involving only a single hydrogen bond (**39**). This carbonyl activation is strikingly similar to Shudo and Ohwada's superelectrophilic activation of 1,3-diphenylpropan-1-one (**40**) and benzaldehyde (**41**).[14,56] Compounds capable of double hydrogen bonding to carbonyl groups belong to a promising new class of catalysts. While these catalysts clearly do not generate a full di-positive charge at the carbonyl group, the multi-dentate interaction significantly lowers the energy of the carbonyl LUMO (the $C=O$ π^* orbital), enhancing the electrophilic reactivities of the carbonyl compounds.

As discussed previously, superelectrophilic activation in biological systems has been found even with a metal-free hydrogenase enzyme found in methanogenic archea, an enzymatic system that converts CO_2 to methane.[57] It was found that $N^5.N^{10}$–menthyl tetrahydromethanopterin (**42**) undergoes an enzyme-catalyzed reaction with H_2 by hydride transfer to the pro-R position and release of a proton to form the reduced product (**43**; eq 36).

$$(36)$$

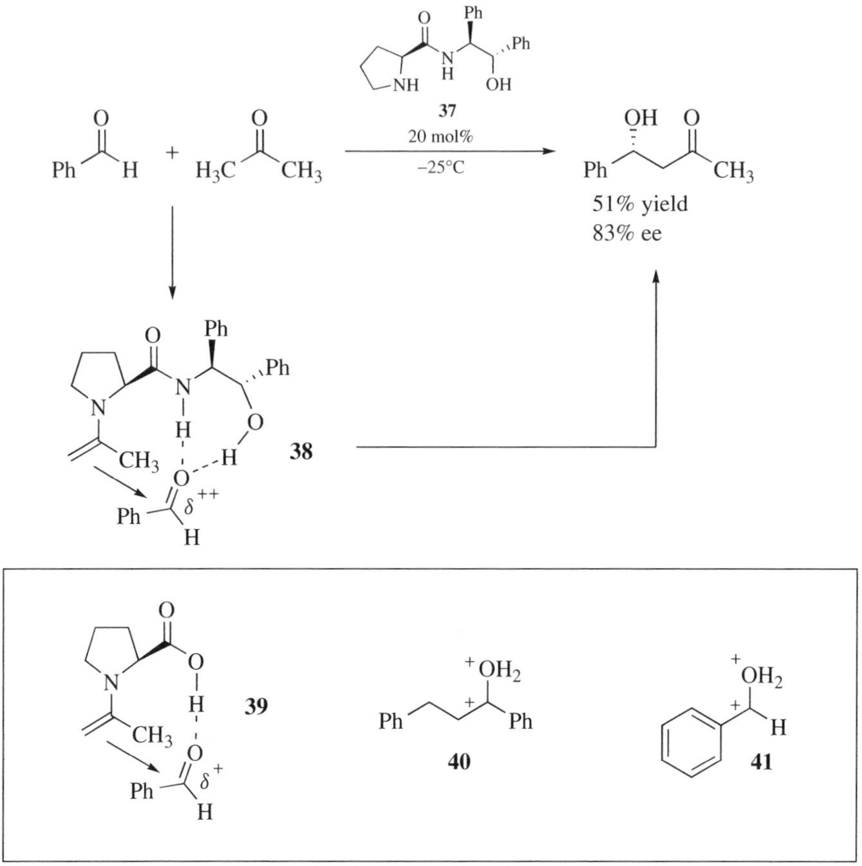

Scheme 1.

It was suggested that protonation of one or both of the adjacent nitrogen sites generates an enhanced electrophilic system. This superelectrophilic system would be the result of decreasing stabilization by neighboring groups and increasing electron deficiency at the electrophilic site, enabling cation **42** to react with molecular hydrogen. As noted by Berkessel and Thauer, a localized, highly acidic enzyme active site is formed by properly positioned acidic functional groups and a nonbasic environment.

The vast majority of examples reported so far of superelectrophilic activation have been found in studies involving liquid superacids with acidities in the range of $H_0 - 12$ to -26. Most often, these acids have been fluorinated sulfonic acids (CF_3SO_3H and FSO_3H, $CF_3SO_3H:SbF_5$, $2\ CF_3SO_3H\text{-}B(O_3SCF_3)_3$), HF-based Lewis conjugates (HF-BF_3 and HF-SbF_5), and systems with excess Lewis acids. Nevertheless, several examples of superelectrophilic activation were reported in weaker acid systems, such as zeolites and perfluorinated resinsulfonic acid, such as Nafion-H. Multi-dentate interaction with weaker acidic sites can extend the realm of superelectrophilic activation to physiological conditions. In principle, superelectrophiles (doubly electron deficient systems) may also be generated from two-electron oxidation of neutral substrates. A number of these systems are discussed in subsequent chapters.

Besides acidity, the involved temperature has also been found to be a factor in the chemistry of superelectrophiles. A number of super-electrophilic reactions have been shown to require more elevated temperatures in order to accomplish reactions with weak nucleophiles (Table 3).[7,12,14,23,54,58–60] In cases, lower temperatures fail to allow the reactions. However, these same electrophilic systems will often react with more nucleophilic substrates at lower temperatures. For example, fluorobenzene gives p-chlorofluorobenzene in 95% yield at 25°C by reaction with N-chlorosuccinimide and $H_2O–BF_3$,[54] 3-pyridinecarboxaldehyde condenses with benzene in 99% yield by reaction with CF_3SO_3H at 25°C,[59] and 5-hydroxy-1-methyl-2-pyrrolidinone reacts with o-dimethoxybenzene in CF_3SO_3H giving the arylated product in 67% yield at 0°C (compared with entries 1, 5, and 6, respectively).[60] The need for more elevated temperatures, however, can sometimes be avoided through the use of a different acid system, as shown in the cyclization of the cinnamoyl amide (entry 3).[12] There may be several reasons for the required higher temperatures in some superelectrophilic reactions. Since the superelectrophiles are often higher-energy species, it may be necessary to use more elevated temperatures to generate their appreciable concentrations. Even with the formation of a dicationic superlectrophile, the higher temperature may also be necessary to overcome the significant activation energy barrier associated with their electrophilic attack on weak nucleophilic substrates. Nitrobenzene, for example, itself is almost completely protonated in superacidic media (protonated nitrobenzene, $pK_a - 11.3$), and, therefore, electrophilic reactions with nitrobenzene may require that the superelectrophilic reagent collide with a protonated nitrobenzene. Higher temperatures may be necessary to overcome the electrostatic repulsive effects in such reactions.

While many superelectrophilic reactions are accomplished at ambient or somewhat elevated temperatures, a significant number of conversions are found to work better at lower temperatures. As some of these reactions employ gaseous reagents such as HF, BF_3, and low molecular weight alkanes, the lower temperatures may be necessary to help to keep the reagents in the condensed phase. In other reactions, the lower temperatures are used to control the excessive reactivities of the superelectrophiles

Table 3. Superelectrophilic reactions requiring more elevated temperatures.

Entry	Substrate	Acid	Temperature	Nucleophile	Proposed Superelectrophile	Product	Yield
(1)		$H_2O–BF_3$	105°C				69%
(2)		$AlCl_3$	110°C				70%
(3)		HUSY	130°C	phenyl			90%
		$H_3PO_4–P_2O_5$	130°C	phenyl			90%
		CF_3SO_3H	25°C	phenyl			85%
(4)	HNO_3	$2CF_3SO_3H–$ $B(O_3SCF_3)_3$	65°C				96%
(5)		CF_3SO_3H	130°C				10%
(6)		CF_3SO_3H	80°C				60%
(7)		CF_3SO_3H	80°C	phenyl			84%
(8)		CF_3SO_3H	150°C	phenyl			69%

and increase selectivity. For example, a number of superelectrophilic systems are based upon halogenated methyl cations, such as CCl_4-$nAlCl_3$, CCl_4-$nSbF_5$, and CBr_4-$nAlBr_3$. These systems tend to form very reactive trihalomethylcation-Lewis acid complex (**44**) and have been used in a variety of reactions of alkanes and cycloalkanes at lower temperatures.

The effect of temperature is clearly seen in the iodination of alkanes and cycloalkanes with CCl_4-AlI_3-I_2, where only monoiodination is observed at $-20°C$, but further products are also obtained at $0°C$ (eq 37).[61]

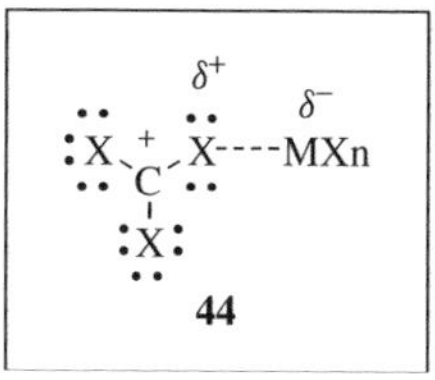

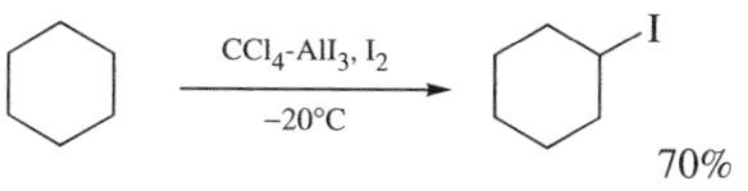

$$(37)$$

In other conversions, better yields (Scheme 2) and product selectivity (Scheme 3) are obtained at lower temperatures.[62] In the case of the Koch-Haaf butane carbonylation, it was noted the importance of temperature on the equilibria between the *sec*-butyl and *tert*-butyl cations and their derived acyl cations. Besides providing better yields and product

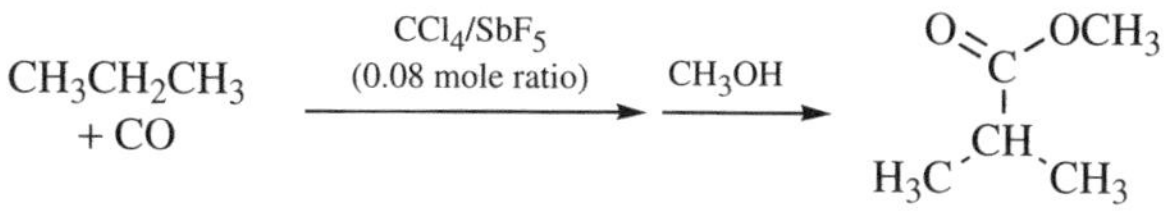

Temp	Yield
$-30°C$	50%
$-10°C$	28%

Scheme 2.

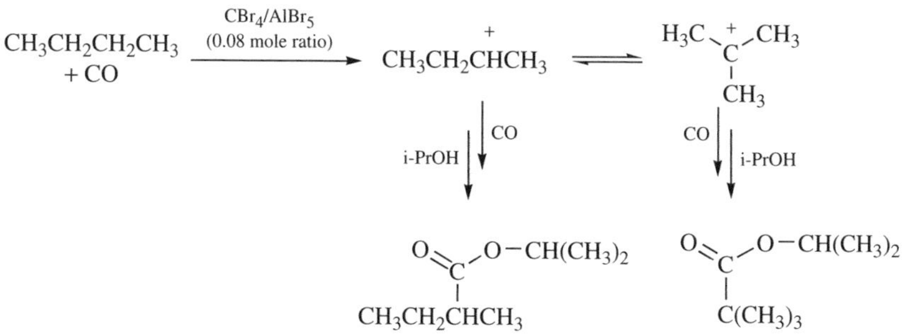

Temp	Product Ratio	
$-20°C$	5.25	1.00
$0°C$	1.29	1.00

Scheme 3.

selectivity, the lower temperatures also prevent cracking and oligomerization reactions which are known to occur with trihalomethyl cation-Lewis
acid complexes and saturated hydrocarbons. Protosolvated nitronium salts
(**6**) are exceptionally reactive electrophiles. A number of synthetic conversions with superelectrophilic nitronium salts have been carried out at lower
temperatures, often to prevent polynitration of the aromatic substrates.
Nitrobenzene and 1-nitronaphthalene can both be prepared effectively at
low temperatures using superelectrophilic nitrations (eqs 38–39).[63] Good
positional selectivity is seen in the low temperature nitration of naphthalene. With the use of a single equivalent of nitronium salt (or its precursor),
di and trinitration is suppressed by lower temperatures. This is also seen
in the superelectrophilc nitration of the trityl cation (eq 40).[46]

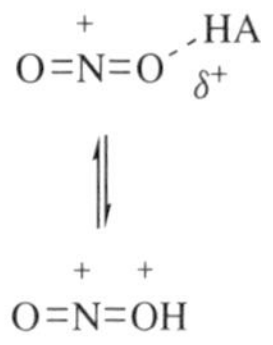

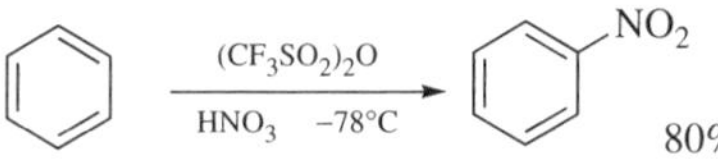

(38)

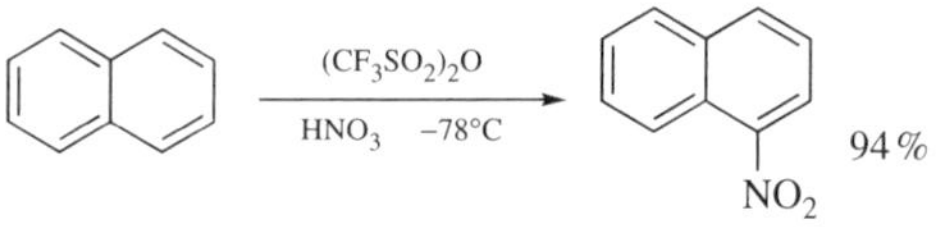

(39)

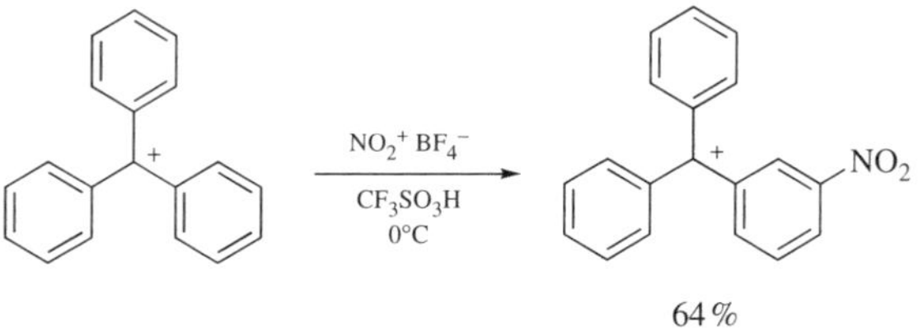

(40)

Several conversions based on superelectrophilic hydroxycarbenium ions
(such as **45**) were also carried out at lower temperatures, including the
condensation reaction (eq 41) and electrocyclization (eq 42).[8, 13] A similar electrocyclization reaction with a benzilic acid methyl ester was also
found to proceed in high yield at $-40°C$ (eq 43).[22]

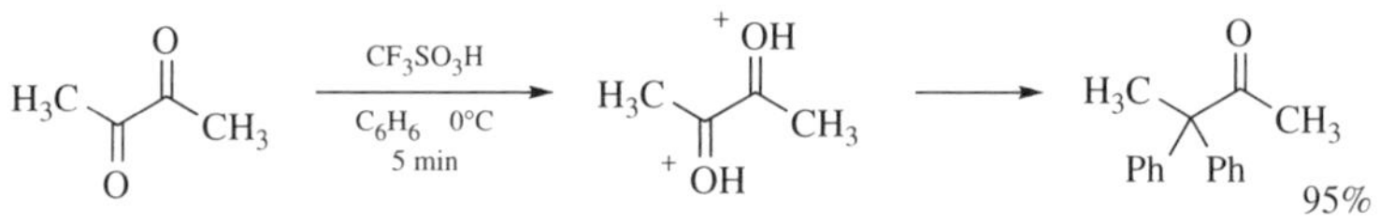

(41)

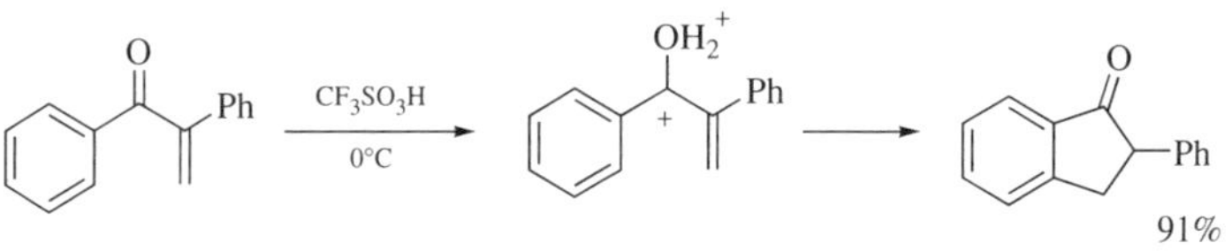

(42)

(43)

46

The reactivity of superelectrophile **46** precludes its direct observation by NMR, but as discussed in Chapter 2, stabilization of the carbenium ion center with *para*-methoxyphenyl groups allowed the dicationic species (**5**) to be observed at low temperature.

As discussed previously, several types of reactive dications and super-electrophiles have been directly observed using NMR spectroscopy. These experiments have all used low temperatures ($-100°C$ to $-30°C$) and superacidic conditions to generate the observable reactive dications and superelectrophiles. Some reactive dications and superelectrophiles are stable at low temperatures and can be directly observed by NMR, but at higher temperatures they readily cleave and decompose. The low temperatures also slow down proton exchange reactions and enable the ions to be observed as static species.

Lower temperatures were also an important aspect of other studies of superelectrophilic chemistry. For example, Olah and co-workers studied the role of superelectrophiles in the acid-catalyzed cleavage of esters.[34] One of the key experiments was carried out under highly acidic conditions and at $-40°C$ to prevent nucleophilic attack of monocationic intermediates (eq 44).

47   **48**

(44)

Under these conditions, it is highly unlikely that exchange of the $-CH_3$ group for $-CD_3$ could occur by nucleophilic attack on the monocation (**47**). The superelectrophile (**48**) is indicated to be involved in the exchange reaction. In studies related to the protosolvation of carbocations, very high acidities and low temperatures were used to probe the proton-deuterium exchange of carbocationic electrophiles (Scheme 4 and *vide supra*).[64] Low temperature "stable ion" conditions are important to prepare the 2-propyl cation (**49**). Despite the low temperatures, isotopic exchange was observed, indicating involvement of the protosolvated superelectrophile

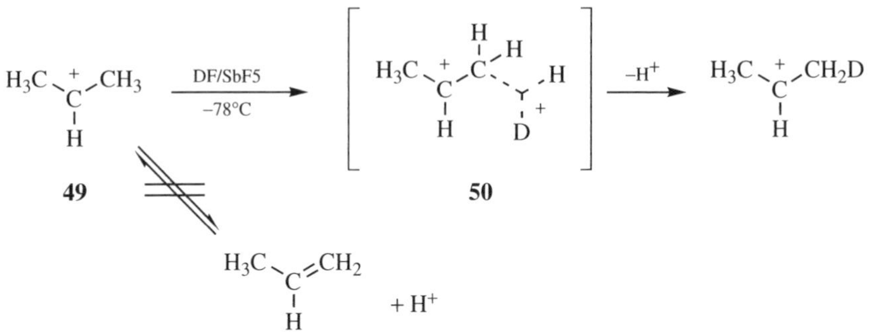

Scheme 4.

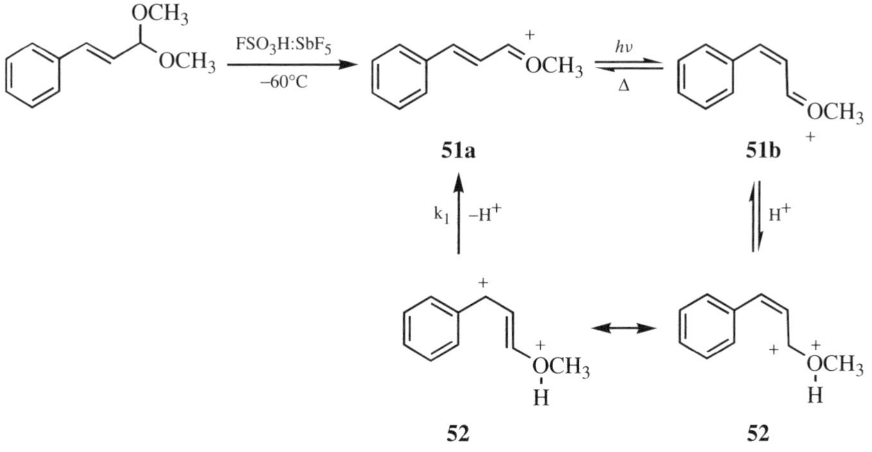

Scheme 5.

(**50**; see also related study of *tert*-butyl cation[44]). The very high acidities and low temperatures prevent any deprotonation-deuteration equilibrium involving propylene. As discussed briefly in Chapter 2, low tempera-ture was also important in the study of a protosolvated carboxonium ion (Scheme 5).[33] At $-60°C$, a photostationary state is found for the conju-gated oxonium ion **51**. As the temperature is increased, the Z (**51b**) -> E (**51a**) isomerization rate is increased. The isomerization is also found to occur at an increasing rate at higher acidities. In FSO_3H at $-5°C$, the stereomutation rate constant is found to be 2.0×10^{-4} s^{-1}, while in FSO_3H-SbF_5 (4:1) at $-5°C$, the rate constant is measured to be 6.9×10^{-3} s^{-1}. This indicates the involvement of dicationic intermediate (**52**). Delocalization of the positive charge in the dicationic structure is thought to enhance the isomerization rate.

In summary, superelectrophilic systems are usually generated with an excess of Brønsted and Lewis superacids, although examples are known in which under suitable conditions superelectrophiles are formed by the multidentate interaction of less strong acids. Multidentate interactions can also lead to doubly electron deficient systems exhibiting superelectrophilic chemistry. Besides the highly acidic conditions, solvents for the superelectrophilic reactions must be themselves of low nucleophilicity. Superelectrophiles are typically generated from the same precursors and functional groups that form the parent electrophiles. Indeed, the discussed superelectrophiles are usually in equilibrium with their corresponding parent electrophiles. Lower temperatures may be needed in superelectrophilic reactions to control excessive, unselective reactivities. Higher temperatures are found important, however, in the reactions of superelectrophiles with weak nucleophiles.

REFERENCES

(1) G. A. Olah; K. K. Laali; Q. Wang; G. K. S. Prakash *Onium Ions*, Wiley, New York, **1998**.

(2) G. A. Olah; G. K. S. Prakash; J. Sommer, in *Superacids*; Wiley, New York, **1985**.

(3) P. J. Stang; M. R. White *Aldrichimica Acta* **1983**, *16*, 15.

(4) G. A. Olah; K. K. Laali; G. K. S. Sandford *Proc. Nat. Acad. Sci. USA* **1992**, *89*, 6670.

(5) (a) T. Ohwada; A. Itai; T. Ohta; K. Shudu *J. Am. Chem. Soc.* **1987**, *109*, 7036. (b) T. Ohwada; K. Okabe; T. Ohta; K. Shudo *Tetrahedron* **1990**, *46*, 7539.

(6) T. Ohta; K. Shudo; T. Okamoto *Tetrahedron Lett.* **1984**, *25*, 325.

(7) Yokoyama, A.; T. Ohwada; K. Shudo *J. Org. Chem.* **1999**, *64*, 611.

(8) T. Yamazaki; S.-i. Saito; T. Ohwada; K. Shudo *Tetrahedron Letters* **1995**, 36, (32), 5749.

(9) D. A. Klumpp; M. Garza; S. Lau; B. Shick; K. Kantardjieff *J. Org. Chem.* **1999**, *64*, 7635.

(10) D. A. Klumpp; K. Y. Yeung; G. K. S. Prakash; G. A. Olah *J. Org. Chem.* **1998**, *63*, 4481.

(11) T. Ohwada; N. Yamagata; K. Shudo *J. Am. Chem. Soc.* **1991**, *113*, 1364.

(12) K. Y. Koltunov; S. Walspurger; J. Sommer *Chem. Comm.* **2004**, 1754.

(13) T. Suzuki; T. Ohwada; K. Shudo *J. Am. Chem. Soc.* **1997**, *119*, 6774.

(14) S. Saito; Y. Sato; T. Ohwada; K. Shudo *J. Am. Chem. Soc.* **1994**, *116*, 2312.

(15) D. A. Klumpp; D. N. Baek; G. K. S. Prakash; G. A. Olah *J. Org. Chem.* **1997**, *62*, 6666.

(16) D. A. Klumpp; S. Fredrick; S. Lau; G. K. S. Prakash; K. K. Jin; R. Bau;
      G. A. Olah *J. Org. Chem*. **1999**, *64*, 5152.

(17) R. Rendy; Y. Zhang; A. McElrea; A. Gomez; D. A. Klumpp *J. Org. Chem*.
      2004, *69*, 2340.

(18) D. A. Klumpp; R. Rendy; Y. Zhang; A. Gomez; A. McElrea *Org. Lett*.
      2004, *6*, 1789.

(19) D. A. Klumpp; R. Rendy; A. McElrea *Tetrahedron Lett*. **2004**, *45*, 7959.

(20) J. P. Hwang; G. K. S. Prakash; G. A. Olah *Tetrahedron* **2000**, *56*, 7199.

(21) Y. Sato; M. Yato; T. Ohwada; S. Saito; K. Shudo *J. Am. Chem. Soc*. **1995**,
      117, 3037.

(22) T. Ohwada; T. Suzuki; K. Shudo *J. Am. Chem. Soc*. **1998**, *120*, 4629.

(23) G. A. Olah; A. Orlinkov; A. B. Oxyzoglou; G. K. S. Prakash *J. Org. Chem*.
      1995, *60*, 7348.

(24) G. A. Olah; G. Rasul; A. Burrichter; G. K. S. Prakash *Proc. Nat. Acad. Sci.
      USA* **1998**, *95*, 4099.

(25) O. Farooq; M. Marcelli; G. K. S. Prakash; G. A. Olah *J. Am. Chem. Soc*.
      1998, *120*, 4629.

(26) G. A. Olah; A.-H. Wu *Synlett* **1990**, 599.

(27) S. Saito; S. Saito; T. Ohwada; K. Shudo *Chem. Pharm. Bull*. **1991**, *39(10)*,
      2718.

(28) (a) G. A. Olah; H. C. Lin *J. Am. Chem. Soc*. **1976**, *96*, 549. (b) G. A. Olah;
      H. C. Lin *Synthesis* **1974**, 444.

(29) S. Walspurger; A. V. Vasilyev; J. Sommer; P. Pale *Tetrahedron* **2005**, *61*,
      3559.

(30) (a) G. A. Olah; J. R. De Member; Y. K. Mo; J. J. Svoboda; P. Schilling;
      J. A. Olah *J. Am. Chem. Soc*. **1974**, *96*, 884. (b) K. Laali; H. Y. Chen;
      R. J. Gerzina *J. Organomet. Chem*. **1988**, *348*, 199.

(31) G. A. Olah; G. K. S. Prakash; M Barzaghi; K. Lammertsma; P. v. R.
      Schleyer; J. A. Pople *J. Am. Chem. Soc*. **1986**, *108*, 1032.   (b) G. A. Olah;
      G. K. S. Prakash; M. Marcelli; K. Lammertsma *J. Phys. Chem*. **1988**, *92*,
      878.

(32) I. B. Repinskaya; K. Y. Koltunov; M. M. Shakirov; V. A. Koptyug *Zhur.
      Org. Khim*. **1992**, *28*, 1013.

(33) C. Blackburn; R. F. Childs *J. Chem. Soc., Chem. Commun*. **1984**, 812.

(34) G. A. Olah; N. Hartz; G. Rasul; A. Burrichter; G. K. S. Prakash *J. Am.
      Chem. Soc*. **1995**, *117*, 6421.

(35) G. Rasul; G. K. S. Prakash; G. A. Olah *J. Org. Chem*. **1994**, *59*, 2552.

(36) (a) D. M. Brouwer; A. A. Kiffen *Recl. Trav. Chim. Pays-Bas* **1973**, *92*,
      809. (b) D. M. Brouwer; A. A. Kiffen *Recl. Trav. Chim. Pays-Bas* **1973**,
      92, 906.

(37) Olah, G. A.; Germain, A.; Lin, H. C.; Forsyth, D. A. *J. Am. Chem. Soc*.
      1975, *97*, 2928

(38) D. M. Brouwer; A. A. Kiffen *Recl. Trav. Chim. Pays-Bas* **1973**, *92*, 689.

(39) (a) G. A. Olah; G. K. S. Prakash; T. Mathew; E. R. Marinez *Angew. Chem., Int. Ed.* **2000**, *39*, 2547. (b) G. A. Olah; T. Mathew; E. R. Marinez; P. M. Esteves; M. Etzkorn; G. Rasul; G. K. S. Prakash; *J. Am. Chem. Soc.* **2001**, *123*, 11556.

(40) A. Bagno; J. Bukala; G. A. Olah *J. Org. Chem.* **1990**, *55*, 4284.

(41) G. A. Olah; G. K. S. Prakash; M Barzaghi; K. Lammertsma; P. v. R. Schleyer; J. A. Pople *J. Am. Chem. Soc.* **1986**, *108*, 1032.

(42) G. K. S. Prakash; G. Rasul; A. Burrichter; K. K. Laali; G. A. Olah *J. Org. Chem.* **1996**, *61*, 9253.

(43) I. B. Repinskaya; M. M. Shakirov; K. Y. Koltunov; V. A. Koptyug *Zhur. Org. Khim.* **1988**, *24*, 1907.

(44) G. A. Olah; N. Hartz; G. Rasul; G. K. S. Prakash *J. Am. Chem. Soc.* **1993**, *115*, 6985.

(45) (a) J. C. Culman; M. Simon; J. Sommer *J. Chem. Soc., Chem Commun.* **1990**, 1098. (b) A. Martin; M.-P. Jouannetaud; J.-C. Jacquesy *Tetrahedron Lett.* **1996**, *37*, 2967. (c) J. Sommer; J. Bukala *Acc. Chem. Res.* **1993**, *26*, 370.

(46) (a) G. A. Olah; Q. Wang; A. Orlinkov; P. Ramaiah *J. Org. Chem.* **1993**, *58*, 5017. (b) G. A. Olah; A. Orlinkov; P. Ramaiah; A. B. Oxyzoglou; G. K. S. Prakash *Russ. Chem. Bull.* **1998**, *47*, 924.

(47) M. Vol'pin; I. Akhrem; A. Orlinkov *New J. Chem.* **1989**, *13*, 771.

(48) G. A. Olah; G. Rasul; A. K. Yudin; A. Burrichter; G. K. S. Prakash; A. L. Chistyakov; I. V. Stankevich; I. S. Akhrem; N. P. Gambaryan; M. E. Vol'pin *J. Am. Chem. Soc.* **1996**, *118*, 1446.

(49) G. A. Olah; Q. Wang; G. Neyer *Synthesis*, **1994**, 276.

(50) G. A. Olah; B. Torok; J. P. Joschek; I. Bucsi; P. M. Esteves; G. Rasul; G. K. S. Prakash *J. Am. Chem. Soc.* **2002**, *124*, 11379.

(51) K. Y. Koltunov; I. B. Repinskaya *Zhur. Org. Khim.* **1995**, 31, 1723.

(52) (a) K. Y. Koltunov; S. Walspurger; J. Sommer *Tetrahedron Lett.* **2004**, *45*, 3547. (b) K. Y. Koltunov; S. Walspurger; J. Sommer *Eur. J. Org. Chem.* **2004**, 4039.

(53) K. Y. Koltunov; L. A. Ostashevskaya; I. B. Repinskaya *Russ. J. Org. Chem. (Engl. Transl.)* **1998**, *34*, 1796.

(54) G. K. S. Prakash; T. Mathew; D. Hoole; P. M. Esteves; Q. Wang; G. Rasul; G. A. Olah *J. Am. Chem. Soc.* **2004**, *126*, 15770.

(55) Z. Tang; F. Jiang; L.-T. Yu, X. Cui; X. Cui; L.-Z. Gong; A.-Q. Mi; Y.-Z. Jiang; Y.-D. Wu *J. Am. Chem. Soc.* **2003**, *125*, 5262.

(56) S. Saito; T. Ohwada; K. Shudo *J. Am. Chem. Soc.* **1995**, *117*, 11081.

(57) A. Berkessel; R. K. Thauer *Angew. Chem. Int. Ed. Engl.* **1995**, *34*, 2247.

(58) K. Y. Koltunov; G. K. S. Prakash; G. Rasul; G. A. Olah *Tetrahedron* **2002**, *58*, 5423.

(59) D. A. Klumpp; S. Lau *J. Org. Chem.* **1999**, *64*, 7309.

(60) D. A. Klumpp, unpublished results.

(61) I. Akhrem; A. Orlinkov; S. Vitt; A. Chistyakov *Tetrahedron Lett.* **2002**, *43*, 1333.

(62) I. Akhrem; A. Orlinkov; L. Afanas'eva, P. Petrovskii; S. Vitt *Tetrahedron Lett.* **1999**, *40*, 5897.

(63) G. A. Olah; V. Prakash Reddy; G. K. S. Prakash *Synthesis* **1992**, 1087.

(64) G. A. Olah; N. Hartz; G. Rasul; G. K. S. Prakash; M. Burkhart; K. Lammertsma *J. Am. Chem. Soc.* **1994**, *116*, 3187.

4

GITONIC GEMINAL SUPERELECTROPHILES

4.1 STRUCTURAL CONSIDERATIONS

As noted in the first chapter, superelectrophiles are divided into two basic categories: *gitonic* (close) and *distonic* (distant) superelectrophiles. The distonic superelectrophiles are characterized by structures having two or more carbon or hetero atoms separating the positive charge centers, while the gitonic superelectrophiles have the two charges in close proximity. Gitonic superelectrophiles thus can be characterized according to the distance between the two (or more) charge centers. This chapter discusses systems in which the charges are located on (or around) the same atom (i.e., H_4O^{2+}). These are referred to as geminal systems. Chapter 5 deals with vicinal superelectrophiles having greater charge separation corresponding to 1,2-dicationic systems. Distonic superelectrophiles are discussed in Chapter 7.

Although gitonic superelectrophiles may be viewed as closely located di- or polycationic systems, in fact the positive charges are often strongly delocalized through resonance, conjugative, and inductive interactions with involved ligands. For example, in the case of water, natural bond analysis (NBO) derived atomic charges from DFT calculations show charges of -0.916 and $+0.458$ on the oxygen and hydrogen atoms,[1]

Superelectrophiles and Their Chemistry, by George A. Olah and Douglas A. Klumpp
Copyright © 2008 John Wiley & Sons, Inc.

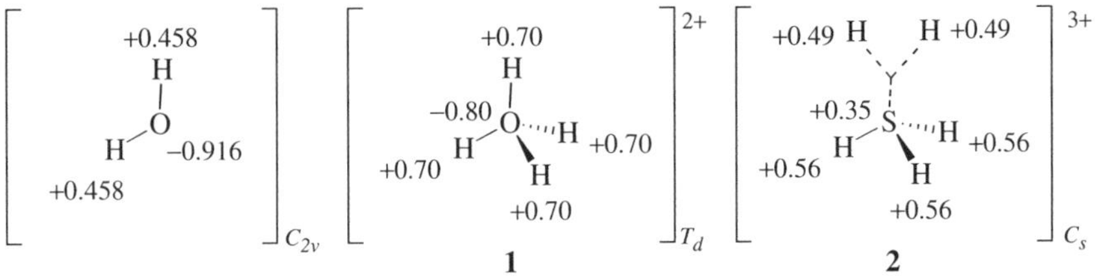

Figure 1. Calculated NBO charges on water, diprotonated water (**1**), and pentahydrido-sulfonium trication (**2**).

respectively, while similar calculations show charges of -0.80 and $+0.70$ on the oxygen and hydrogen atoms for the gitonic superelectrophilic, diprotonated water (H_4O^{2+}, **1**, Figure 1).[2] Similarly, theoretical studies of the pentacoordinate sulfonium trication H_5S^{3+} (**2**) have shown that most of the positive charge is delocalized on the hydrogen atoms.[3] It has also been noted that in systems like H_4O^{2+} and H_5S^{3+}, the distribution of the positive charge on the periphery atoms leads to a large decrease in the quantum mechanical Coulombic repulsion energy.[4]

In a similar sense, charge delocalization through resonance-type contributions is important in the nature of some superelectrophiles and in the understanding of their chemistry. For example, the superelectrophilic Nazarov-cyclization is also considered to involve delocalization of positive charge in a gitonic superelectrophile (Scheme 1).[5] Although double protonation of the carbonyl group can be represented as the geminal-type dication **4a**, this is only a minor contributing resonance structure. The *de facto* best representation of this superelectrophile involves the 1,2-dication **4b** and delocalized structure **4c**, which leads directly to the transition state structure of the 4π-electrocyclization (**5**). Compared with a cyclization

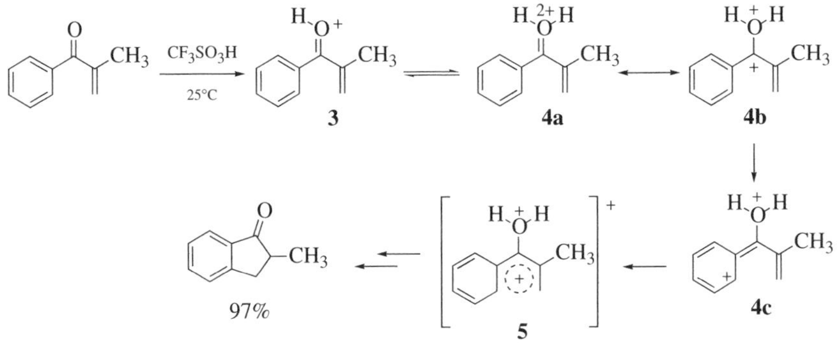

Scheme 1.

reaction involving the monoprotonated intermediate (**3**), it was shown (by *ab initio* calculations) that the reaction involving the superelectrophile has a significantly lower energy barrier to the cyclization (*vide supra*).[5]

It should be clear when considering superelectrophilic chemistry that fully formed dicationic superelectrophiles are only the limiting case. There is always a varying degree of protolytic (or Lewis acids complexating) interaction involved in superelectrophilic condensed phase reactions (Figure 2).[6] While much evidence has been accumulated for the *de facto* role of varied superelectrophiles, their direct observation (in most cases) has not yet been accomplished in the condensed phase. Besides being energetic, high lying and therefore by necessity, short-lived species, this could also suggest either incomplete protonation to a superelectrophilic state (i.e., **6**), formation of low concentrations of fully formed dicationic superelectrophiles (**7**), or an equilibium system involving both structures. Based on the results of kinetic studies, Shudo and Ohwada have found evidence to suggest the varying degree of protonation of superelectrophiles (*vide infra*). In one of the studied systems, it was proposed that proton transfer at the transition state is about 50%.[7]

In the following, we describe geminal gitonic superelectrophiles, including carbo, azo, oxo and sulfo, and halodicationic gitonic superelectrophiles. There is even evidence for the protonation and alkylation of noble gases and thus superelectrophiles based on them are a possibility.[8] Though doubly ionized noble gases (i.e., Xe^{2+}) are well known, there have been no reports yet describing geminal gitonic superelectrophiles centered at noble gases, although they have been studied by *ab initio* calculations.[4] As noted in Chapter 1, metal atoms in high oxidation states are not considered geminal gitonic superelectrophiles within the context of this discussion.

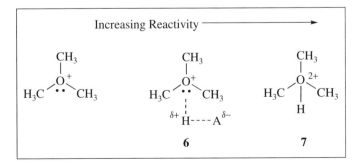

Figure 2. Degree of protosolvation of the oxonium cation.

4.2   GEMINAL SYSTEMS

4.2.1   Geminal Carbodications

One of the foundations of organic chemistry is Kekulé's concept (suggested independently by Couper) of the tetravalency of carbon. The structure of higher bonded (coordinated) carbon systems, such as alkyl-bridged organometallics $[Al(CH_3)_2]_2$ or carbonium ions of which CH_5^+ is the parent, however, cannot be explained by Kekulé's four-valent concept involving only two electron-two center bonding (2e-2c). Extensive experimental and theoretical studies have provided evidence for the involvement of two-electron three center (2e-3c) bonding in such systems showing that carbon atoms can be simultaneously coordinate to more than four groups or atoms.[9] Diprotonated methane, CH_6^{2+} (**8**) or its analogues can be considered as gitonic superelectrophiles with multiple 2e-3c bonding. *Ab initio* calculations at the HF/6-31G* level have shown CH_6^{2+} to have a C_{2v} symmetrical structure, with two orthogonal 2e-3c interactions. It was shown to be kinetically stable, with a calculated barrier of 63 kcal/mol to deprotonation.[10] Despite this kinetic stability and the fact that CH_5^+ is readily observed in gas-phase studies, CH_6^{2+} has not yet been observed in condensed phase studies. However, a remarkably stable analogous gold complex (**9**) has been prepared and even isolated as a crystalline compound allowing X-ray structure determination by Schmidbaur and co-workers.[11] Compound **9** may be considered an isolobal analog of the CH_6^{2+} ion. In addition to CH_6^{2+}, the CH_7^{3+} ion (**10**, triprotonated methane) has been studied computationally at the MP2/6-31G** level.[12]

The *ab initio* calculations determined the C_{3v} structure to be a kinetically stable minimum.

Varied one carbon geminal dications have been generated in the gas-phase and studied by theoretical methods. For example, mass spectroscopic studies of methane have detected dicationic species, CH_4^{2+}, $CH_3\bullet^{2+}$, CH_2^{2+}, and $CH\bullet^{2+}$, while other studies have examined the structures of halogenated one carbon geminal dications.[13] The halogenated systems

(**11a-b**) have been studied by both theoretical and experimental methods.[14]

$$\left[\text{H--C--F} \right]^{2+} \qquad \left[\text{H--C--Cl} \right]^{2+} \qquad \left[\text{H--C--Br} \right]^{2+} \qquad \left[\text{H--C--I} \right]^{2+}$$

$$\qquad \textbf{11a} \qquad\qquad\qquad \textbf{11b} \qquad\qquad\qquad \textbf{11c} \qquad\qquad\qquad \textbf{11d}$$

In gas-phase studies, these dicationic species were found to react by competing electron and proton transfer routes.[15] These studies found that for ion-molecule reactions involving CHX^{2+} and nonpolar molecules (N_2 and O_2) or rare gas atoms, the electron transfer reaction dominates, at the expense of proton transfer, if the electron transfer process is significantly exothermic ($>2\,eV$). However, in reactions involving CHX^{2+} species and polar molecules (H_2O, CO, and HCl), proton transfer becomes far more important. The difference in reaction paths was attributed to an increasing stabilization of the encounter complexes involving the CHX^{2+} species and polar molecules. This stabilization tends to favor the formation of the thermodynamic products from proton transfer, versus the kinetic products from electron transfer reaction.

In other theoretical and experimental studies, CF_2^{2+} and CF_3^{2+} have been examined for their gas-phase chemistry.[16] The gas-phase reaction of CF_2^{2+} with molecular hydrogen has been extensively studied as was briefly mentioned in Chapter 2.[16a] In chemistry that is characteristic of several superelectrophilic systems (*vide infra*), the gas-phase reaction of CF_2^{2+} with hydrogen leads to the formation CHF_2^+ and H^+, products in which the two formal positive charges are separated by a fragmentation process. Other reaction pathways have been described for gas-phase collisions involving CF_2^{2+} and diatomic molecules, including non-dissociative electron transfer, dissociative electron transfer, and collision induced charge separation.[16a] The carbodication CCl_2^{2+} (**13a,b**) may be involved in the conversion of trichlorofluoromethane to methylene chloride by the action of SbF_5 in the presence of a hydride donor.[17] Such a reaction would involve superelectrophilic solvation of the trichloromethyl cation by SbF_5 (Scheme 2), producing an equilibrium involving the incipient trichloromethyl dication (**12**) and the geminal dication (**13a,b**). Subsequent reactions with the hydride donor then give the reduced product. The CCl_2^{2+} ion (**13a,b**) has also been studied by theoretical calculations and mass-spectroscopic experiments.[18] Calculations have estimated the carbon-halogen bond length to be quite short, indicating a significant amount of charge delocalization (i.e., **13b**). Interestingly, superelectrophile **13** is isoelectronic with carbon disulfide (S=C=S).

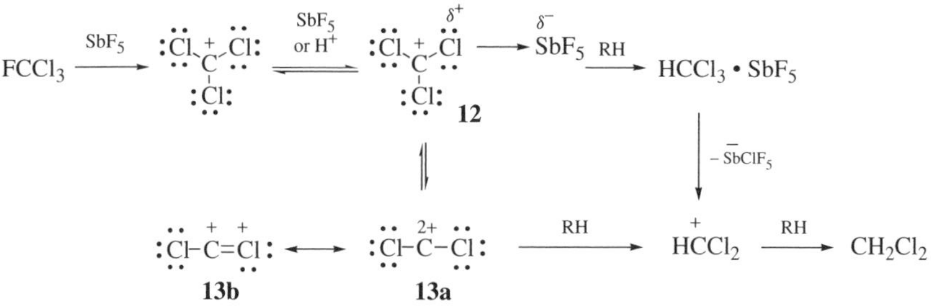

Scheme 2.

Table 1. Observed gas-phase one-carbon dications and trications

Dicationic Species	Tricationic Species
$CCl_2 \cdot Cl_2^{2+}$, CF_3^{2+} $CHCl_3^{2+}$, CCl_2^{2+}, CBr_2^{2+}, CI_2^{2+}, $CH_2Cl_2^{2+}$, CH_3Cl^{2+}, CO^{2+}, CO_2^{2+}, CS_2^{2+}, CF^{2+}, CS_2^{2+}	COS^{3+}, CS_2^{3+}, CSe_2^{3+}, CS^{3+}

Gas-phase experiments have been used to detect a variety of doubly and triply-charged one carbon species (Table 1).[19] While many of these species presently do not have a direct analog in condensed phase chemistry, studies of their generation, stability (including determinations of electronic states, structures, and energies), and gas-phase reactivities may provide further insight towards the behavior of gitonic superelectrophiles, especially geminal-type systems.

4.2.2 Geminal Azodications

The formal expansion of the valence octet involving hypervalent and hypercoordinate first row elements has been a subject of considerable interest. Some of these species can be directly observed (i.e., BH_5 and CH_5^+). The gitonic superelectrophile NH_5^{2+} (**14**; doubly protonated ammonia) has been studied by experimental and theoretical methods.[20] At both MP2(fu)/6-31G** and QCISD/6-311G** levels of theory, structure **14** (C_{4v} symmetry) is found to be a minimum on the potential energy surface. The dissociation of **14** into NH_4^+ and H^+ is estimated to be exothermic by 99.1 kcal/mol, however, the gitonic species (**14**) has a considerable kinetic barrier to dissociation, estimated to be about 25 kcal/mol. Interestingly, the same study suggests the possibility of directly observing NH_5^{2+} in mass-spectrometry experiments, as theoretical calculations indicate that the NH_3^{2+} species (**16**) reacts with molecular hydrogen in a highly exothermic reaction (eq 1).

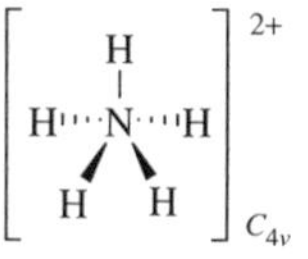

$$NH_3^{2+} + H_2 \xrightarrow{\Delta H = -143 \, kcal/mol} NH_5^{2+} \tag{1}$$
$$\mathbf{16} \qquad\qquad\qquad\qquad\qquad \mathbf{14}$$

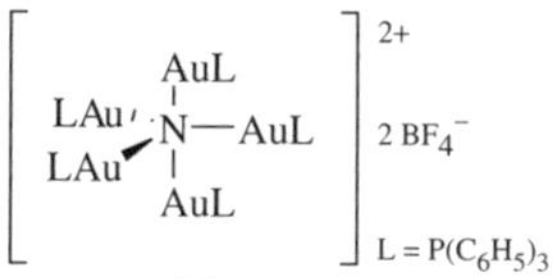

No evidence for the gitonic superelectrophile **14** has yet been obtained from condensed phase superacidic experiments. In experimental studies, ammonium chloride $NH_4^+Cl^-$ was reacted with $DSO_3F:SbF_5$ in solution at $100°C$ for 14 days, but no hydrogen/deuterium exchange was observed. However, a gold (I) stabilized species (**15**) analogous to **14** has been prepared and characterized by X-ray crystallography in work by Schmidbaur and associates.[21] It was found that the central nitrogen atom is coordinated to five $(C_6H_5)_3PAu$ ligands in a trigonal-bipyramid geometry.

Besides the hypervalent or hypercoordinate gitonic ammonium dications such **14**, there is considerable experimental and theoretical evidence for trivalent, dicationic nitrogen species, of which NH_3^{2+} (**16**) can be considered the parent system. The NH_3^{2+} dication (**16**) itself has been observed in the gas-phase by charge-transfer mass spectroscopy.[22] Related gitonic superelectrophiles have also been generated in the condensed phase by superacid-promoted reactions of hydroxylanilines, aniline *N*-oxides, and nitrosobenzene.[23] For example, when *N*-phenylhydroxylamine (**17**) is reacted in superacid with benzene, diphenylamine (**18**) and the amino-biphenyls (**19**) are formed as products (Scheme 3).[23a] With CF_3CO_2H, the major product is diphenylamine (**18**), while in CF_3SO_3H the major

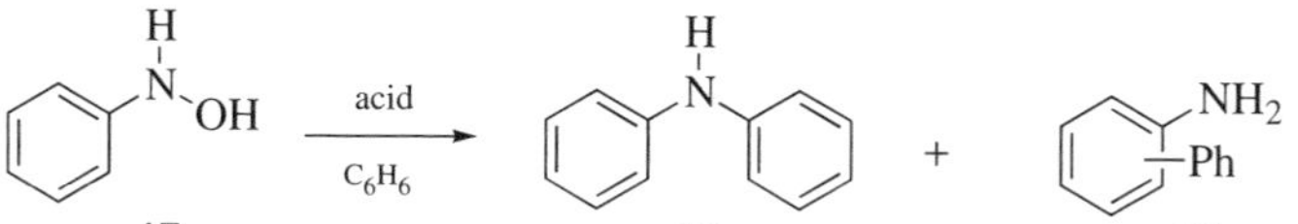

CF_3CO_2H:	56%	9% para
		8% ortho
CF_3SO_3H:	trace	43% para
		23% ortho

Scheme 3.

products are the aminobiphenyls (**19**). In the reaction with CF_3CO_2H ($H_0 -2.7$), it is proposed that monocationic intermediates (**20**, **21**, or **22**) lead to diphenylamine (**19**) (Scheme 4). Although the divalent anilenium ion (**21**) cannot be ruled out, the authors favor either an S_N2-like bimolecular substitution (A2) mechanism, or one that involves a cationic species having partial anilenium ion character (**22**). In the reaction with CF_3SO_3H ($H_0 -14.1$), dicationic intermediates are thought to be involved and lead to the biphenyl products **19** (Scheme 5). Protonation of both the N and O-sites provides the diprotonated species **23**, which gives a gitonic superelectrophile (or iminium-benzenium dication) **24** upon loss of H_2O. Electrostatic repulsive effects make structure **24b** the more important resonance contributor to the overall structure of **24**.The reaction of dication **24** with benzene produces intermediates **25** and **26**, and eventually leads to the aminobiphenyl products (**19**). Similarly, aniline N-oxide (**27**) and

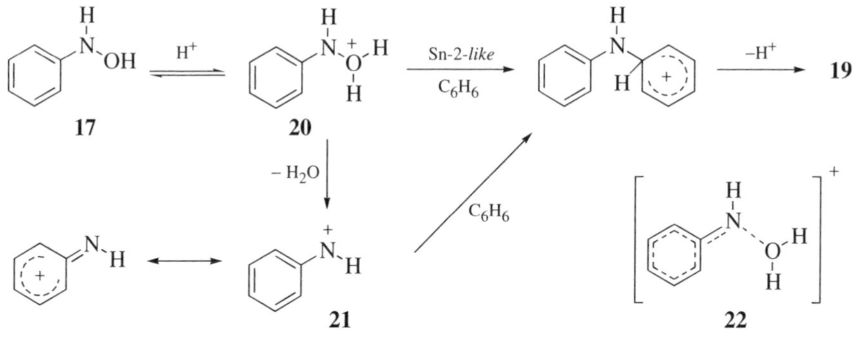

Scheme 4.

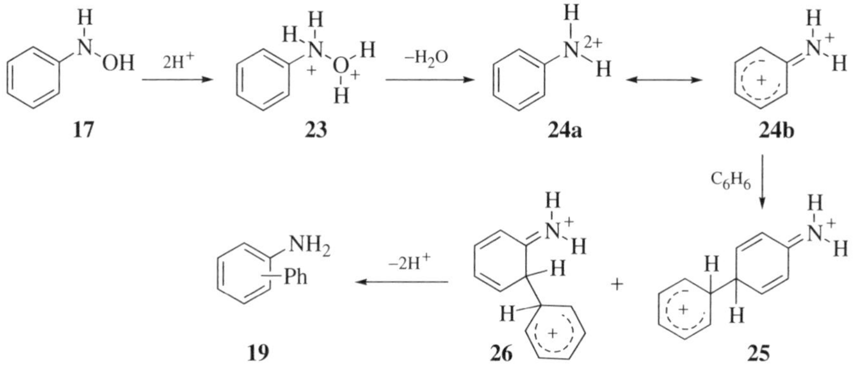

Scheme 5.

the nitrosobenzene (**29**) react in superacid to form superelectrophiles **28** and **30**, respectively (eqs 2–3).[23b] Both superelectrophiles (**28** and **29**) have been shown to react with benzene in electrophilic aromatic substitution reactions. Analogous chemistry was used in the superacid-promoted reaction of the *N*-oxide (**31**) in the preparation of the natural product alkaloid (**32**, eq 4).[23c]

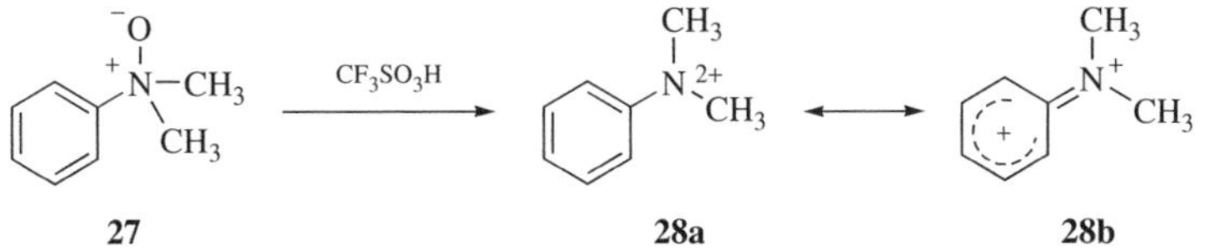

(2)

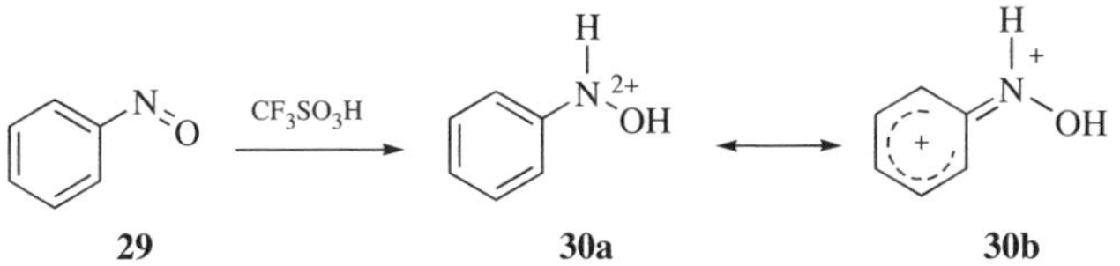

(3)

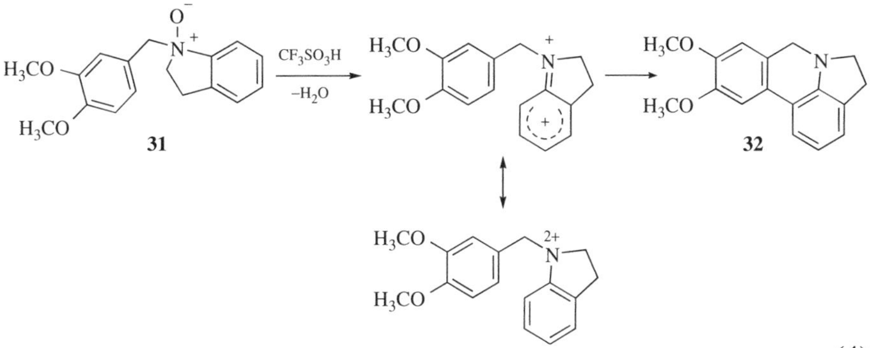

(4)

4.2.3   Geminal Oxo and Sulfodications

There are several types of oxygen-based geminal dications that have been studied by experiment and theory. A fundamentally important system is diprotonated water, i.e., H_4O^{2+}. In superacidic solution, the hydronium ion has been shown to be extremely stable by ^1H and ^{17}O NMR spectroscopy.[24] Isotopic hydronium ions have also been prepared from $HSO_3F{:}SbF_5-D_2O/SO_2ClF$.[25] The ions H_3O^+, H_2DO^+, and HD_2O^+, can be generated and characterized by ^1H and ^2H NMR spectroscopy, and there is no evidence for any deprotonation or hydrogen/deuterium exchange when the hydronium ions are formed in the superacid $HSO_3F{:}SbF_5$ (H_0 −21). However in even stronger superacidic medium (HF/DF–SbF₅,

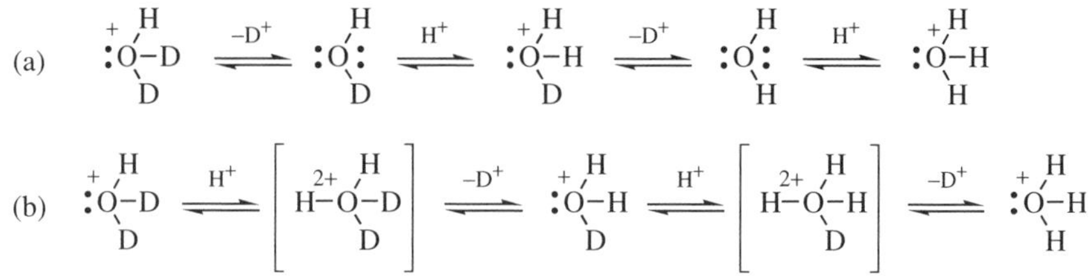

Scheme 6. Two possible mechanisms for H/D exchange in superacids.

$H_0 - 25$), hydrogen/deuterium exchange is observed.[26] ^{1}H, ^{2}H, and ^{17}O NMR spectroscopy was used to study the system indicating isotopomeric ions including $D_2H^{17}O^+$ and $DH_2^{17}O^+$ exchange through $D_2H_2^{17}O^{2+}$. Two mechanisms are possible for the observed H/D exchange (Scheme 6): (a) deprotonation-protonation equilibria involving neutral molecules and monocations or (b) protonation-deprotonation equilibria involving monocations exchanging H and D through H_4O^{2+} dications. Because H/D exchange only occurs at the highest acidities (i.e., with $HF/DF-SbF_5$ but not with $HSO_3F:SbF_5$) and deprotonation to water is not taking place in these systems, the results suggest that isotopic exchange takes place via isotopomeric H_4O^{2+} or corresponding protosolvated species ($H_3O^+..HA$) involving the nonbonded electron pair of the oxygen atom.

The observed H/D is in accord with theoretical studies by Schwarz et al. and Olah et al., respectively, which indicating H_4O^{2+} to be a kinetically stable species.[27] At the MP4SDTQ/6-311++G**//MP/6-31G** level, calculations show that H_4O^{2+} lies in a potential energy well with a barrier for proton loss estimated to be 39.6 kcal/mol in the gas-phase. Calculations at the HF/6-31G* level likewise indicate a deprotonation barrier of 39.4 kcal/mol. The optimized geometry is shown to possess T_d symmetry (isoelectronic with NH_4^+, CH_4, and BH_4^-) and much of the charge is located on the hydrogen atoms (calculated charges: hydrogen atoms +0.70, oxygen atom −0.80).[2] The GIAO-MP2 derived δ ^{17}O for H_4O^{2+} has also been calculated and it is estimated to be 39.2 ppm, which is 29 ppm deshielded from the experimentally observed ^{17}O chemical shift from H_3O^+.[27a] Despite the predicted kinetic stability of H_4O^{2+}, the ion has not yet been generated in gas-phase experiments, although $H_3O^{2+•}$ has been observed from charge stripping of the hydronium ion, H_3O^+.[27b] Other computational studies have investigated the structures and energies of the fluorooxonium dications, FOH_3^{2+} and $F_2OH_2^{2+}$.[2] Schmidbaur and co-workers have remarkably succeeded in preparing dipositive tetrahedral gold complexes of O^{2+} (**33**).[28]

$$\left[\begin{array}{c} \text{Ar}_3\text{PAu} \cdots \text{AuPAr}_3 \\ \text{O} \\ \text{Ar}_3\text{PAu} \cdots \text{AuPAr}_3 \\ \textbf{33} \end{array}\right]^{2+}$$

$$\text{Ar} = \text{C}_6\text{H}_5 \text{ or } (o\text{–CH}_3\text{C}_6\text{H}_4)$$

The gold complexes were characterized by X-ray crystallography and represent isolobal analogs of H_4O^{2+}.

Other superelectrophilic oxonium dication systems have also been studied. For example, trialkyloxonium salts (Meerwein salts, R_3O^+ X^-) in the presence of strong acids show substantial increase in their reactivities, indicating superelectrophilic activation.[29] Meerwein salts are excellent alkylating agents for nucleophiles containing heteroatoms, but in the absence of superacids, they are not capable of C-alkylating aromatic or aliphatic hydrocarbons. When superacids like FSO_3H-SbF_5 (Magic Acid, $H_0 -21$ to -24) or $CF_3SO_3H-B(O_3SCF_3)_3$ (triflatoboric acid, $H_0 -20$ to -21) are added, trialkyloxonium ions (**34**) readily alkylate aromatics such as benzene, toluene or chlorobenzene (eq 5).[30]

$$\underset{\textbf{34}}{H_3C \overset{\cdot\cdot}{\underset{\underset{CH_3}{|}}{\overset{+}{O}}} CH_3} \underset{}{\overset{FSO_3H-SbF_5}{\rightleftharpoons}} H_3C \overset{\overset{H}{|}}{\underset{\underset{CH_3}{|}}{O}} CH_3 \overset{2+}{} \overset{C_6H_6}{\longrightarrow} C_6H_5CH_3 + H_3C \overset{\cdot\cdot}{\underset{\underset{H}{|}}{\overset{+}{O}}} CH_3$$

(5)

Weaker acids do not catalyze the methylation. Despite that trialkyloxonium ions have a formal positive charge, the oxygen atom can still act as a Lewis base resulting in an interaction between the lone pair electrons on the oxygen and the highly acidic protosolvating system.

Protolytic activations can also be achieved in cases of primary and secondary oxonium ions. Primary alcohols, even methanol, and ethers form acidic oxonium ions with strong acids. Methanol is completely protonated in superacids and forms the methyloxonium ion (**35**) as a stable, well-defined oxonium ion.[31] The methyloxonium ion requires forcing conditions to C-alkylate arenes such as phenols. The methylation of aromatics including benzene is greatly facilitated by the use of solid or liquid superacids.[32] This is indicative of the activation of the methyloxonium ion through protosolvation of the nonbonded electron pair of the oxygen atom (**36** or **37**, eq 6), or alternatively it may suggest a protosolvated species (**39**) arising from the dimethyloxonium ion (**38**, eq 7).

$$H-\overset{..}{\underset{..}{O}}-CH_3 \;\underset{}{\overset{H^+}{\rightleftharpoons}}\; \underset{H}{\overset{H}{\underset{+}{\overset{..}{O}}}}-CH_3 \;\underset{}{\overset{HA}{\rightleftharpoons}}\; \underset{H}{\overset{H}{\underset{+}{\overset{\overset{\overset{\delta^-}{H--A}}{}}{\overset{.\,.\,\delta^+}{O}}}}}-CH_3 \;\rightleftharpoons\; \underset{H}{\overset{H}{H-\overset{2+}{O}}}-CH_3 \quad (6)$$

$$\qquad\qquad\qquad\quad \mathbf{35} \qquad\qquad\qquad\qquad \mathbf{36} \qquad\qquad\qquad \mathbf{37}$$

$$2\,CH_3OH \;\xrightarrow{-H_2O}\; CH_3OCH_3 \;\underset{}{\overset{H^+}{\rightleftharpoons}}\; \underset{H}{\overset{+}{CH_3OCH_3}} \;\underset{}{\overset{HA}{\rightleftharpoons}}\; \underset{H}{\overset{H}{\overset{2+}{CH_3OCH_3}}} \quad (7)$$

$$\qquad\qquad\qquad\qquad\qquad\qquad\qquad\qquad\qquad \mathbf{38} \qquad\qquad\quad \mathbf{39}$$

The methyloxonium ion $CH_3OH_2{}^+$ in superacidic media also readily undergoes ionic hydrogenation to give methane (eq 8),[33]

$$CH_3OH \;\xrightarrow[{-H_2}]{CF_3SO_3H-B(OSO_2CF_3)_3}\; CH_4 \quad (8)$$

again suggesting the formation of superelectrophilically activated species (**36** or **37**). Under forcing conditions in varied superacids (even polyphosphoric acid) methanol undergoes self-condensation to give C_3-C_9 alkanes, toluene, and C_8-C_{10} aromatics (Scheme 7).[34,35] The carbon-carbon bonds are initially formed by a reaction step involving the methyloxonium ion $CH_3OH_2{}^+$ and protosolvated dication $CH_3OH_3{}^{2+}$. Due to the superelectrophilic nature of $CH_3OH_3{}^{2+}$ (**37**), it is capable of insertion into the carbon-hydrogen bond of $CH_3OH_2{}^+$ (**35**) with concomitant loss of hydronium ion. Subsequent reactions of protonated ethanol readily yield the observed hydrocarbon products (by further alkylation or through dehydration to ethylene). A similar mechanism has been proposed in the formation of hydrocarbon products from dimethyl ether in superacidic media.[35] Protosolvation is also suggested in the superacid-catalyzed carbonylation of dimethyl ether (or methanol) to form methyl acetate (eq 9).[36]

$$CH_3OCH_3 \;\xrightarrow[{CO}]{HF-BF_3}\; \underset{\underset{\mathbf{38}}{BF_4^-}}{\overset{+}{\underset{H}{CH_3OCH_3}}} \;\underset{\underset{O}{\overset{|||}{C}}}{\rightleftharpoons}\; \underset{\underset{2BF_4^-}{H}}{\overset{H}{\underset{|}{\overset{2+}{CH_3OCH_3}}}} \;\xrightarrow{-2HF}\; \overset{+}{CH_3CO}\;\overset{-}{BF_3OCH_3}$$

$$\qquad\qquad\qquad\qquad\qquad\qquad\qquad\qquad\qquad\qquad\qquad\qquad\qquad\quad \Big\downarrow {\scriptstyle -BF_3}$$

$$\qquad\qquad\qquad\qquad\qquad\qquad\qquad\qquad\qquad\qquad\qquad\qquad\qquad CH_3CO_2CH_3$$

$$\quad (9)$$

The dimethyloxonium ion (**38**) is itself not reactive towards CO. However, superelectrophilic activation enables it to react with carbon monoxide.

A considerable amount of work has also been done to show that superelectrophilic sulfonium dications can be generated. For example, like the

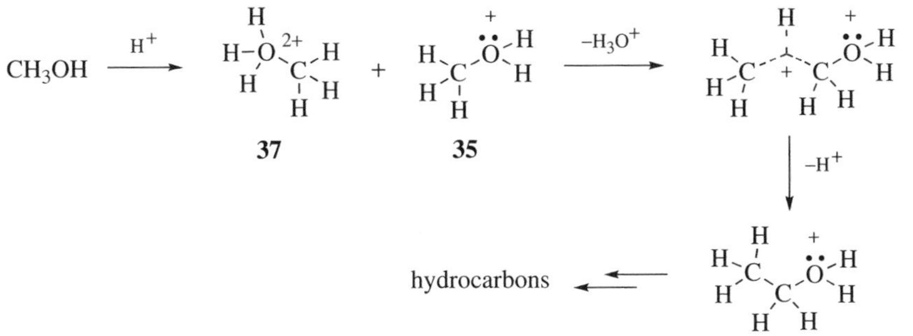

Scheme 7.

protosolvated (protonated) hydronium ion (H_4O^{2+}), evidence has been obtained for protosolvation of H_3S^+ involving H_4S^{2+}.[37] When HD_2S^+ is generated in $FSO_3H:SbF_5(4:1, H_0 -18)$ solution, it also forms the isotopomeric products (i.e., H_2DS^+) by H/D exchange (eq 10).

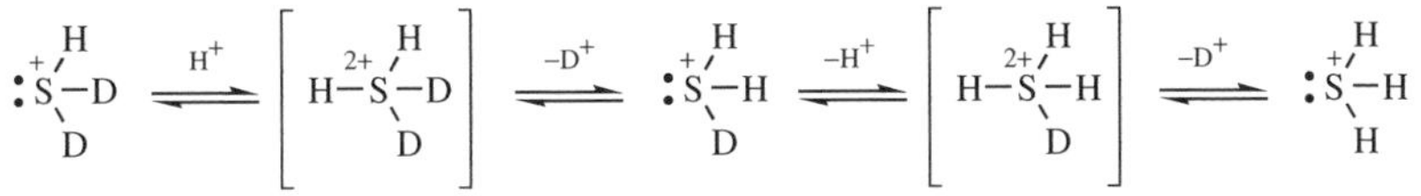

Moreover, the exchange rate increases when $FSO_3H:SbF_5(1:1, H_0 -21.5)$ is used as the reaction medium. The increasing rate of H/D exchange at higher acidities is in accord with exchange occurring by a protosolvation mechanism involving isotopomeric H_4S^{2+}. *Ab initio* calculations have been done on the H_4S^{2+} ion, and like H_4O^{2+}, it was shown to be thermodynamically unstable but kinetically stable. The global minimum is found at the tetrahedral structure (T_d symmetry, isoelectronic with AlH_4^-, SiH_4, and PH_4^+) with a significant barrier to gas-phase dissociation (59.2 $kcal^{-1}$ at the MP4SDTQ/6-31G**//HF/6-32G* level). Interestingly, calculations also suggest that the H_4S^{2+} ion may be more readily obtained than H_4O^{2+}. Although the calculations were done on isolated gas-phase species, the proton transfer from H_4O^{2+} to H_3S^+ is estimated to be thermodynamically very favorable at the MP4SDTQ/6-31G** level (eq 11).[37]

$$\Delta H° = -27.7 \text{ kcal} \cdot \text{mol}^{-1}$$

The increased stability of the sulfonium dication may be due to the larger size of sulfur and thus greater ability to disperse the positive charge.

Other stable, organosulfurane (IV) dications (**40–42**) were reported.[38]

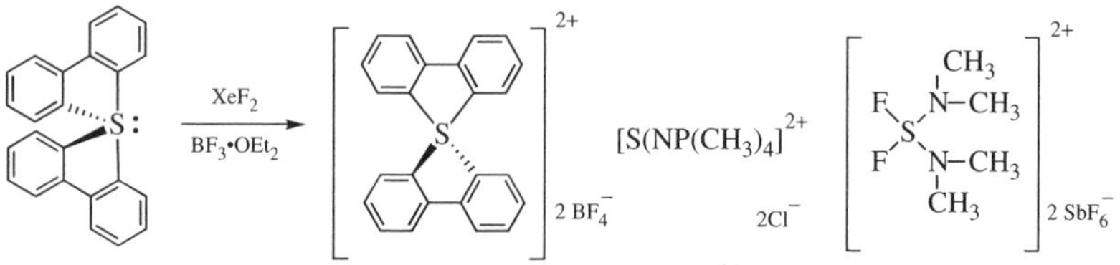

40   **41**   **42**

Salt **40** is prepared by the oxidation of the neutral organosulfurane with XeF$_2$ to obtain **40** as a crystalline solid.[38a] X-ray diffraction studies of **33** reveal a tetrahedral geometry at the sulfur and C-S bond length of 1.753 Å, which is comparable to C-S single bond length. Laguna and co-workers were able to prepare and study by X-ray crystallography the perchlorate salt of $[(C_6H_5)_3PAu]_4S^{2+}$ (**43**) an analog of H$_4$S^{2+}.[39] The salt was prepared by the reaction of [(Ph$_3$PAu)$_2$S] with two equivalents of [AuPPh$_3$ClO$_4^-$] and exhibits an approximately tetragonal pyramidal framework.

Little experimental or theoretical work has been done on the related H$_4$Se^{2+} and H$_4$Te^{2+} dications (**44** and **45**).

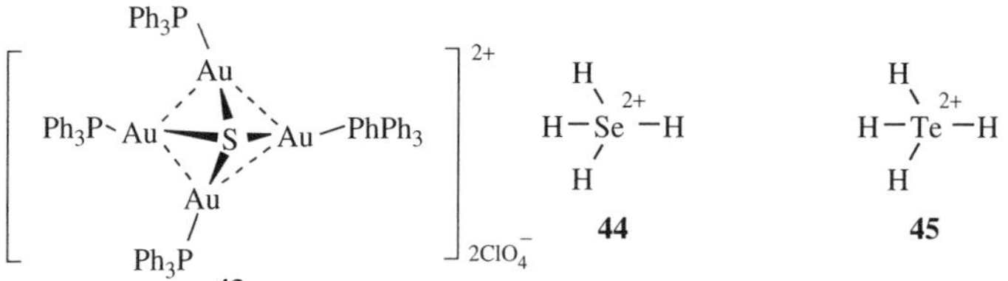

43   **44**   **45**

The tetraaurated species $[(C_6H_5)_3PAu]_4Se^{2+}$ has been prepared and characterized, however.[40]

Like the trialkyloxonium superelectrophiles, the salts of trimethyl sulfonium (CH$_3$)$_3$S$^+$, selenonium (CH$_3$)$_3$Se$^+$, and telluronium (CH$_3$)$_3$Te$^+$ ions have also been shown by Laali et al. to undergo superelectrophilic activation.[41] These onium salts methylate toluene in FSO$_3$H-SbF$_5$, but with the weaker Bronsted superacid CF$_3$SO$_3$H (triflic acid, H_0 −14.1), no methylation takes place (eq 12).

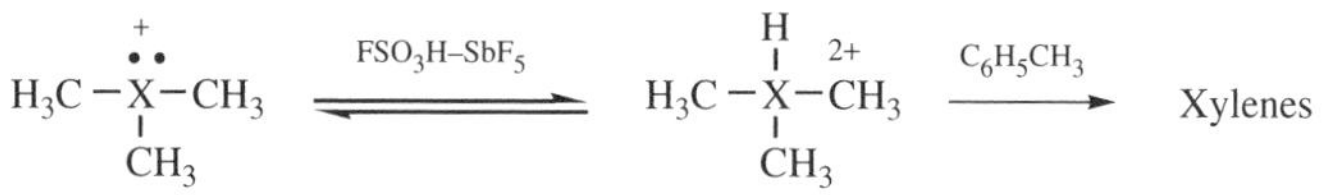

$$X = S, Se, Te$$

(12)

The need for extremely high acidity is in accord with the protosolvation of the onium salts.

There has been some experimental and theoretical work related to even higher coordinate polycationic geminal superelectrophiles, such as H_5O^{3+} and H_5S^{3+}.[42] When triprotonated water, H_5O^{3+}, was studied at the MP2/6-31G** level of calculation, the entire potential energy surface is found to be repulsive. The trication H_5O^{3+} dissociates into H_4O^{2+} and H^+ upon optimization. However, with the larger sulfur atom, the tricationic species (**2**, H_5S^{3+}) is found at a stable minimum. At the MP2/6-31G** level, the C_s-symmetric form of **2** is determined to be the only stable structure, one that resembles a complex between SH_3^{3+} and a hydrogen molecule (eq. 13).

$$\left[\begin{array}{c} \text{H} \diagdown \diagup \text{H} \\ | \\ \text{S} \\ \text{H}^{\text{\tiny\textbullet}} \diagup \text{\textbackslash} \text{H} \\ \text{H} \end{array} \right]^{3+}_{C_s} \qquad SH_3^{3+} + H_2 \longrightarrow SH_5^{3+} \quad \Delta H = -124 \text{ kcal} \bullet \text{mol}^{-1}$$

$$\hspace{2cm} \mathbf{2} \hspace{5cm} \mathbf{2} \hspace{4cm} (13)$$

The structure **2** is isostructural with the C_s-symmetric structure of the pentacoordinate carbonium ion CH_5^+, involving a 3 center-2 electron bond. Interestingly, theoretical calculations at the CCSD(T)/cc-pVTZ//QCISD (T)/6-311G** + ZPE level indicate that the reaction of SH_3^{3+} and H_2 may be a viable gas-phase reaction as it is shown to be highly exothermic (eq 13). Calculations also indicate that SH_3^{3+} itself is located at a potential energy minima. Although SH_5^{3+} (**2**) has not yet been thus far generated experimentally, the isolobal gold complex [$S(AuPR_3)_5^{3+}$] has been reported in the solution phase.[39] The same study also reports evidence for the [$S(AuPR_3)_6^{4+}$] species. However, neither of the salts, [$S(AuPR_3)_5^{3+}$] or [$S(AuPR_3)_6^{4+}$], could be characterized by crystallography. These sulfur complexes represent isolobal analogues of the SH_5^{3+} and SH_6^{4+}, respectively.

4.2.4 Geminal Halodications

Halonium ions are an important class of onium ions.[43] The dialkylchloro, bromo, and iodohalonium ions can be prepared and even isolated as stable salts (i.e., **46**), as shown by Olah et al. by reacting an excess of haloalkane with strong Lewis acid halides in solvents of low nucleophilicity (eq 14). In superacid solution, dialkylhalonium ions show enhanced alkylating reactivity.[44] It is considered that this enhanced reactivity is due to further protolytic (or electrophilic) activation involving the non-bonded electron pairs of the halogen atom leading to gitonic superelectrophiles (**47** or **48**, eq 15).

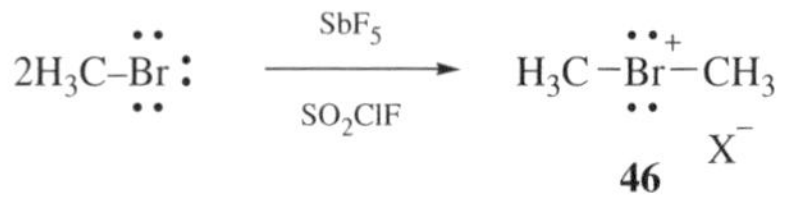

$$2H_3C-\overset{\cdot\cdot}{\underset{\cdot\cdot}{Br}}: \quad \xrightarrow[\text{SO}_2\text{ClF}]{\text{SbF}_5} \quad H_3C-\overset{\cdot\cdot+}{\underset{\cdot\cdot}{Br}}-CH_3 \qquad X^- \qquad (14)$$

$$\underset{\substack{46}}{H_3C-\overset{\cdot\cdot+}{\underset{\cdot\cdot}{Br}}-CH_3 \\ X^-} \quad \underset{\text{HX}}{\rightleftharpoons} \quad \underset{47}{H_3C-\overset{\cdot\cdot+}{Br}-CH_3 \\ \overset{\delta^+}{H}---X^{\delta^-}} \quad \rightleftharpoons \quad \underset{48}{H_3C-\overset{\cdot\cdot 2+}{Br}=CH_3 \\ H \quad 2\,X^-} \qquad (15)$$

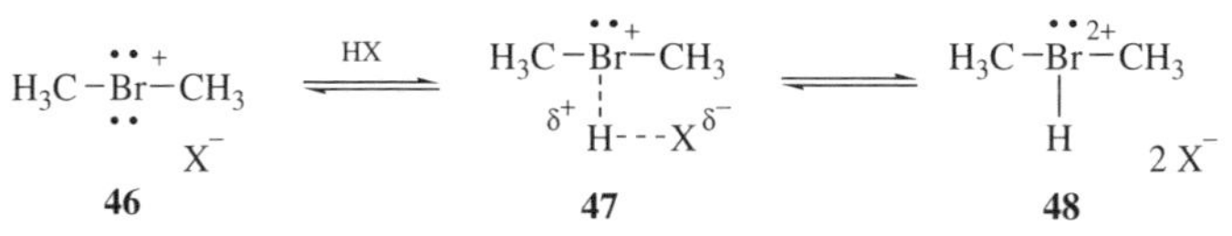

There is further evidence for the involvement of the halogen-centered gitonic superelectrophiles from isotopic exchange experiments.[45] When bromonium ion **46** (X=SbF$_6^-$) is generated in a solution containing the isotopomeric CD$_3$F–SbF$_5$, methyl interchange is observed (Scheme 8). Under conditions of low nucleophilicity and high acidity, it is unlikely that demethylation of **46** occurs to give free bromomethane (and subsequently the exchange product **50**). Thus, methyl exchange is best explained through the methylation of **46** to give the geminal dication **49**. Protosolvated dialkylhalonium ions such as **48** have also been studied by theoretical methods.[44a]

In studies of fluoro, chloro, bromo, and iodonium ions and their superelectrophilic, H$_3$X^{2+} ions were investigated by density functional theory calculations at the B3LYP/6-31G**//B3LYP/6-31G** level. The H$_3$I^{2+}, H$_3$Br^{2+}, and H$_3$Cl^{2+} ions were found to be stable minima (C_{3v} symmetry), with significant energy barriers to dissociation, 35.8, 37.9, and 27.4 kcal/mol, respectively. In all cases, however, gas-phase dissociation into H$_2$X$^+$and H$^+$ is shown to be exothermic. The H$_3$F^{2+}species is also located at a potential energy minimum, however, the optimized geometry shows D_{3h} symmetry and a significantly lower barrier to dissociation,

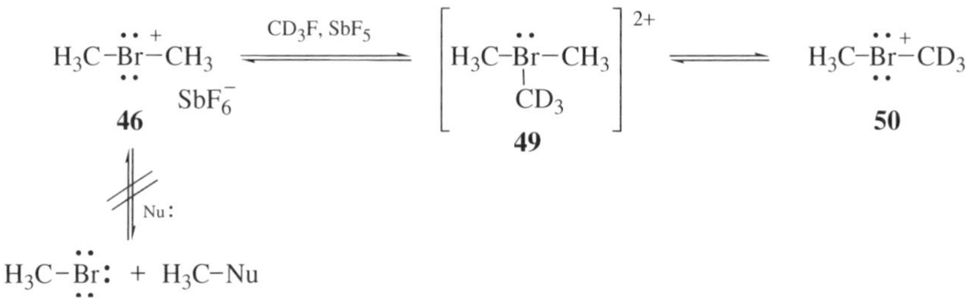

$$\underset{46}{H_3C-\overset{\cdot\cdot+}{\underset{\cdot\cdot}{Br}}-CH_3 \\ SbF_6^-} \quad \underset{\text{CD}_3\text{F, SbF}_5}{\rightleftharpoons} \quad \left[\underset{49}{H_3C-\overset{\cdot\cdot}{Br}-CH_3 \\ CD_3}\right]^{2+} \quad \rightleftharpoons \quad \underset{50}{H_3C-\overset{\cdot\cdot+}{\underset{\cdot\cdot}{Br}}-CD_3}$$

$$\downarrow \text{Nu:}$$

$$H_3C-\overset{\cdot\cdot}{\underset{\cdot\cdot}{Br}}: \; + \; H_3C-Nu$$

Scheme 8.

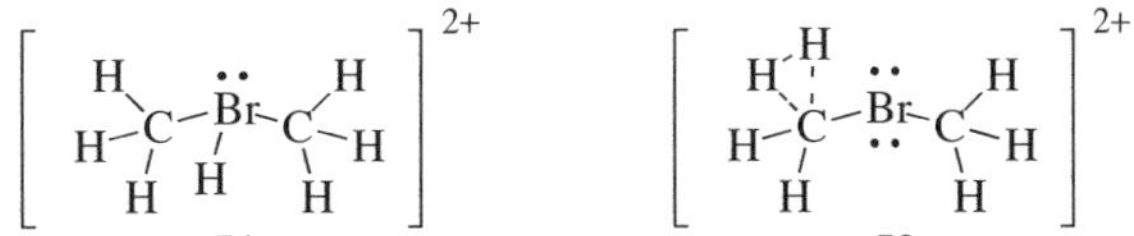

Relative Energy: 0.0 kcal•mol^{-1} 21.7 kcal•mol^{-1}

Figure 3.

8.6 kcal/mol. These calculations agree well with earlier calculations done at the MP2(FU)/6-31G** level.

In the HF–SbF$_5$ superacid-catalyzed carbonylation reactions of alkanes, Sommer and co-workers reported that addition of bromide ion to the reaction mixture leads to greatly increased reaction rates.[46] This can be best interpreted to be a consequence of the protolytic activation of H$_2$Br$^+$ via the *gitonic* trihydrobromonium dication H$_3$Br^{2+}, though the authors suggested that the enhanced reactivity may be due to the formation of "Br$^+$" for which, however, there is no evidence in the condensed state.[6a] Thermodynamically, H$_3$Br^{2+} is expected to be more stable in solution than in the gas phase, due to interaction with the involved counter ions. In comparing H$_3$Br^{2+} to H$_4$O^{2+}, it has been noted that the calculated proton affinity of H$_2$Br$^+$ (-42.9 kcal/mol) is greater than the calculated proton affinity of H$_3$O$^+$ (-60.4 kcal/mol). Yet, the calculated dissociation barriers are comparable (H$_3$Br^{2+}, 37.9 kcal/mol; H$_4$O^{2+}, 38.2 kcal/mol). Given the discussed experimental evidence for the formation of H$_4$O^{2+} in liquid superacid, it seems likely that H$_3$Br^{2+} is a viable gitonic superelectrophile.

In the case of the dimethylbromonium cation **46**, *ab initio* calculations have found two stable minima for the protosolvated superelectrophile, the bromine and carbon protonated forms, **51** and **52**, respectively (Figure 3).[44a] Dication **51** is estimated to be more stable than **52** by about 21 kcal/mol at the B3LYP/6-31G**//B3LYP/6-31G** level. When dimethylbromonium cation (**46**) is reacted in DF/SbF$_5$ in SO$_2$ at $-78°$C, no deuterium incorporation is observed into the methyl groups. This observation is in accord with the theoretical calculations in that Br-protonation is expected to be the preferred process.

The protonated methyl halides have also been studied by *ab initio* computational methods.[44a] Whereas initial protonation occurs on the halogen atom, the second protonation occurs primarily on the C-H bonds. The B3LYP/6-31G** potential energy surfaces (PES) for CH$_3$XH$_2$$^{2+}$ (X=F, Cl, Br, I) are found to be repulsive.

REFERENCES

(1) D. Liu; T. Wyttenbach; M. T. Bowers *Int. J. Mass Spectrom.* **2004**, *236*, 81.

(2) V. Prakash Reddy; E. Sinn; G. A. Olah; G. K. S. Prakash; G. Rasul *J. Phys. Chem. A.* **2004**, *108*, 4036.

(3) G. A. Olah; G. Rasul; G. K. S. Prakash *Chem Eur. J.* **1997**, *3*, 1039.

(4) A. I. Boldyrev; J. Simons *J. Chem. Phys.* **1992**, *97*, 4272.

(5) T. Suzuki; T. Ohwada; K. Shudo *J. Am. Chem. Soc.* **1997**, *119*, 6774.

(6) G. A. Olah; D. A. Klumpp *Acc. Chem. Res.* **2004**, *37*, 211.

(7) Y. Sato; M. Yato; T. Ohwada; S. Saito; K. Shudo *J. Am. Chem. Soc.* **1995**, *117*, 3037.

(8) (a) A. J. Heck; L. J. d. Koning; N. M. N. Nibbering *J. Phys. Chem.* **1992**, *96*, 8870. (b) G. A. Olah; J. Shen *J. Am. Chem. Soc.* **1973**, *95*, 3582.

(9) G. A. Olah; G. K. S. Prakash; R. E. Williams; L. D. Field; K. Wade *Hypercarbon Chemistry*, Wiley: New York, 1987.

(10) (a) G. A. Olah; G. K. S. Prakash; G. Rasul *J. Org. Chem.* **2001**, *61*, 2907. (b) G. A. Olah; G. Rasul *J. Am. Chem. Soc.* **1996**, *118*, 12922. (c) G. A. Olah; G. Rasul *Acc. Chem. Res.* **1997**, *30*, 245.

(11) F. Scherbaum; A. Grohmann; B. Huber; C. Krüger; H. Schmidbaur *Angew. Chem. Int. Ed. Engl.* **1988**, *27*, 1544.

(12) G. A. Olah; G. Rasul *J. Am. Chem. Soc.* **1996**, *118*, 8503.

(13) K. Lammertsma; P. v. R. Schleyer; H. Schwarz *Angew. Chem. Int. Ed. Engl.* **1989**, *28*, 1321.

(14) (a) M. W. Wong; B. F. Yates; R. H. Nobes; L. Radom *J. Am. Chem. Soc.* **1987**, *109*, 3181. (b) J. Roithova; J. Hrusak; Z. Herman *Int. J. Mass Spectrom.* **2003**, *228*, 497.

(15) (a) J. Roithova; J. Zabka; Z. Herman; R. Thissen; D. Schröder; H. Schwarz *J. Phys. Chem. A.* **2006**, *110*, 6447. (b) J. Roithova; Z. Herman; D. Schröder; H. Schwarz *Chem. Eur. J.* **2006**, *12*, 2465.

(16) (a) Z. Herman; J. Zabka; Z. Dolejsek; M. Farnik *Int. J. Mass Spectrom.* **1999**, *192*, 191. (b) S. D. Price; M. Manning; S. R. Leone *J. Am. Chem. Soc.* **1987**, *109*, 3181. (c) S. D. Price; *J. Chem. Soc., Faraday Trans.*, **1997**, *93*, 2451.

(17) G. A. Olah *Angew. Chem. Int. Ed. Engl.* **1993**, *32*, 767.

(18) (a) K. Leiter; K. Stephean; E. Mark; T. D. Mark *Plasma Chem. Plasma Processes* **1984**, *4*, 235. (b) C. J. Porter; C. J. Proctor; T. Ast; J. H. Beynon *Croat. Chem. Acta* **1981**, *54*, 407. (c) Y.-Y. Lee; S. R. Leone *J. Phys. Chem. A.* **1995**, *99*, 15438. (d) I. S. Akhrem; A. L. Chistyakov; N. P. Gambaryan; I. V. Stankevich; M. E. Vol'pin *J. Organomet. Chem.* **1997**, *536–537*, 489.

(19) D. Schröder; H. Schwarz *J. Phys. Chem. A.* **1999**, *103*, 7385.

(20) (a) G. A. Olah; A. Burrichter; G. Rasul; G. K. S. Prakash *J. Am. Chem. Soc*. **1997**, *119*, 4594. (b) G. Rasul; G. K. S. Prakash; G. A. Olah *J. Am. Chem. Soc*. **1997**, *119*, 12984.

(21) A. Schier; A. Grohmann; J. M. Lopez-de-Luzuriaga; H. Schmidbaur *Inorg. Chem*. **2000**, *39*, 547.

(22) M. Hamden; A. G. Brenton *Int. J. Mass Spectrom*. **1988**, *84*, 211.

(23) (a) K. Shudo; T. Ohta; T. Okamoto *J. Am. Chem. Soc*. **1981**, *103*, 645. (b) T. Ohta; R. Machida; K. Takeda; Y. Endo; K. Shudo; T. Okamoto *J. Am. Chem. Soc*. **1980**, *102*, 6385. (c) K. Shudo; T. Ohta; Y. Endo; T. Okamoto *Tetrahedron Lett*. **1977**, 105.

(24) (a) K. O. Christe; R. D. Schack; R. D. Wilson *Inorg. Chem*. **1975**, *14*, 2224. (b) G. D. Mateescu; G. M. Benedikt *J. Am. Chem. Soc*. **1979**, *101*, 3959.

(25) V. Gold; J. L. Grant; K. P. Morris *J. Chem. Soc., Chem. Commun*. **1976**, 397.

(26) G. A. Olah; G. K. S. Prakash; M Barzaghi; K. Lammertsma; P. v. R. Schleyer; J. A. Pople *J. Am. Chem. Soc*. **1986**, *108*, 1032.

(27) (a) G. A. Olah; A. Burrichter; G. Rasul; R. Gnann; K. O. Christe; G. K. S. Prakash *J. Am. Chem. Soc*. **1997**, *119*, 8035. (b) W. Koch; N. Heinrich;
H. Schwarz; F. Maquin; D. Stahl *Int. J. Mass Spec. Ion Proc*. **1985**, *67*, 305. (c) Reference 4.

(28) H. Schmidbaur; S. Hofreiter; M. Paul *Nature* **1995**, *377*, 503.

(29) V. G. Granik; B. M. Pyatin; R. G. Glushkov *Russ. Chem. Rev*. **1971**, *40*, 747.

(30) (a) G. A. Olah; J. R. De Member; Y. K. Mo; J. J. Svoboda; P. Schilling; J. A. Olah *J. Am. Chem. Soc*. **1974**, *96*, 884. (b) G. A. Olah; G. Rasul; A. Burrichter; G. K. S. Prakash *Proc. Nat. Acad. Sci. USA* **1998**, *95*, 4099.

(31) G. A. Olah; A. M. White; D. H. O'Brien *Chem. Rev*. **1970**, *70*, 561.

(32) J. Kaspi; D. D. Montgomery; G. A. Olah *J. Org. Chem*. **1978**, *43*, 3147.

(33) G. A. Olah; A.-H. Wu *Synlett*, **1990**, 599.

(34) D. E. Pearson *J. Chem. Soc. Chem. Commun*. **1974**, 397.

(35) G. A. Olah; K. A. Laali; Q. Wang; G. K. S. Prakash *Onium Ions*; Wiley: New York, NY, **1998**, p. 435.

(36) A. Bagno; J. Bukala; G. A. Olah *J. Org. Chem*. **1990**, *55*, 4284.

(37) G. A. Olah; G. K. S. Prakash; M. Marcelli; K. Lammertsma *J. Phys. Chem*. **1988**, *92*, 878.

(38) (a) S. Sato; H. Ameta; E. Horn; O. Takahashi; N. Furukawa *J. Am. Chem. Soc*. **1997**, *119*, 12374. (b) H. Henle; R. Hoppenheit; R. Mews *Angew. Chem., Int. Ed. Engl*. **1984**, *23*, 507. (c) H. Folkerts; W. Hiller; M. Herker; S. F. Vyoishchikov; G. Frenking; K. Dehnicke *Angew. Chem., Int. Ed. Engl*. **1995**, *34*, 1362.

(39) F. Canales; C. Gimeno; A. Laguna; M. D. Villacampa *Inorg. Chim. Acta* **1996**, *244*, 95.

(40) S. Canales; O. Crespo; M. C. Gimeno; P. G. Jones; A. Laguna *Chem. Comm.* **1999**, 679.

(41) K. Laali; H. Y. Chen; R. J. Gerzina *J. Organomet. Chem.* **1988**, *348*, 199.

(42) G. A. Olah; G. Rasul; G. K. S. Prakash *Chem Eur. J.* **1997**, *3*, 1039.

(43) G. A. Olah; *Halonium Ions*, Wiley: New York, **1975**.

(44) (a) G. A. Olah; G. Rasul; M. Hachoumy; A. Burrichter; G. K. S. Prakash *J. Am. Chem. Soc.* **2000**, *122*, 2737. (b) Reference 30a.

(45) G. A. Olah et. al., unpublished result.

(46) J. Bukala; J.-C. Culmann; J. Sommer *J. Chem. Soc., Chem. Commun.* **1992**, 481.

5

GITONIC *VICINAL* SUPERELECTROPHILES

5.1 STRUCTURAL CONSIDERATIONS

Similar to the geminal-type superelectrophiles described in the previous chapter, *vicinal* systems may be formally depicted with adjacent positive charges, but these charges are often delocalized by induction and conjugation. In the case of the ethylene dication ($C_2H_4^{2+}$, **1**), theoretical calculations have shown the perpendicular structure (D_{2d}) to be the most stable geometry.[1] The two formally vacant orbitals are orthogonal and capable of interacting with the vicinal CH_2 groups by hyperconjugation (Figure 1). This leads to a remarkably short carbon-carbon bond (estimated to be 1.432 Å), as well as positive charge residing on the hydrogen atoms. Similarly, protosolvation of the *tert*-butyl cation (**2**) gives the protio-*tert*-butyl dication (**3**). Theoretical calculations (MP2/6-311+G** level) of the NBO charges show that much of the positive charge is located on the methyl groups.[2]

There have been many studies related to tetraaryl-1,2-ethylene dications, formally gitonic superelectrophiles, which show extensive charge delocalization into the aryl rings (Scheme 1). Low temperature ^{13}C NMR studies have shown that the *para* carbons are deshielded, from $\delta^{13}C$

Superelectrophiles and Their Chemistry, by George A. Olah and Douglas A. Klumpp
Copyright © 2008 John Wiley & Sons, Inc.

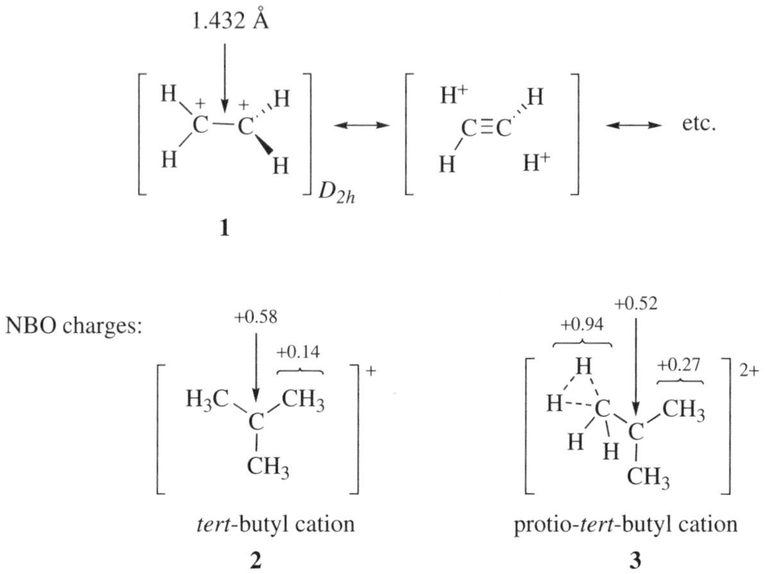

Figure 1. Ethylene dication (**1**) and calculated NBO charges for the *tert*-butyl cation (**2**) and the proto-*tert*-butyl cation (**3**).

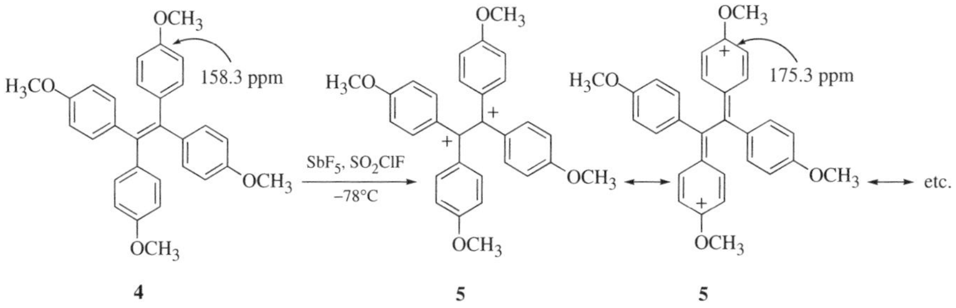

Scheme 1.

158.3 in the tetrakis(*p*-methoxyphenyl)ethylene (**4**), to $\delta^{13}C$ 175.3 in the corresponding dication (**5**).[3] This large downfield shift is best explained by the delocalization of the positive charge, caused by the repulsive interaction of the two positive charge centers. Inductive effects have likewise been shown to increase the acidic and electrophilic character of adjacent groups. For example, the bicyclic hydrazinium dications react with nucleophiles at the α-carbon instead of at one of the nitrogen centers (eq 1).

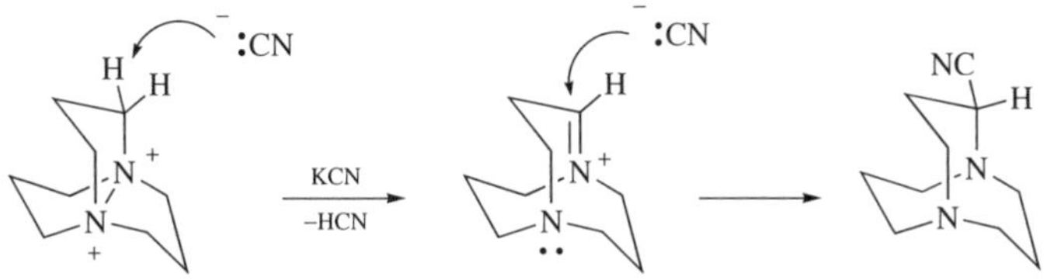

(1)

As described previously, a fully formed dicationic superelectrophile is generally considered the limiting case, and there are likely varying degrees of superelectrophilic activation. In Chapter 1, the 1,2-dicationic systems, such as the protoacetyl dication (CH_3COH^{2+}) and protonitronium dication (NO_2H^{2+}), were considered in terms of varying degrees of protosolvation. Superelectrophilic activation in these systems may involve only partial protonation or partial Lewis acid coordination to generate the reactive species. The kinetic studies of Shudo and Ohwada were described in Chapter 2 wherein the cyclizations of 4-phenylbutyronitrile (**6**) and 1,3-diphenyl-1-propanone (**9**) were compared (eqs 2 and 3).[4]

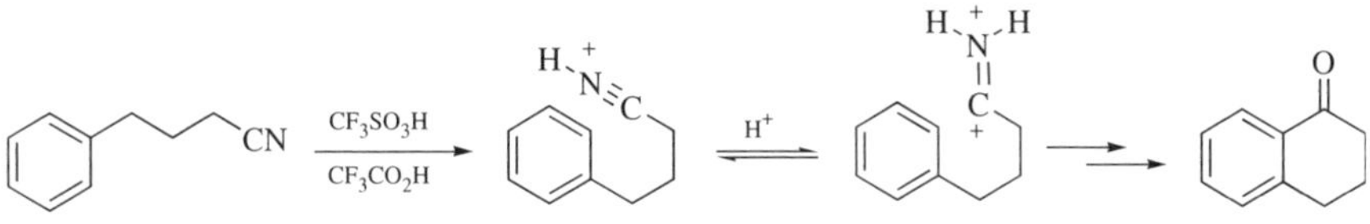

(2)

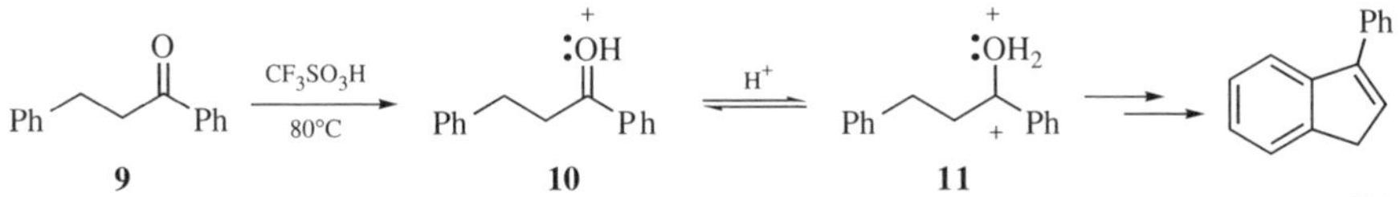

(3)

While both 4-phenylbutyronitrile (**6**) and 1,3-diphenyl-1-propanone (**9**) cyclizations show a dependence on acid strength, the kinetic data suggest differing degrees of protonation at their transition states, or differing protosolvation of the activated complexes.

Koltunov and Sommer have noted that an important consideration is the *timing* of the second protonation in superelectrophilic reactions.[5] For example, in the superacid-catalyzed condensation of benzaldehyde with benzene (*vida infra*), two studies have proposed superelectrophilic diprotonated benzaldehyde (**12** or **13**) as the reactive electrophile leading to condensation products (Scheme 2).[6,7] An alternative explanation for this

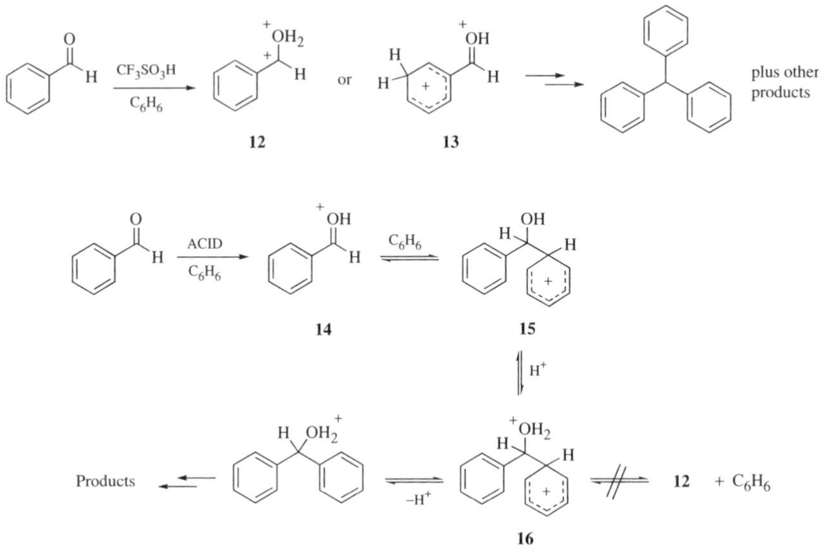

Scheme 2.

conversion, without invoking superelectrophiles **12** or **13**, involves mono-protonated benzaldehyde (**14**) forming the σ-complex with benzene (**15**) followed by protonation of the resulting hydroxy group to give the distonic superelectrophile (**16**). This proposal suggests another, potentially important, aspect of superelectrophilic chemistry: formation of diprotonated (or other multiply charged species) intermediates may lead to favorable shifts in equilibria leading to products.

The *vicinal*-dicationic intermediates comprise a large group of super-electrophilic species. We are subsequently discussing sections on boron, aluminum, carbon, nitrogen, oxygen, halogen, and even noble-gas based systems. Each of these sections is further divided as necessary. For example, the carbon-based systems include, *vicinal*-carbon-carbon, carbon-nitrogen, carbon-oxygen, and carbon-halogen dicationic systems.

5.2 *VICINAL* SYSTEMS

5.2.1 Boron and Aluminum-Centered Systems

There are relatively few examples known of doubly electron-deficient gitonic superelectrophiles involving boron or aluminum. In studies by Schwartz and associates, the NBH_4^{2+} dication has been generated in gas-phase studies using charge stripping mass spectroscopy, and its potential energy surface has been studied by theoretical calculations.[8] The

NBH_4^{2+} dication is formed by the collision of the $NBH_4^{\bullet+}$ ion with O_2, and experimental detection of the NBH_4^{2+} ion indicates a lifetime greater than 5×10^{-6} s. This is in accord with computational studies that find significant energy barriers to decomposition or fragmentation. At the MP4/6-311G(d,p)//6-31G(d)+ZPE level of theory, two minima are located, the ammoniaborene dication NH_3BH^{2+} (**17**) and its isomer $NH_2BH_2^{2+}$ (**18**).

The ammoniaborene dication NH_3BH^{2+} (**17**) is estimated to be about 60 kcal/mol more stable than $NH_2BH_2^{2+}$ (**18**). Much of the interest in activated boron and aluminum electrophiles arises from their isoelectronic relationships to related carbon-centered superelectrophiles (Table 1).[9] For example, a variety of superelectrophilic trihalomethylcations (i.e., **19**) have been proposed in superacid-catalyzed conversions and these species have been shown to have significance in the functionalization of alkanes (*vide infra*). Isoelectronic boron and aluminum species **20** and **21** are thought to be involved in some reactions with weak nulceophiles: the reaction of BCl_3 with arenes in the presence of $AlCl_3$ gives phenylborondichloride (eq 4).[10]

$$C_6H_6 + BCl_3 \xrightarrow{\text{AlCl}_3} C_6H_6BCl_2 + HCl \qquad (4)$$

The activity of aluminum chloride as a strong Lewis acid (including its ability to crack or isomerizes alkanes) may be related to the formation of its mono-bridged dimer (**21**).[11] The carbodication CCl_2^{2+} (**22**) has been suggested as a possible intermediate in condensed phase reactions,[12] while the carbodication CH_2^{2+} (**25**) has been studied by high level *ab initio* calculations and directly observed by mass-spectrometric techniques.[13] The isoelectronic boron and aluminum species (**23–24**) have likewise been studied by theoretical methods,[14] while cation **26** has also been observed in gas-phase experimental studies.[15] Carbodication **27** has also been studied extensively by experimental and theoretical work, and computations have shown **27** to have planar structure (C_{2v}).[16] Similarly, the isoelectronic boron species **28** was studied theoretically and it was proposed as an intermediate in the condensed phase.[17] Evidence for the boron analog (**29**) to the protosolvated tert-butyl dication (**3**) has also been suggested from condensed phase superacid chemistry.[18] Diprotonated carbonic acid (**30**) may

Table 1. Carbon superelectrophiles and their isoelectronic boron and aluminum systems.

Superelectrophile	Isoelectronic Boron or Aluminum Systems

be involved in the superacidic cleavage of carbonates or bicarbonates.[19] The isoelectronic boron species, protonated boric acid (**31**), has been observed in gas-phase studies and examined by theoretical methods.[20] Even the methonium dication CH_6^{2+} (**32**, diprotonated methane) has an analogous boron species (BH_6^+, **33**) which has been observed in gas-phase studies and been examined by computational methods.[21] The BH_6^+ (**33**) ion is generated in a flowing afterglow-selected ion flow tube apparatus by the gas-phase reaction of BH_2^+ (**26**; previously generated from molecular oxygen and B_2H_6) with two dihydrogen molecules. Calculations at the MP2(fu)/6-311G9 d,p)//MP2(fu)/6-311G(d,p) level of theory indicate that BH_2^+ binds to two H_2 molecules in reaction steps that are exothermic by 14.5 kcal/mol (first reaction) and 17.6 kcal/mol (second reaction).[17b] The reactions of BH_2^+ (**26**) and BCl_3-$AlCl_3$ (**20**) with weak nucleophiles (H_2 and C_6H_6) indicate that the isoelectronic boron and aluminum species possess a high level of electrophilic reactivity. Although these boron and aluminum systems are monocationic, they can be good models for carbon-centered gitonic superelectrophiles, or other doubly electron-deficient systems.

5.2.2 Carbon-Centered Systems

Vicinal 1,2-carbodicationic systems are some of the most important and thoroughly studied superelectrophiles. They include *inter alia* 1,2-ethylene dications, related carbon-nitrogen superelectrophiles (diprotonated imines and nitriles), protosolvated carboxonium ions, and superelectrophilic tri-halomethyl cations. It is understood that many of these systems involve extensive charge delocalization, and in a sense may not be considered formal 1,2-dicationic systems. For example, diprotonated 2,3-butanedione (**34**) may be represented formally as a 1,2-ethylene dication (**35a**, a gitonic superelectrophile), but even in monocationic carboxonium ions, there is a significant amount of double-bond character retained in the carbon-oxygen bond (eq 5).[22]

Charge-charge repulsive effects increase the importance of the resonance form (**35b**) having dione-type structure (a 1,4-dication and representing a distonic superelectrophile). Despite the importance of the charge separated structure **35b**, the system is included here with other 1,2-ethylene dications and gitonic superelectrophiles.

5.2.2.1 Carbon-Carbon Vicinal-Dications
Ethylene dications are a major group of carbon-centered superelectrophilic systems. The parent ethylene dication ($CH_2CH_2^{2+}$, **1**) has been studied experimentally in the gas-phase, as well as in several theoretical studies.[1] Two electron

ionization of ethylene in the gas-phase produces the ethylene dication **1**. Calculations at the HF/6-31G(d) and MP2/6-31G(d) levels indicate the D_{2h} structure as the most stable geometry. Other two carbon systems studied include the acetylene dication (**36**), the ethane dication (**37**, MP4/6-311G** optimized structure), and diprotonated ethane (**38**).[13,23]

Halogenated ethylene dications have also been studied in theoretical and gas-phase studies.[24] Although the ethylene dication (**1**) can be detected in mass spectrometric experiments by charge stripping of the ethylene radical cation with molecular oxygen, **1** could not be generated yet in the condensed phase. On the other hand, substitution by good n- or π-donor groups leads to ethylene dications that can be readily generated in solution.

Among the stabilized ethylene dications produced, the aryl-substituted ethylene dications have received most attention (Table 2). These have been prepared by the direct oxidation of the ethylene derivatives or by ionization of suitable functional groups. In the case of tetraanisylethylene (**4**), oxidation by $SbCl_5$, Br_2, or I_2/CCl_4 produces the remarkably stable dark green ethylene dication **5**, which has been characterized by X-ray crystallography.[25] Other organic oxidants, such as anthracene-based radical cations, have also been used to generate **5**. The tetraanisylethylene dication (**5**) can itself oxidize hydrocarbon substrates (octamethylbiphenylene and cyclooctatetraenes) to their radical cations by electron-transfer reactions.[26] The reversible oxidation of the chiral tetraaryl ethylene (**50**) provides a chiral tetraaryldication (**51**), which may be useful in chiroptical switching and memory devices (eq 6).[27]

(6)

The (R)-alkyl groups induce left-handed helicity in dication **51** which gives an intense Cotton effect in the circular dichroism (CD) spectrum. As in the case of the tetraanisylethylene dication (**5**), the X-ray crystal

Table 2. Generation of aryl-substituted 1,2-ethylene dications.

Substrate	Reagent	Dication
Ar₂C=CAr₂ **4** Ar = *p*-methoxyphenyl	SbCl₅, Br₂, or I₂	Ar₂C⁺–⁺CAr₂ **5**
Ar₂C=CAr₂ **39** Ar = *p*-dimethylaminophenyl	SbCl₅ or AgNO₃	Ar₂C⁺–⁺CAr₂ **40**
Ph₂C=CPh₂ **41**	SbF₅	Ph₂C⁺–⁺CPh₂ **42**
(Ar)(CH₂CH₃)C=C(CH₃CH₂)(Ar) **43** Ar = *p*-methoxyphenyl	Electrochemical Oxidation	(Ar)(CH₂CH₃)C⁺–⁺C(CH₃CH₂)(Ar) **44**
(Ph)(HO)C–C(=O)(Ph) Ph	CF₃SO₃H	(Ph)C⁺–⁺C(OH) Ph Ph **45**
(Ph)(Cl)C–C(=O)(OCH₃) Ph	Ag (I), CF₃SO₃H	(Ph)C⁺–⁺C(OH) Ph OCH₃ **46**
(Ar)(HO)C–C(=O)(OCH₃) Ar Ar = *p*-methoxyphenyl	CF₃SO₃H-SbF₅	(Ar)C⁺–⁺C(OH) Ar OCH₃ **47**
(Ph)(HO)C–C(=O)(OH) Ph	CF₃SO₃H	(Ph)C⁺–⁺C(OH) Ph OH **48**
(Ar)(O=)C–C(=O)(Ar) Ar = *p*-methoxyphenyl	CF₃SO₃H or FSO₃H	(HO)C⁺–⁺C(OH) Ar Ar **49** Ar = *p*-methoxyphenyl

structure of the chiral dication **51** shows extensive delocalization (and thus stabilization) of the positive charges. Another related and important system is the tetrakis-(*p*-dimethylaminophenyl)-ethylene dication (**40**) which can be readily prepared by the chemical oxidation (by Ag (I) salts or SbCl$_5$) of the olefin (eq 7).[28]

39   40

$$(7)$$

The stability of dication **40** is seen in the fact that it can be studied in DMSO solution and isolated with counter-ions having somewhat higher nucleophilic nature. The X-ray crystal structure of the bis-triiodide salt of dication **40** reveals that the C=C bond length increases from 1.35 Å in starting compound **39** to 1.50 Å for the dication (**40**), reflecting a charge delocalized structure and roughly a single bond between the two central carbons.[28] While the stabilized tetraarylethylene dications such as **5**, **51**, and **40**, do not possess the high electrophilic reactivities typical of gitonic superelectrophiles, the reactivities of these systems are greatly increased in structures having substituents that are less electron donating.

The tetraphenylethylene dication (**42**) can be generated by the direct oxidation of the olefin, or by other routes (Scheme 3).[29] Ionization of benzopinacol, tetraphenylethylene oxide, and 2,2,2-triphenylacetophenone in triflic acid gives the ethylene dication **42** by dehydration of the carbeniumoxonium dication (**52**).[30] Although dication **42** can be directly observed under stable ion conditions at low temperatures, warming of the solutions containing dication **42** leads to cyclization with proton loss giving quantitatively 9,10-diphenylphenanthrene. The same 9,10-diphenylphenanthrene is also produced from the reaction of 1,2-dichloro-1,1,2,2-tetraphenylethane with SbCl$_5$, a reaction that likely involves dication **42**. For dication **42**, the ^{13}C NMR spectrum ($-75°$C, SbF$_5$-SO$_2$ClF) shows δ^{13}C resonances at C$^+$,198.6; C$_{ipso}$,140.1; C$_{ortho}$,143.0, 148.4; C$_{metal}$,143.3; C$_{para}$,152.9. When compared to the ^{13}C NMR spectrum of tetraphenylethylene, it is clear that the positive charge is delocalized into the phenyl rings **42**. It is, however, also apparent that the charge delocalization is considerably less in **42** than in the tetraanisylethylene dication (**5**) or the tetrakis-(*p*-dimethylaminophenyl)ethylene dication (**40**).

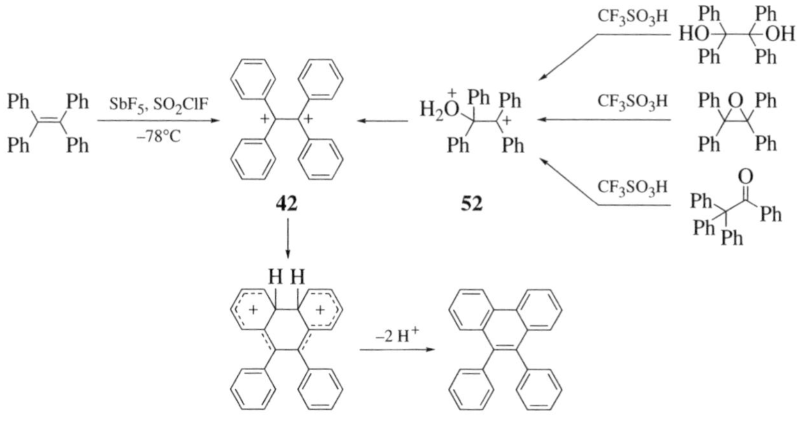

Scheme 3.

The dication **44** was generated by the reversible electrochemical oxidation of the corresponding olefin (**43**).[26] It was also prepared by chemical
oxidation using the ethanoanthracene derivative (**53**, eq 8).

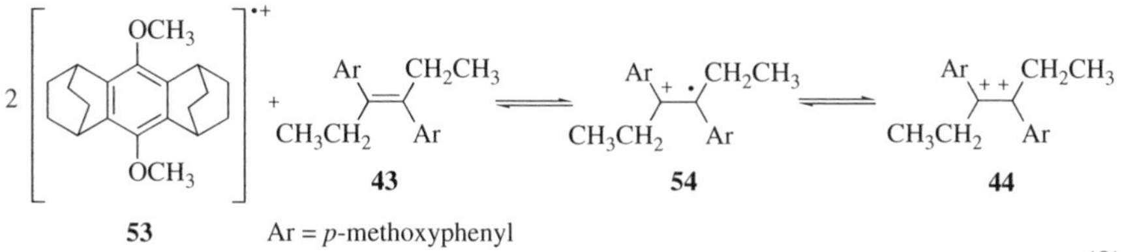

Ar = *p*-methoxyphenyl

(8)

Dication **44** decomposes at room temperature. The poor stability is thought
to arise from the liable α-protons of the ethyl groups of dication **44**. A
series of 1,2-ethylene dications have been generated and studied involving
both aryl-stabilized and oxygen-stabilized carbocationic centers (**45–49**;
(Table 2). When the hydroxyketone is reacted with a substantial excess of
CF_3SO_3H at $-50°C$, two cyclization products (**55–56**) are obtained (eq
9).[31] The reaction is thought to occur through the dication **45**. Similarly,
the fluorene products (**59–60**) are obtained from the hydroxyester (**57**)
and acid (**58**), via the dications **46** and **48** (eq 10).

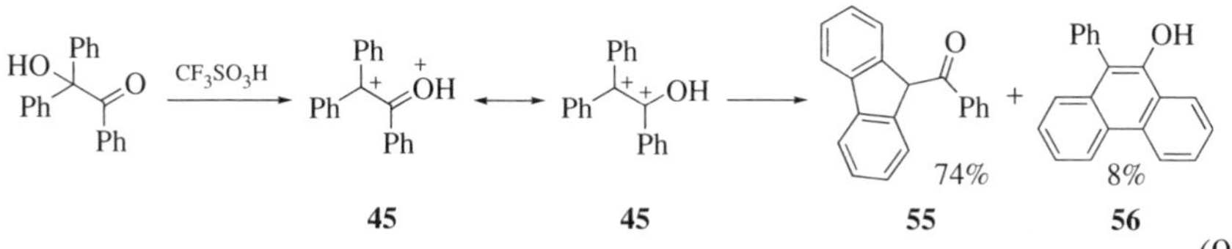

(9)

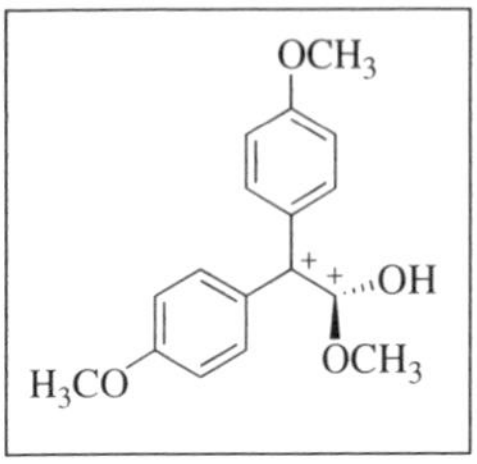

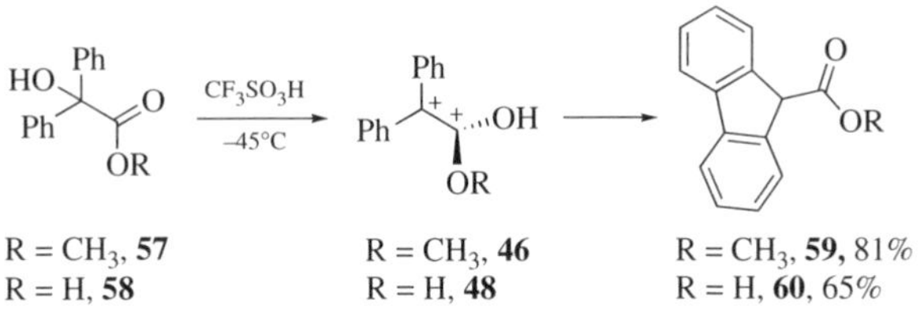

(10)

Several lines of evidence in the studies of Ohwada and Shudo support the involvement of these gitonic superelectrophiles. An analagous dicationic species (**47**) has been directly observed by ^{1}H and ^{13}C NMR at low temperature in both CF$_3$SO$_3$H–SbF$_5$ and CF$_3$SO$_3$H solutions. With dication **47**, the NMR spectra show an equivalence of the aromatic rings suggesting a perpendicular relationship between the two cationic centers. The dications having the 4-methylphenyl substituents (**61–62**) have likewise been directly observed by NMR at low temperature, and upon warming the solutions, the expected cyclization products are formed (eqs 11–12).

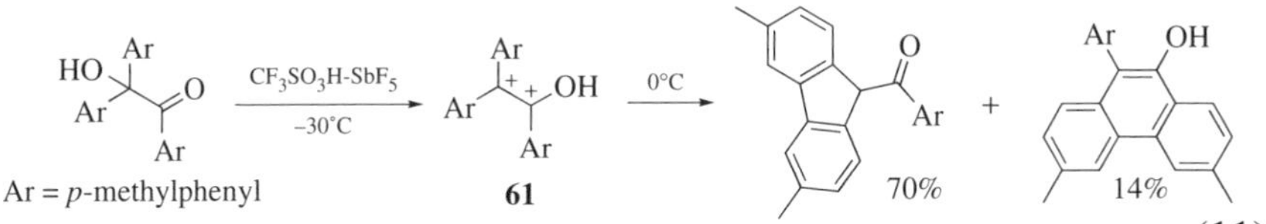

(11)

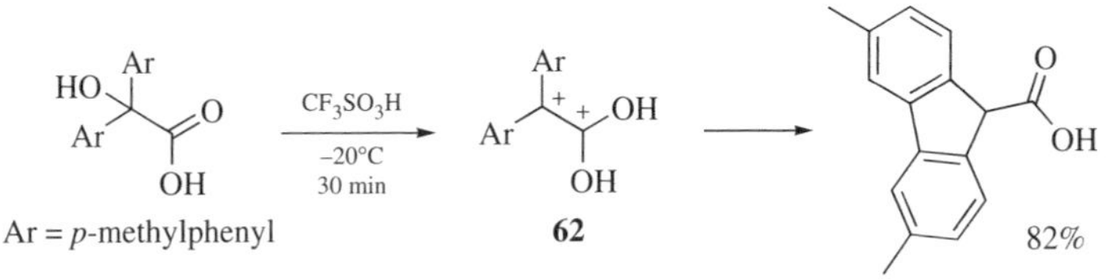

(12)

In the case of dication **62**, the cyclization reaction was found to exhibit first order kinetics (k $= 4.03 \times 10^{-3}$ s^{-1}). The activation parameters were also calculated (ΔH$^{\ddagger} = 11.1$ kcal/mol and ΔS$^{\ddagger} = -28.7$ eu).[32a]

Evidence for the dicationic intermediates was also obtained from kinetic experiments involving the hydroxyester **57**.[32a] The yield of the cyclization product (**59**) increases considerably with the acidity of the reaction media (Scheme 4). The fluorene product **59** is formed in appreciable quantities only in superacids $H_0 \leq -12$). Following this observation, the kinetics of the fluorene cyclization was also studied in solutions of varying acidity. When compound **57** is reacted in solutions with acidity in the H_0 -11 to -13 range, the cyclization rate is found to increase linearly with acidity.

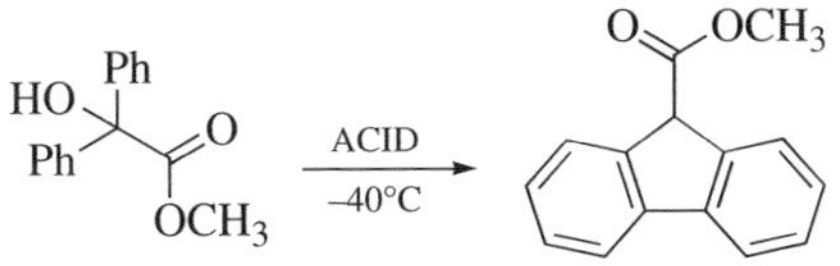

H_0:	−2.7	−11.8	−13.2
% YIELD:	0%	20%	94%

Scheme 4.

The linear rate profile provides strong evidence for the second protonation of the monocationic intermediate, and a rate-determining step involving the dicationic superelectrophile **46**.

In addition to the above kinetics studies, the fluorene cyclization was studied using *ab initio* computational methods.[32a] It was found that the theoretically predicted barriers to the cyclizations for the dicationic intermediates agree well with the values obtained from the kinetic experiments. For example, geometry optimization and energy calculations at the B3LYP/6-31* level estimated that the activation energy (E_a) is 14.0 kcal/mol for the 4π-electron conrotatory electrocyclization reaction involving compound **57** and the diprotonated intermediate (**46**, eq 13).

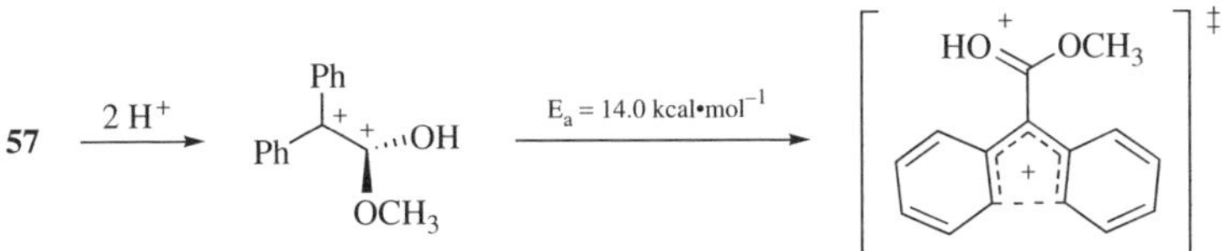

$$(13)$$

This is reasonably close to the experimentally determined value of $\Delta H^{\ddagger} = 11.1$ kcal/mol for the superelectrophilic cyclization of **62** (eq 12). Another computational study showed that the energy barriers dramatically decrease for the electrocyclization when the monocations are protonated to form superelectrophiles. In the case of **63**, cyclization provides the acetyl-substituted fluorene in 70% yield from CF_3SO_3H (Scheme 5). At the B3LYP/6-31* level of theory, dication **64** is estimated to have a cyclization barrier to fluorene of 8.5 kcal/mol, compared to a value of 25 kcal/mol for the cyclization of monocation **65**.

The described superelectrophilic activation and fluorene-cyclization is thought to involve a lowered energy of the LUMO and concomitant delocalization of positive charge into the aryl ring(s).[32b] Calculations at the 4-31G//STO-3G level on a model system (Figure 2) have shown that the amount of positive charge in the phenyl ring increases upon formation of the dication (**67**) when compared to the monocation (**66**) and the benzyl cation (calculations are based on fully planar structures). It is well known

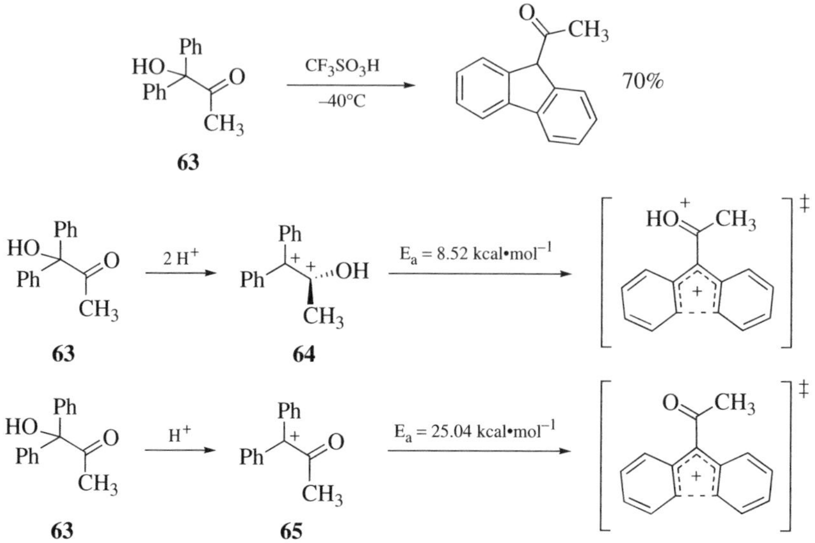

Scheme 5.

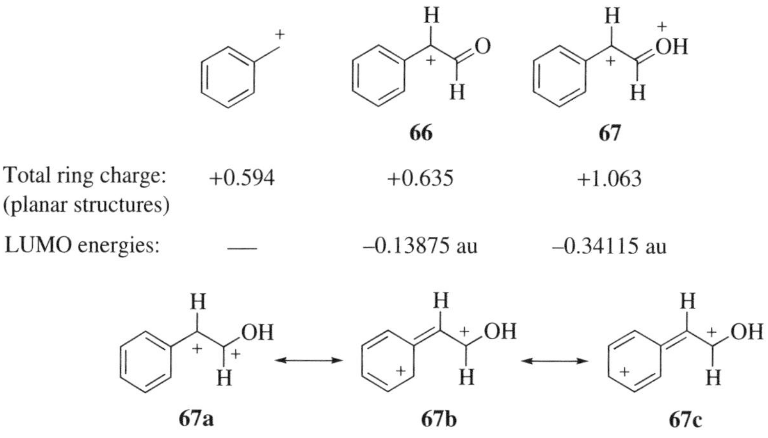

Total ring charge: (planar structures)	+0.594	+0.635	+1.063
LUMO energies:	—	−0.13875 au	−0.34115 au

Figure 2.

from experimental studies that charge-charge repulsive effects can lead to separation of the charge centers. The calculations also show that the LUMO energy decreases considerably upon formation of the dicationic system (−0.13875 au for the monocation **66** and −0.34115 au for the dication **67**). Moreover, calculations of the LUMO coefficients indicate that resonance forms such as **67b** and **67c** are important to the description of the dicationic structure. These two aspects, LUMO lowering and charge delocalization, are critical factors in the superelectrophilic activation of these 4π-electron controtatory electrocyclizations.

Further experimental evidence for gitonic superelectrophiles comes from the catalytic activation of the monocationic species.[32a] When the α-chloro ester (**68**) is reacted with $AgBF_4$ or $Ag(O_2CCF_3)$, the monocationic intermediate (**69**) is cleanly formed. This is evident from the observed trapping product (**57**, eq 14).

$$(14)$$

With the monocationic species, no fluorene cyclization is observed. However upon addition of CF_3SO_3H, the cyclization occurs almost quantitatively. This is consistent with formation of the protonated, dicationic intermediate (**46**) leading to the cyclization product (**59**). In this same study, it is noted that other stable monocationic 1,1-diarylethyl cations (i.e., the 1,1-diphenylethyl cation) do not readily form the fluorene ring system, indicating the importance of superelectrophilic activation.

Besides the fluorene cyclization, the 1,2-ethylene dications are known to undergo another type of electrocyclization reaction to produce the phenanthrene ring system (see eqs 9 and 11 and Scheme 3).[32a] Although derived carbenium-carboxonium dications (**70**) will give the phenanthrene-type products (albeit in low yield; eq 15), the bis-carboxonium dication **71** does not. Moreover, the fluorene cyclization occurs readily in 1,1-dihydroxylic systems (eq 10).

$$(15)$$

$$(16)$$

In order to understand this apparent difference in reactivities, two isomeric dications (**72** and **73**) were studied calculationally.

$$
\begin{array}{cc}
\underset{\textbf{72}}{
\overset{\displaystyle
\begin{array}{c}
\text{H-O} \quad \text{H} \\
{}^{+}\;\;{}^{+} \\
\text{C-C} \\
\text{H} \qquad \text{O-H}
\end{array}
}{}
}
&
\underset{\textbf{73}}{
\overset{\displaystyle
\begin{array}{c}
\text{H} \qquad \text{O-H} \\
{}^{+}\;\;{}^{+} \\
\text{C-C} \\
\text{H} \qquad \text{O} \\
\qquad \text{H}
\end{array}
}{}
}
\end{array}
$$

Relative Energies: 0 kcal•mol^{-1} 25 kcal•mol^{-1}

Calculations at the 4-31G level of theory found that the 1,2-bis-carboxonium ion (**72**) is significantly more stable than the carbenium-carboxonium dication (**73**). Thus, when compared with the 1,1-dihydroxylic structure **73**, the 1,2-dihydroxylic structure **72** is more effective at stabilizing the 1,2-ethylene dication. This effect leads to the stabilization of dication **71** (eq 16), and not to phenanthrene cyclization.

Other oxygen stabilized ethylene dications have been prepared and studied using stable ion conditions. For example, Olah and White were able to observe as early as 1967 diprotonated oxalic acid (**74**) in FSO$_3$H–SbF$_5$ solutions.[33] This species and related dications were also studied subsequently by theoretical methods. Diprotonated α-keto acids and esters have likewise been generated in superacidic media, such as **75–77**.[34]

74   **74**   **75**   **76**

77   **77**

For each of these ions, the ^{1}H and ^{13}C NMR spectra are consistent with dicationic species. In the case of dication **77**, the ^{13}C signal for the *para*-carbon is found at δ^{13}C 177.5 (CF$_3$SO$_3$H–SbF$_5$ solution at $-20°$C). This again indicates a significant amount of positive charge delocalization arising from charge-charge repulsive effects.

Similar oxygen stabilized ethylene dications were proposed in several types of superacid-catalyzed condensation reactions involving 1,2-dicarbonyl compounds. For example, 2,3-butanedione condenses in high yield with benzene and the superelectrophile (**35**) is considered to be the key intermediate, because the monoprotonated species (**78**) is not sufficiently electrophilic to react with benzene (eq 17).[35] Several biologically important α-ketoacids were also found to generate superelectrophiles

$$(17)$$

$$(18)$$

(**79–82**) in CF_3SO_3H, all of which condense with benzene, (eq 18 and *vide infra*).[36] In the case of **81** and **82**, it is likely that these superelectrophiles are in equilibrium with diprotonated species.

Dication **79** has been observed by low temperature NMR from $FSO_3H–SbF_5$ solutions.[34] Although diprotonated quinones (**83–84**) are expected to have extensive delocalization of the positive charges, several have been reported, and their condensation chemistry suggests a relatively high level of electrophilic reactivity (eq 19).[37] Diprotonated acenaphthenequinone and aceanthrenequinone (**83** and **84**) have been proposed as intermediates in superacid-catalyzed condensation reactions. High molecular weight polymers have also been prepared by superelectrophilic condensations involving **83**.[37b,c] Both superelectrophiles **83** and **84** were studied by calculations, while **83** has been directly observed by spectroscopic methods. In each case, calculations at the B3LYP/6-311G**//B3LYP/6-311G** level show that the global minima involve protonation at both carbonyl oxygen atoms.[38] Aceanthrenequinone and acenaphthenequinone have been shown to condense with moderately deactivated arenes in the superacid-promoted reactions (eq 19).

$$(19)$$

This indicates that strong electrophilic species (i.e., **84**) are generated from the quinones.

Besides oxygen and aryl-stablized ethylene dications, there have been several reports of nitrogen-stabilized systems. For example, Olah and co-workers described the oxamide dication (**85**) using theoretical and experimental studies (eq 20).[39] When oxamide is dissolved in FSO_3H–SbF_5 solution at $-78°C$, the ^{13}C NMR of dication **85** is obtained with a single ^{13}C signal at $\delta^{13}C$ 157.9. This experimental result agrees reasonably well with the IGLO-calculated chemical shift of $\delta^{13}C$ 165.7 for the dication **85**.

$$(20)$$

Calculations at the MP2/6-31G*//MP2/6-31G* level indicates that the C_{2h} fully planar structure is about 8 kcal/mol more stable than the C_{2v} perpendicular form (a transition state). Likewise, condensation reactions with isatins have suggested the involvement of the superelectrophile **86** (eq 21).[40] This superelectrophilic chemistry has been successfully applied in the preparation of hyperbranched polymers and other macromolecules.[40b–d] Other synthetic studies have used the isatin condensations to prepare biologically active products.[40e] The monooxime of 2,3-butanedione is protonated twice in CF_3SO_3H to generate the dication **87** (eq 22).[41]

$$(21)$$

$$(22)$$

Both of these superelectrophiles (**86–87**) are capable of reacting with benzene in condensation reactions. Dication **87** can be directly observed by low-temperature NMR, being generated by dissolving the oxime in CF_3SO_3H–SbF_5 solution at $-30°C$. Based on the NMR studies, the pK_a of dication **87** is estimated to be about -17 to -19, suggesting that in CF_3SO_3H ($H_0 - 14$) a low but significant concentration of the dication is present.

Another ethylene dication with significant nitrogen stabilization is the tetrakis(dimethylamino)ethylene dication (**88**). By reaction of the substituted-ethylene with Cl_2 or Br_2 (eq 23),

$$(23)$$

the dication salt (**88**) can be formed and its X-ray crystal structure has been determined.[42] Shortening of the C–N bonds and twisting of the N–C–N sub-units $76°$ relative to each other is consistent with almost complete delocalization of the positive charges. The same dication (**88**) can be prepared as a 1:2 salt by the reaction of tetrakis(dimethylamino)ethylene and tetracyanoethylene. The stability of this dication is evident from the formation of stable salts with counter ions having relatively high nucleophilic nature.

An unusual class of gitonic superelectrophiles (having *vicinal*-dicationic structure) are the protosolvated alkylcarbenium ions (of carbenium-carbonium dication nature). Several studies have shown that alkylcarbenium

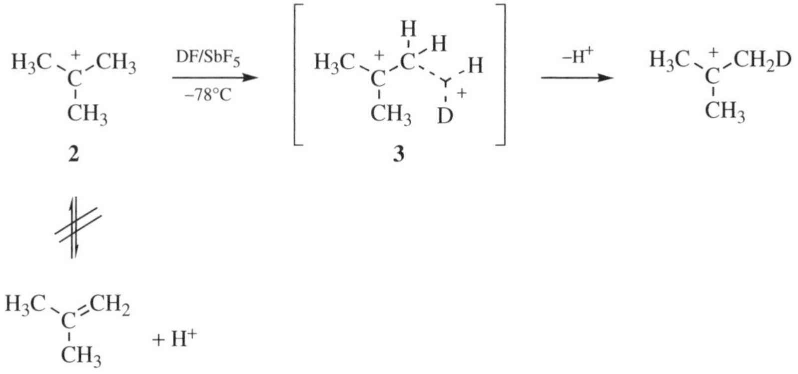

Scheme 6.

ions can undergo protosolvation of the C–H σ-bonds in superacidic media.[43] For example, the *tert*-butyl cation (**2**) is a remarkable stable species in superacids with no deprotonation to isobutylene. However, when it is generated in an SbF_5/SO_2ClF solution and then 1:1 DF/SbF_5 is added, the *tert*-butyl cation undergoes slow but well-characterized (by ^1H and ^2H NMR) hydrogen-deuterium exchange (Scheme 6).[43a] To account for the isotopic exchange in superacid, protonation (deuteration) of the *tert*-butyl cation (**2**) to the protio-*tert*-butyl dication (**3**) or the corresponding protosolvation must occur. Theoretical calculations have estimated that C–H protonation of **2** is more favorable than C–C protonation by about 14 kcal/mol.

In a similar respect, 2-chloropropane can be ionized in SbF_5/SO_2ClF at $-78°C$ to the 2-propyl cation (**89**).[43b] Subsequent addition of 1:1 DF/SbF_5 again leads to slow hydrogen-deuterium exchange. The exchange is envisaged to occur by reactions at the C–H bond via the dication **90** (eq 24).

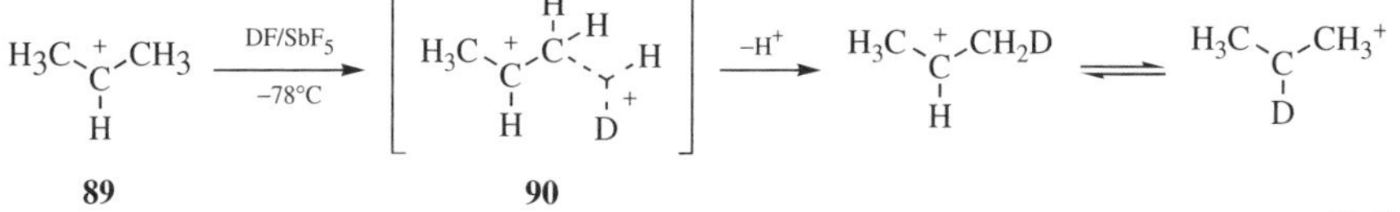

(24)

Alkyl cations like the *tert*-butyl cation (**2**) and 2-propyl cation (**89**) are significantly stabilized by hyperconjugative C–H and C–C σ-back donation into the empty carbocationic p-orbitals. Protosolvation involving σ-bonds can diminish this hyperconjugative stabilization and thus lead to superelectrophilic carbocationic species.

5.2.2.2 Carbon-Nitrogen Vicinal-Dications

There are several types of gitonic superelectrophiles having the 1,2-dicationic structure involving carbon-nitrogen bonds. As described in Chapter 2, diprotonated nitriles and cyanides have been proposed as intermediates in superacid-catalyzed reactions.[44] The Gatterman reaction (the acid-catalyzed reaction of cyanides with arenes) is known to provide improved yields of products with increasing acidity of the catalyst. Shudo and Ohwada studied the superacid catalyzed reaction of trimethylsilyl cyanide with aromatics and found significantly increased reaction rates and yields with stronger superacids (eq 25).

acid system	time	product yield (ca. 25% ortho, 75% para)
100% CF_3SO_3H ($H_0 - 14$)	30 min	2%
95% CF_3SO_3H, 5% SbF_5 ($H_0 < -18$)	10 min	93%

$$(25)$$

Based on these results, the diprotonated superelectrophile (**92**) is suggested to be involved in the reactions (eq 26).

$$(26)$$

Theoretical studies of dication **92** have shown it to reside in a deep potential energy well, with a gas-phase barrier to deprotonation estimated to be about 42 kcal/mol (MP4/6-311G** level).[45]

The nitrilium salt **93** is known to react with activated arenes, but in relatively weak acids (CF_3CO_2H, H_0 -2.7), it does not react with benzene or deactivated arenes. However in 5% SbF_5 : 95% CF_3SO_3H H_0 -18 solution, the nitrilium salt **93** gives with benzene benzophenone in 55% yield (eq 27).

$$(27)$$

Since the reaction only occurs in highly acidic media, it is suggested that the gitonic superelectrophile (**94**) is formed and leads to product.[44]

Kinetic evidence suggests the involvement of gitonic superelectrophiles also in superacid-promoted Houben-Hoesch reactions.[44] As discussed previously, 4-phenylbutyronitrile (**6**) is found to cyclize at an appreciable rate only in solutions more acidic than $H_0 -10$ (eq 28).

$$(28)$$

This is indicative of the formation of the diprotonated intermediate **8** and its involvement in the rate-limiting step of the cyclization.

Protonated iminium dications have also been studied by theoretical and gas-phase experimental studies.[46] For the methyleniminium ion (**95**), the N-protonated species (**96**) was found to be a stable minimum. A second dicationic species was also located as a minimum, corresponding to the nonclassical structure **97**, but it lies 27 kcal/mol above **96** on the potential energy surface (MP2/6-31G** level).

Using charge stripping mass spectrometry, Schwarz and co-workers found evidence for the formation of the gitonic superelectrophile **96** in the gas phase (generated from the ethylamine radical cation, $CH_3NH_2^{+\bullet}$).[47] The ethyleniminium and related dications have also been studied using theoretical methods.[46] Three types of structures have been found at the global energy minima from their respective iminium ions (eqs 29–31):

	98	**99**
Relative Energy, kcal/mol:	0.0	6.0

$$(29)$$

$$(30)$$

100

$$(31)$$

101

the nitrogen-protonated charge-separated dication (**98**), the nitrogen protonated dications (**99** and **100**), and the C–H protonated dication (**101**). The preferred type of structure is evidently determined by the pattern of substitution of the iminium ion. In the case of the isopropyleniminum ion (eq 30), nitrogen protonation leads to a carbenium ion center (**100**) that is stabilized by hyperconjugation with the methyl groups. Similar stabilization occurs with dication **99** (eq 29). However the repulsive effects of the positive charges lead to charge separation and stabilization of structure **98**.

In superacidic reactions, diprotonated imines form gitonic superelectrophiles.[48] As described in Chapter 2, kinetic experiments have shown that diprotonated intermediates are involved in these conversions. Other experiments showed that the reaction provides higher yields in stronger acid systems (eq 32),

102   **103**   **104**

acid system	time	temperature	product yield
100% CF_3SO_3H ($H_0 - 14$)	2 h	60°C	94%
100% CF_3CO_2H ($H_0 - 2.7$)	26 h	72°C	50%

$$(32)$$

indicating the involvement of *vicinal*-superelectrophiles (i.e., **104**). In order to further probe the superelectrophilic activation, compound **105** was reacted in deuterated triflic acid (CF_3SO_3D; 23°C for 24 hrs) and deuterium incorporation was only seen in the phenethyl group (Figure 3). This H/D exchange must occur through intermediates such as **106**. Thus, superelectrophilic ring closure does not occur via dications **107** or **108**, but rather involves an *N,N*-diprotonated species (such as **104** or **109**).

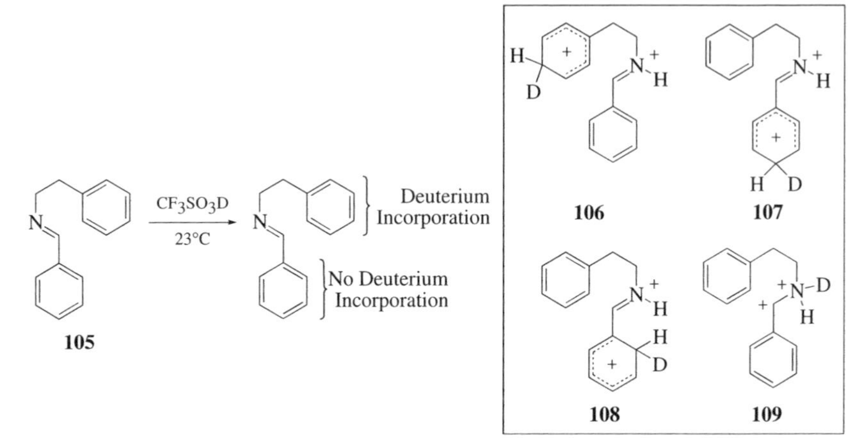

Figure 3.

Another example of a carbon-nitrogen gitonic superelectrophile is the protonated guanidinium ion.[49] The guanidinium ion (**110**) can be protonated in $FSO_3H:SbF_5$ to give the gitonic superelectrophile (**111**; eq 33),

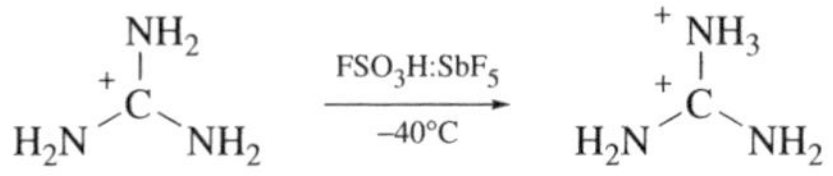

$$(33)$$

which is a stable dication in the superacidic medium. Dication **111** was characterized with ^{13}C, ^{1}H, and ^{15}N NMR spectroscopy.

As described in Chapter 2, a unique gitonic superelectrophile is considered to be involved in an enzyme system that converts CO_2 to methane. Berkessel and Thauer have studied this metal-free hydrogenase enzyme from methanogenic archaea and a mechanism is proposed involving activation through a *vicinal*-superelectrophilic system (eq 34).[50]

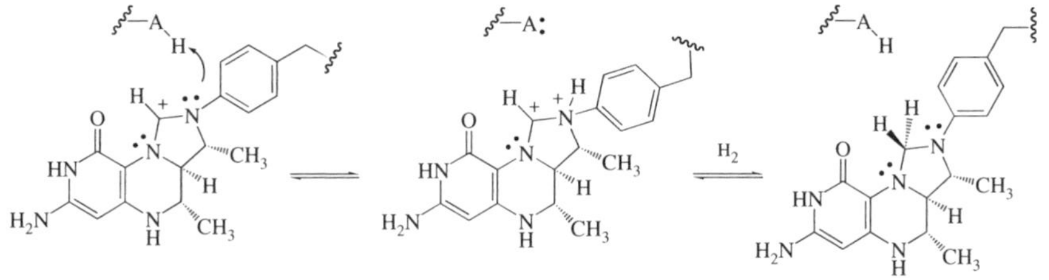

$$(34)$$

Although the exact nature of the superelectrophilic intermediate is not known, it was suggested that protosolvation of one or both of the iminium nitrogen atoms leads to the superelectrophile, such as **113**. This results in a decreased neighboring group stabilization and enables reaction with the weakly nucleophilic molecular hydrogen. It was also noted that localized, highly acidic realms (such as those necessary to produce a superelectrophile) may be possible in an enzyme's active site given properly positioned acidic functional groups and a nonbasic local environment. In accord with the proposed superelectrophilic activation, model studies have shown that the carbenium ion salts themselves are unreactive towards molecular hydrogen, even in the presence of added base (eq 35).[51]

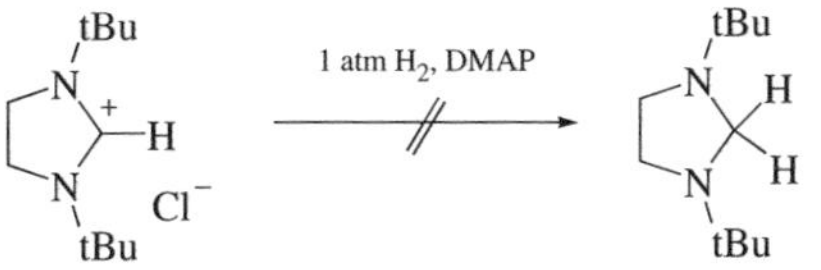

$$
\text{(35)}
$$

The metal-free hydrogenase enzyme system was also studied by theoretical calculations.[52] This study involved the reactions of two model systems (**115** and **116**) with molecular hydrogen (Table 3). Cation **115**

Table 3. Calculated model reactions of the amidinium ions 115 and 116 with H_2 in the absence and presence ofa base ((BLYP/6-311G(2 d, 2p) level).

	ΔH (kcal•mol^{-1})
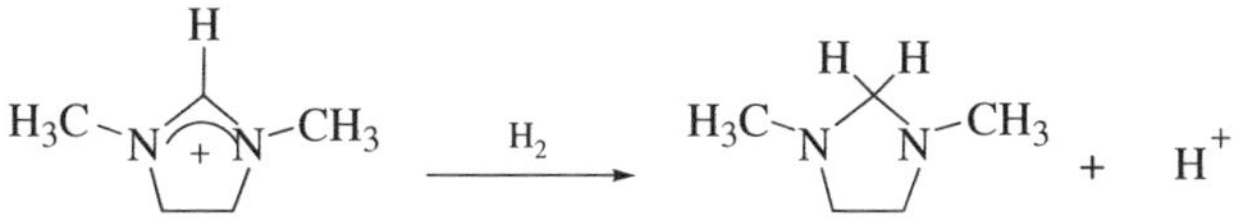 **115** $\xrightarrow{H_2}$ **117** + H^+	244
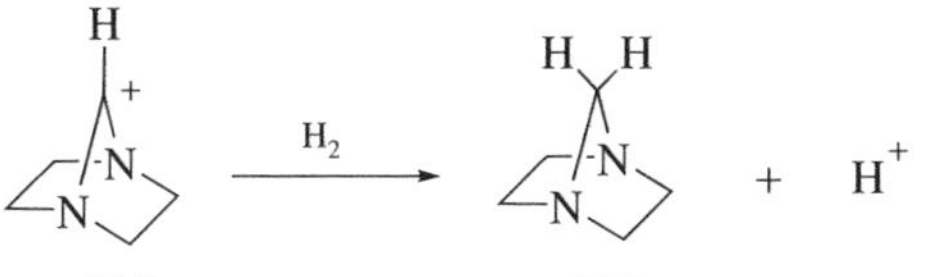 **116** $\xrightarrow{H_2}$ **118** + H^+	149
115 $\xrightarrow[H_2O]{H_2}$ **117** + H_3O^+	33
116 $\xrightarrow[H_2O]{H_2}$ **118** + H_3O^+	−57
115 $\xrightarrow[NH_3]{H_2}$ **117** + NH_4^+	4
116 $\xrightarrow[NH_3]{H_2}$ **118** + NH_4^+	−87

represents the resonance stabilized amidinium ion of $N^5.N^{10}$–methenyl tetrahydromethanopterin (**112**), while cation **116** serves as a model having resonance stabilization diminished or eliminated. In Berkessel and Thauer's mechanistic proposal, loss of planarity is suggested as one of the factors that makes the system (**112**) more electrophilic and reactive towards hydrogen. Using the BLYP/6-311G(2 d, 2p) level of theory and solvation modeling using SCI-PCM (self-consistent isodensity polarized continuum model), it was found that reactions with molecular hydrogen were energetically more favorable with cation **116** compared with cation **115**. When the calculations included reaction of base with the proton generated from the electrophilic reaction, the reaction energetics show exothermic or slightly endothemic values (4 kcal/mol, **115**, H_2, and NH_3). The authors suggest that these data indicate the thermodynamic feasibility of the Berkessel-Thauer mechanism in the presence of a base. However, these studies did not address the possibility of superelectrophilic activation, and ascribed reactivtity to decreased neighboring group stabilization by a folding or bending of the imidazolidine ring, rather than due to protonation of the nitrogen lone pair electrons.

Pnictogenocarbenium ions $[H_2C{=}XH_2]^+$ (**119–121**, X = P, As, Sb) and their superelectrophilc dications have been studied using high-level *ab initio* calculations.[53] Cations **119–121** can be protonated at either the C- or X-sites (Figure 4), and studies of proton affinities (PA) show an increasing PA in the order P < As < Sb (for both C-protonation and X-protonation). For cation **119**, calculations indicate that protonation at the

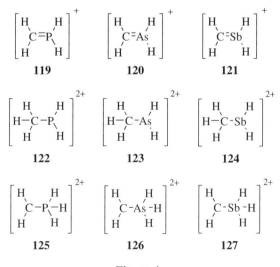

Figure 4.

P-site (**125**) is favored by 3.3 kcal/mol over protonation at the C-site (**122**). However, protonation at the carbon is favored for cations **120** and **121**. All three dicationic species **123**, **124**, and **125**, are predicted to be kinetically stable with significant energy barriers towards homolytic or heterolytic fragmantation reactions. In studies of isodesmic reactions with methane, the dicationic species are shown to be far more reactive (eqs 36–37).

$$
\begin{bmatrix} \text{H} & \text{H} \\ \ & \text{C=X} \\ \text{H} & \text{H} \end{bmatrix}^{+} + \text{H}-\overset{\text{H}}{\underset{\text{H}}{\text{C}}}-\text{H} \xrightarrow{\text{X = P, As, Sb}} \text{H}-\overset{\text{H}}{\underset{\text{H}}{\text{C}}}-\text{X}\overset{\text{H}}{_{\text{H}}} + \text{H}-\overset{\text{H}}{\underset{\text{H}}{\text{C}}}^{+} \qquad (36)
$$

119–121

$$
\begin{bmatrix} \text{H} & \text{H} \\ \text{C}-\text{X}-\text{H} \\ \text{H} & \text{H} \end{bmatrix}^{2+} + \text{H}-\overset{\text{H}}{\underset{\text{H}}{\text{C}}}-\text{H} \xrightarrow{\text{X= P, As,Sb}} \begin{bmatrix} \text{H} & \text{H} \\ \text{H}-\text{C}-\text{X}-\text{H} \\ \text{H} & \text{H} \end{bmatrix}^{+} + \text{H}-\overset{\text{H}}{\underset{\text{H}}{\text{C}}}^{+} \qquad (37)
$$

125–127

For the reactions of cations **119–121** with methane, all the conversions are endothermic (eq 36). Hydride abstractions involving the dications (**125–127**) are, however, strongly exothermic (eq 37). Thus, the isodesmic reactions go from being moderately endothermic to strongly exothermic upon changing involvement of monocationic species (**119–121**) to the superelectrophilic, dicationic species (**125–127**).

5.2.2.3 *Carbon-Oxygen Vicinal-Dications*

A parent system of these *vicinal*-dications is formed upon diprotonation of carbon monoxide. Carbon monoxide can be protonated at carbon to yield the formyl cation (**128**, HCO^+) or at oxygen to yield the isoformyl cation (COH^+), though the latter is estimated to be around 38 kcal/mol less stable than the formyl cation.[54] The formyl cation has been observed in the gas phase using infrared, microwave, and mass spectroscopy. In the condensed phase, the direct observation of the formyl cation has been reported by Gladysz and Horvath in $HF–SbF_5$ solution using high-pressure NMR spectroscopy.[55] While the formyl cation is considered to be sufficiently electrophilic to react with arenes (Gatterman-Koch formylation), a more activated protosolvated formyl cation (or the protoformyl dication **129**) is needed in reactions with saturated hydrocarbons (Scheme 7). For example, Olah et al. found that isoalkanes are converted to branched ketones in high yields by the reaction of carbon monoxide with $HF–BF_3$.[56] The carbonylation-rearrangement of isoalkanes to branched ketones is fundamentally different from the Koch-Haaf type carbonylative carboxylation which exclusively gives branched carboxylic acids. A mechanism is suggested

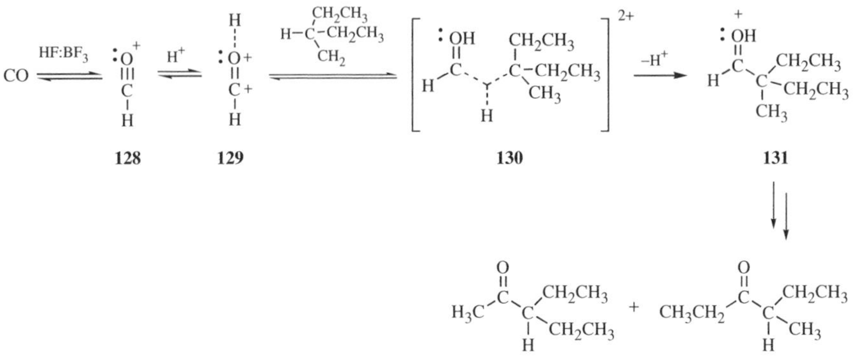

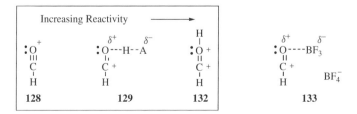

Scheme 7. Proposed mechanism for the formylation of isoalkanes and rearrangement to ketone.

involving the superelectrophilic formyl dication **129** (or its protosolvated form) and reaction with the σ-donor *tert* C–H bonds of the isoalkanes through a 3 center-2 electron bond (**130**). Rearrangement of the intermediate products (i.e., **131**) then gives the branched ketones, potential high octane gasoline additives. As described in Chapter 2, superelectrophilic carbonylation has also been demonstrated with adamantane.

Further evidence for the protoformyl dication comes from aromatic formylation with CO in superacids.[57] It has been shown in the formylation of toluene/benzene that the k_T/k_B rate ratio is sensitive to the strength of acidic catalyst. Formylation in HF–SbF$_5$ gives a low k_T/k_B rate ratio of 1.6 and a significant amount of *ortho* substitution, while with less acidic systems, formylation gives a k_T/k_B rate ratio of up to 860 with predominantly *para* substitution. These data indicate that the formylating agent in HF–SbF$_5$ is an extreme reactive electrophilic species as it does not discriminate between toluene and benzene. It has been proposed that the protoformyl dication (**132**) or a protosolvated species (**129**) are involved in these superacid-catalyzed formylations (Scheme 8). Olah and co-workers have also shown that HCOF with excess BF$_3$ is also an exceedingly reactive formylating agent, having a significantly enhanced

Scheme 8. Protosolvation of the formyl cation (**128**).

reactivity over the 1:1 complex HCOF:BF$_3$.[58] The Lewis acid-complexed 1:2 superelectrophile (**133**) was suggested as the reacting species.

In work by Stahl and Maquin, the protoformyl dication was observed in the gas-phase by charge-stripping mass spectrometry.[59] Ionization of methanol and reaction with O$_2$ leads to detectable HCOH^{2+} (or DCOH^{2+} when CD$_3$OH is used). Several theoretical studies have examined the structure and stability of the protoformyl dication (**132**) and its isomers.[60] At the MP4/6-31G** level, the protoformyl dication (**132**) is found to be in a fairly deep potential energy well. The barriers for proton loss are 20 kcal/mol (O–H protonation, **134**) and 47 kcal/mol (C–H deprotonation, **135**) above the energy minimum for **132** (Figure 5). In structural studies, the protoformyl dication is found to have a relatively short C–O bond length, suggesting considerable π-bonding.

Related classes of gitonic superelectrophiles are the previously mentioned protoacetyl dications and activated acyl cationic electrophiles. The acyl cations themselves have been extensively studied by theoretical and experimental methods,[22] as they are intermediates in many Friedel-Crafts reactions. Several types of acyl cations have been directly observed by spectroscopic methods and even were characterized by X-ray crystal structure analysis. Acyl cations are relative weak electrophiles as they are effectively stabilized by resonance. They are capable of reacting with aromatics such as benzene and activated arenes, but do not generally react with weaker nucleophiles such as deactivated arenes or saturated alkanes.

There are several reports of activation of acyl cations by superacids, suggesting the involvement of gitonic superelectrophiles.[61] As discussed in Chapter 2, hydride transfer from isobutane to the acetyl cation has been reported when the reaction is carried out in excess HF–BF$_3$. At the same

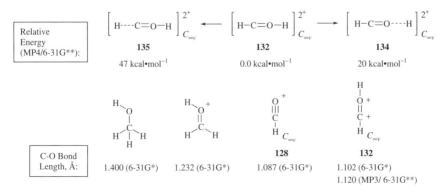

Figure 5. Calculated structures/energies and C-O bond lengths of protonated formyl cations.

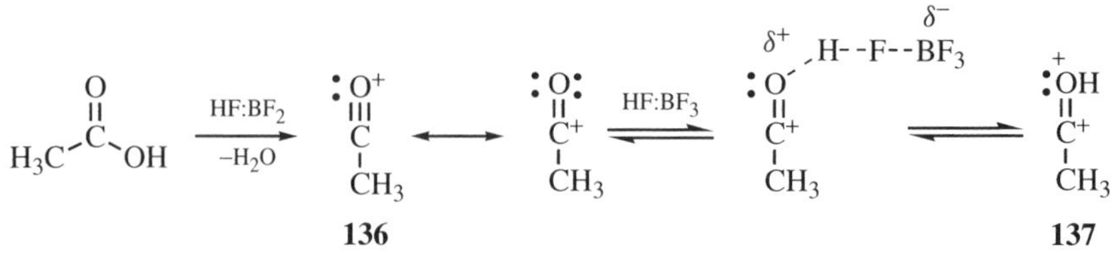

Scheme 9. Protosolvation of the acetyl cation **136** to superelectrophile **137**.

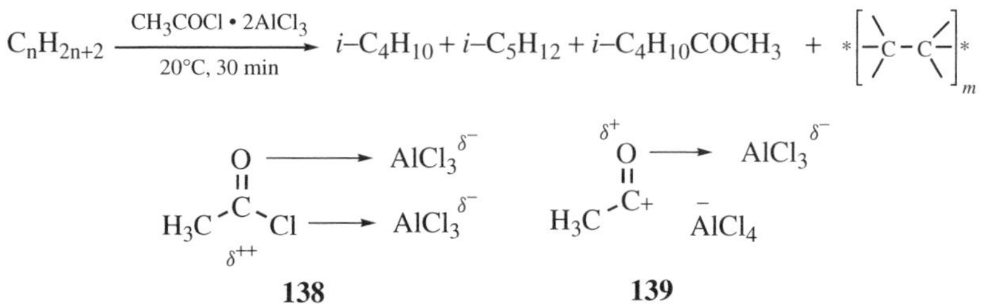

Scheme 10. Electrophilic solvation of the acetyl cation and reactions with alkane.

time, hydride transfer does not occur with acetyl cation salts in aprotic solvents such as SO_2, SO_2ClF, or CH_2Cl_2.[62] These results are consistent with the activation by protosolvation of the acetyl cation (generated from acetic acid) by the Brønsted superacid (Scheme 9). The monocationic acetyl ion (**136**) is unreactive, while the dicationic species (**137**) abstracts hydride. In addition to protosolvation, the activation of the acetyl cation can also be accomplished with Lewis acids. As Vol'pin found, reaction of CH_3COCl in the presence of two or more equivalents of $AlCl_3$ with alkanes leads to ready isomerization and other electrophilic reactions (Scheme 10).[63] Since the reaction does not occur when only one equivalent of $AlCl_3$ is used, the results indicate evidence for electrophilic solvation involving a dicationic complex (**138** or **139**). In a similar respect, pivaloyl chloride was found to give methyl isopropyl ketone in reactions with hydride donors and a large excess of $AlCl_3$ (eq 38).[64]

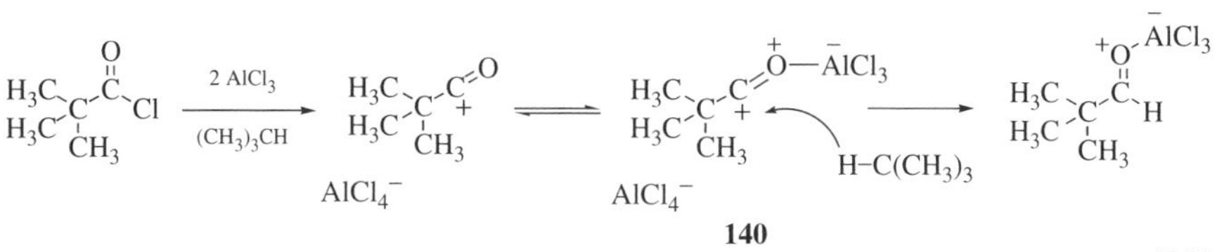

$$(38)$$

A proposed mechanism involves hydride abstraction by the superelectrophilic pivaloyl-$AlCl_3$ dicationic complex (**140**), which in turn undergoes the mentioned rearrangement to methyl isopropyl ketone.

As discussed in Chapter 2, Shudo and Ohwada found that acetyl and benzoyl hexafluoroantimonate salts ($CH_3CO^+SbF_6^-$ and $C_6H_5CO^+SbF_6^-$, respectively) acylate aromatic compounds with increasing reaction rates and greater yields in progressively more acidic media.[44] These data are consistent with the formation of the superelectrophilic species **137** and **141** (eq 39).[65]

$$(39)$$

Like other superelectrophiles, protonation of the acyl cations leads to decrease of neighboring group participation and a lowering of the LUMO energy level, which facilitates reactions with even weak nucleophiles.

Although attempts to directly observe the protoacetyl dication (**137**) by low temperature NMR spectroscopy have not been successful, the stretching frequency of the acetyl cation carbonyl group is significantly shifted in superacids.[63] The protoacetyl dication (**137**) has been observed in the gas phase by charge-stripping mass spectroscopy, obtained by the dissociative ionization of pyruvic acid to generate $CH_3COH^{+\bullet}$ followed by charge stripping via collision with O_2.[66] The gas phase observation of the protoacetyl dication is also consistent with *ab initio* calculations.[67] Despite possessing high thermochemical instability towards deprotonation, calculations indicate a significant kinetic barrier towards such unimolecular dissociation. The protoacetyl dication (**137**, C_{3v} symmetry) was found to be at the global energy minimum for the $C_2H_4O^{2+}$ structures (MP2/6-31G*//4-31G+ZPE level of theory). Based on the high enthalpy of formation of **137** and the exothermicity of model disproportionation reactions, it was predicted that the free protoacetyl dication was unlikely to exist in solution.[66] Nevertheless, superelectrophilic activation of the acyl salts may not require formation of the discrete fully formed dicationic species (**137**), but may involve partial protonation or protosolvation (Scheme 9) in the condensed phase. Moreover, solvation and ion pairing effects may significantly stabilize these doubly-electron deficient species.

Another related class of gitonic superelectrophiles are the superelectrophilic halocarbonyl dications. The halocarbonyl cations (XCO^+, X = F, Cl, Br, I) have been prepared under long-lived stable ion conditions and characterized by ^{13}C NMR spectroscopy.[68] Sommer and co-workers studied the bromine-assisted carbonylation of propane in superacids, and

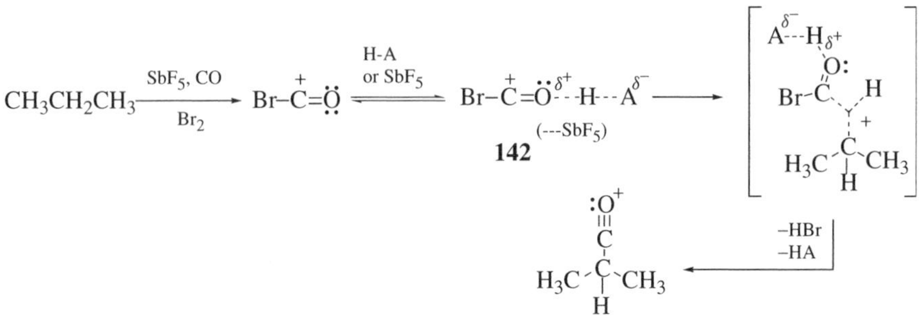

Scheme 11.

the bromocarbonyl cation was proposed as the reactive electrophilic intermediate.[69] Considering the high acidity of the reaction mixture, an alternative mechanistic proposal involves the superelectrophilic protosolvated species (**142**). Reaction with propane, an extremely weak nucleophile is considered to require further activation of the electrophile by coordination with the superacid (Scheme 11). This leads to the carbonylation product.

Carboxonium ions are an important group of electrophiles in chemistry.[22] Carboxonium ions are categorized by the number of oxygen atoms bound to the carbon atom (one, two, or three) and by the groups bonded to the oxygen atoms (as acidic or non-acidic carboxonium ions). All of these types of carboxonium ions have the potential to form *vicinal*-dications by protonation or complexation of the non-bonded electron pairs of oxygen with acids (eq 40).

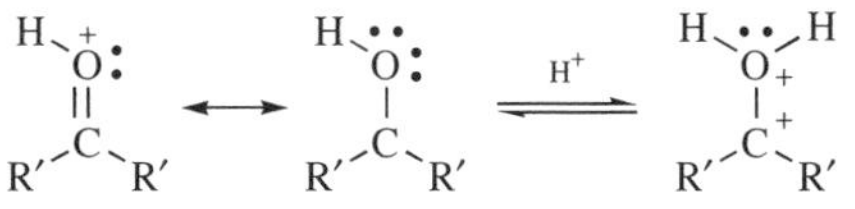

$$R' = \text{alkyl, aryl, alkoxy, OH} \tag{40}$$

A wide variety of these superelectrophilic carboxonium ions have been studied. It has long been recognized that carboxonium ions are highly stabilized by strong oxygen participation and therefore are much less reactive than alkyl cations. For example, trivalent carbocations are efficient hydride abstractors from tertiary isoalkanes (eq 41).

$$\tag{41}$$

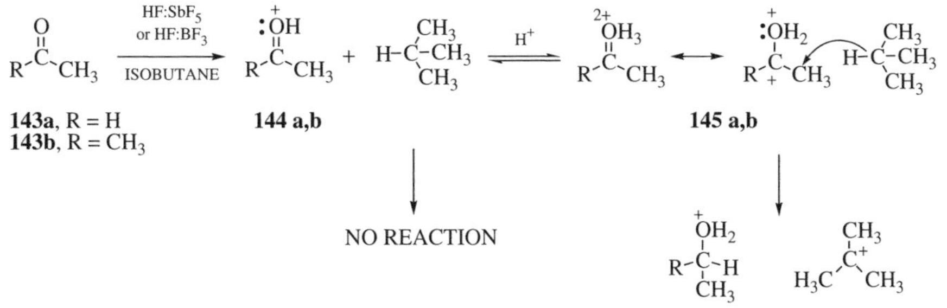

143a, R = H
143b, R = CH₃

144 a,b

145 a,b

NO REACTION

Scheme 12. Proposed mechanism for the superacid-catalyzed hydride transfer involving carboxonium ions and isobutane.

In contrast, carboxonium ions do not hydride abstract from isoalkanes when they are reacted in aprotic solvents. However, as mentioned Brouwer and Kiffin observed that reaction of acetaldehyde or acetone (**143a,b**) with isobutane in $HF:SbF_5$ or $HF:BF_3$, leads to formation of the *tert*-butyl cation (Scheme 12).[70] Under these conditions, the carbonyl group of acetaldehyde or acetone is completely protonated. However, calculations estimate hydride transfer from isobutane to the carboxonium ions (**144a,b**) to be thermodynamically unfavorable. These considerations lead to the early mechanistic proposal by Olah *et al.* in 1975 involving the protosolvation of the carboxonium ions (**144a,b**) forming superelectrophilic species (**145a,b**), enabling hydride abstraction to occur.[62] Similar reactions have also been observed for formaldehyde and its protonated ions.[71] In gas-phase studies, it has been shown that charge-stripping of the radical cation $[CH_2OH_2]^{+\bullet}$ leads to the formation of $CH_2OH_2^{2+}$ (diprotonated formaldehyde) and quantum mechanical calculations also show that $CH_2OH_2^{2+}$ sitting in a considerably deep potential energy well.[72]

Several recent reports have described kinetic studies suggesting protosolvated carboxonium ion intermediates. Shudo, Ohwada, and associates have described kinetic experiments and theoretical studies to show that superelectrophilic carboxonium ions (i.e., **146**) are involved in superacid-catalyzed Nazarov reactions (Scheme 13; see also Chapter 2).[73] Using deuterated superacid, no deuterium was incorporated into the product, suggesting protonation only at the carbonyl group rather than at the olefinic moiety resulting in an intramolecular Friedel-Crafts-type reaction. In *ab initio* calculations at the B3LYP/6-31G* level, diprotonation is shown to significantly lower the energy barriers (compared to the monoprotonated species) leading to the products of the 4π-electrocyclizations.

Similarly, kinetic experiments have shown that superelectrophilic carboxonium dications are involved in the cyclodehydrations of 1,3-diphenyl-1-propanones.[4a] Several examples of the cyclodehydrations were described, including the trifluoromethyl-substituted system (Scheme 14;

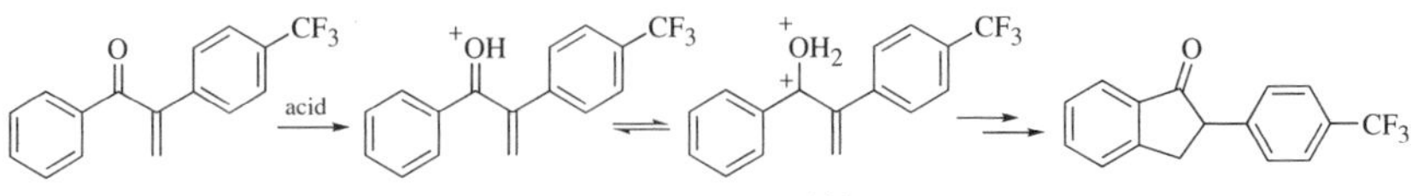

146

acid system	time	temperature	product yield
100% CF₃SO₃H (H₀−14)	5 h	0°C	99%
94% CF₃CO₂H : 6% CF₃SO₃H (H₀−9)	5 h	0°C	43%

Scheme 13.

see also Chapter 2). The reactions exhibit significantly increased yields and reaction rates in more highly acidic media. Along with the results from kinetic studies, these data indicate the involvement of superelectrophilic carboxonium ions (i.e., **148**). Monocationic carboxonium ions such as **147** are in contrast significantly stabilized, and they either do not give the cyclization products or they react very slowly. The activation parameters from the superacid-promoted cyclization (CF₃SO₃H, H_0 −14.1) of 1,3-diphenyl-1-propanone were also determined in this study (ΔH 20 kcal/mol, ΔG 29 kcal/mol, and ΔS − 26 cal/K•mol).

The acid-catalyzed condensation reactions of benzaldehyde with benzene have been well studied. A gitonic superelectrophile has been proposed in one study,[6] while distonic superelectrophiles have been suggested in others.[5,7] When benzaldehyde (**149**) is reacted with CF₃SO₃H and C₆H₆, four products are obtained (**150–153**, Table 4). The relative yields of the products vary with conditions. For example, no anthracene is formed when large excess of C₆H₆ is used in the reaction. In studies using catalytic systems of varying acidities, it was found that overall product formation (% yield) increased with acidity. Given that the pK_{BH+} of benzaldehyde **149** is −6.8, the carbonyl group should be almost completely protonated at H_0 −9 or −10. Since the product yields increase with systems more

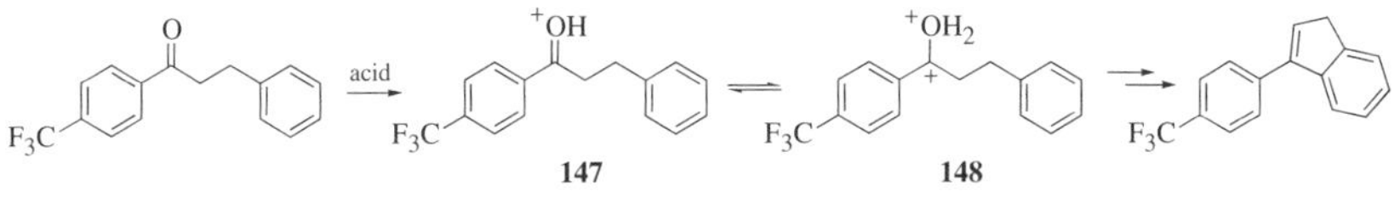

147   **148**

acid system	time	temperature	product yield
100% CF₃SO₃H (H₀−14)	20 min	80°C	68%
95% CF₃CO₂H : 5% CF₃SO₃H (H₀−9)	20 min	80°C	2%

Scheme 14.

Table 4. Products and yields of the reactions of benzaldehyde (149) with benzene at varying acidities.

Acid	Acidity, H_0	Total Yield Products $\left(\begin{array}{c} 150+151+152+153 \end{array} \right)$
100 % CF$_3$CO$_2$H	−2.7	0%
5% CF$_3$CO$_2$H: 95% CF$_3$SO$_3$H	−9.0	4%
5% CF$_3$CO$_2$H: 95% CF$_3$SO$_3$H	−10	15%
5% CF$_3$CO$_2$H: 95% CF$_3$SO$_3$H	−12	40%
5% CF$_3$CO$_2$H: 95% CF$_3$SO$_3$H	−13	77%
100% CF$_3$SO$_3$H	−14	83%

acidic than H_0 −10, it is suggested that diprotonated benzaldehyde is the superelectrophilic intermediate responsible for the condensation chemistry. In considering the mechanism, the first protonation of benzaldehyde occurs at the carbonyl oxygen while the second protonation could be at the ring positions (*ortho*, *meta*, *para*, or *ipso*) or at the carbonyl group. When benzaldehyde is reacted even with a large excess of CF$_3$SO$_3$D at 50°C for 19 h, no deuterium incorporation is observed either on the phenyl ring or aldehyde group of compound **149**. Based on these considerations, it was proposed that the *O,O*-diprotonated species (**12**) is the key intermediate involved (eq 42).

$$(42)$$

In theoretical studies of the benzaldehyde condensation, the diproto-nated structures were calculated at the MP2/6-31G*//MP2/6-31G* level and five structures were found at energy minima (**12–13** and **154–156**). The global minima for the gas-phase $C_7H_8O^{2+}$ structure are found at **13** and **155** (distonic superelectrophiles; same relative energies), both being *C,O*-diprotonated benzaldehyde isomers. The *O,O*-diprotonated species (**12**) is found about 20 kcal/mol^{-1} above **13** and **155** in energy, while **154** and **155** are 14 and 5 kcal/mol^{-1} above the global minima, repec-tively. The condensed phase experiments with CF_3SO_3D and benzalde-hyde argue however against the involvement of **13**. Consequently, an alternative explanation has been suggested involving neither **12** or **13**, based also on the results from analogous chemistry with acidic zeolites.[5] In the latter case, benzaldehyde was found to give triphenylmethane when reacting with benzene in the presence of HUSY zeolite, a solid acid cata-lyst with an acidity estimated at $H_0 \sim -6$. Benzaldehyde is also found to react with cyclohexane (a hydride source) using the same zeolite catalyst. It was argued that the relatively low acidity (and low density of acidic sites) of HUSY makes it improbable to form diprotonated benzaldehyde (**12** or **13**), and a mechanism was proposed involving only monoproto-nated benzaldehyde (**14**, Scheme 2). These considerations assumed the same mechanism to be involved in liquid superacid-promoted condensa-tion. It should be noted however that confinement effects associated with zeolite-catalysis may also have a role in the reactions.

Besides promoting reactions with weak nucleophiles, several molecular rearrangements have been reported with superelectrophilic carboxonium ions suggested as intermediates. This includes the superacid promoted rearrangment of 2,2,2-triphenylacetophenone (**157**) with subsequent for-mation of the 9,10-diphenylphenanthrene (**160**, eq 43).[30]

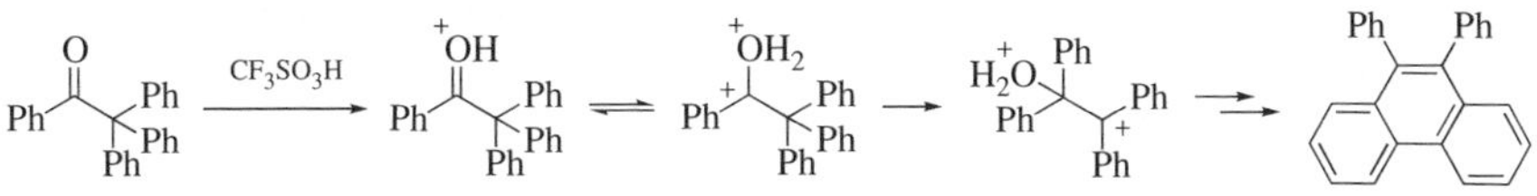

(43)

As described in Chapter 2, the superacid produces the diprotonated gitonic superelectrophile (**159**) which undergoes a retropinacol rearrangement. Phenyl migration is driven by the resulting charge-charge separation. Subsequent reaction steps then give the final cyclization product (**160**).

A related reaction was reported in the discussed acid-catalyzed formylation of isobutane with rearrangement to methyl isopropyl ketone (a potentially significant gasoline additive).[74] Following the insertion of the protosolvated formyl cation into isobutane (*vide supra*), a key step in the formation of the methyl isopropyl ketone involves the superelectrophilic *O,O*-diprotonated species (**162**, Scheme 15). Methyl shift is driven by Coulombic repulsion to give initially the charge-separated species (**163**). Quantum mechanical calculations indicate that the methyl shift is very favorable energetically ($\Delta H = -48$ kcal/mol for the gas-phase reaction). Upon undergoing deprotonation, subsequent hydride shift then gives protonated methyl isopropyl ketone (**166**) in another energetically favorable step (estimated gas-phase $\Delta H = -24$ kcal/mol). Interestingly, calculations also suggest that the hydride shift is unlikely to involve the distonic dication **165**, because such a step would be significantly endothermic ($\Delta H = 24$ kcal/mol) due to charge-charge repulsive effects in **165**. Although, according to calculations, the fully formed gitonic superelectrophile **162** is not a stable minimum in the gas-phase at the B3LYP/6-31G* level of theory, partial proton transfer (protosolvation) is also likely to activate the electrophilic center. Moreover, ion pairing may also stabilize the superelectrophile **162** in the condensed phase.

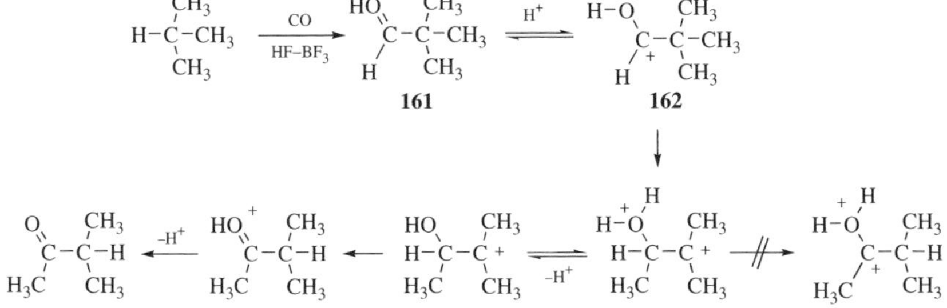

Scheme 15.

Table 5. Results of the $CF_3SO_3H:CF_3CO_2H$ catalyzed isomerization of pivaldehyde.[a]

H_0	Acid System, w/w	Pivaldehyde	Methyl Isopropyl Ketone
-10.9	26.9% CF_3SO_3H	0%	100%
-9.7	11.4% CF_3SO_3H	17%	83%
-9.4	8.0% CF_3SO_3H	29%	71%
-8.4	3.1% CF_3SO_3H	68%	32%
-7.7	0.9% CF_3SO_3H	83%	17%
-2.7	100% CF_3CO_2H	100%	0%

[a]Reaction conditions: 2 h, 25°C, 1:5 pivaldehyde: acid.

Further support for the involvement of the superelectrophile (**162**) was obtained from experimental studies establishing the effect of acidity on the rearrangement. The yield of methyl isopropyl ketone from pivaldehyde is found to increase in proportion to acidity (Table 5) with complete conversion occurring at H_0 -10.9. The pK_a for the protonated pivaldehyde is estimated to be -8.0, indicating that more than half of the aldehyde is in its monoprotonated form (**161**) at $H_0 -10.9$. Thus, protosolvation of the monoprotonated species (**161**) should lead to the formation of the superelectrophile (**162**) albeit in low concentrations.

Another *O,O*-diprotonated gitonic superelectrophile has been proposed in the condensation chemistry of ninhydrin (Scheme 16).[75] It was shown that ninhydrin condenses with arenes in H_2SO_4 to give the 2,2-diaryl-1,3-indanediones (**167**) while reaction in superacidic CF_3SO_3H give the 3-(diarylmethylene)isobenzofuranones (**168**). If the 2,2-diaryl-1,3-indanedione products are isolated and then reacted with CF_3SO_3H, they isomerize cleanly to the 3-(diarylmethylene)isobenzofuranones. As triflic acid (H_0 -14.1) is more than 100 times more acidic than anhydrous H_2SO_4 (H_0 -12), it suggests the involvement of a protosolvated superelectrophile (**169**) in the formation of the 3-(diaryl-methylene)-isobenzofuranones. There may be several other factors contributing to the protosolvated dication **169** undergoing rearrangement. Ring-opening serves to spread out the positive charge between the acyl cation and a protonated enol group (**170**). Moreover, the acyl cation is in conjugation with the diphenylmethylene group as shown by the resonance structure. This stabilizing effect of the *gem*-diphenyl group may be also an important aspect in the rearrangement, because when 2-phenyl-1,3-indanedione is reacted in CF_3SO_3H, no rearrangement is observed and the starting material is recovered quantitatively.

As discussed in Chapter 2, kinetic evidence has been reported to indicate the involvement of the diprotonated product **174** from the unsaturated acetal **171**, formally a superelectrophilic carboxonium dication (eq 44).[76]

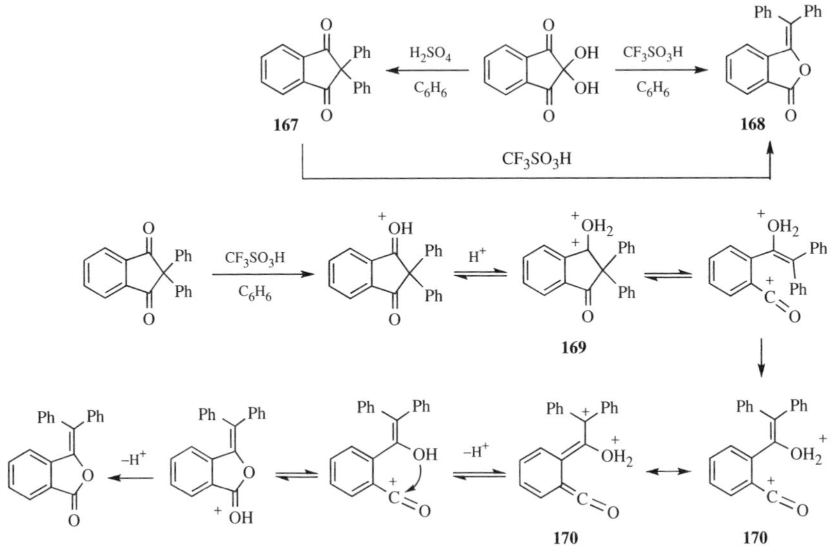

Scheme 16.

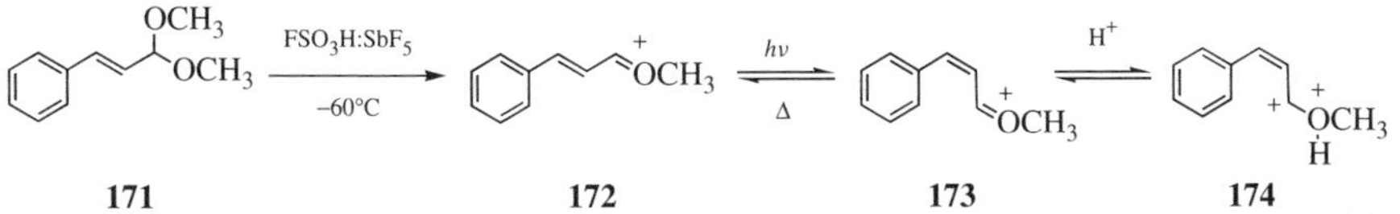

$$(44)$$

Protonated carbonic acid ($H_3CO_3^+$, **175**) is recognized as a stable species in superacids and even a salt was isolated with its X-ray crystal structure obtained. In strong superacids, diprotonated carbonic acid (**30**) was found to undergo ionic hydrogenation to methane (eq 45).[77]

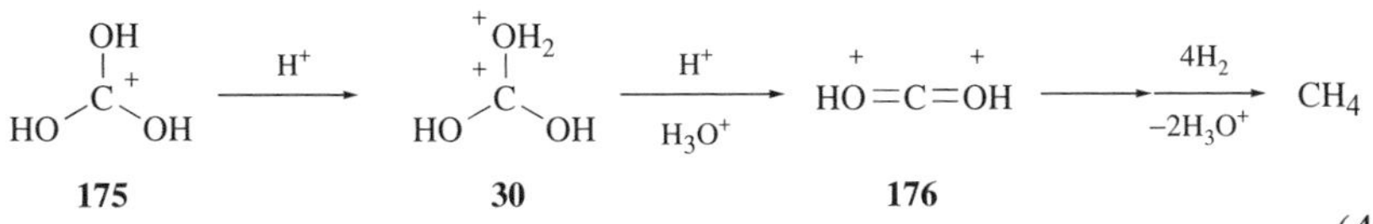

$$(45)$$

This has been interpreted as a protosolvation of **175** to give the superelectrophile **30**, which itself undergoes protolytic cleavage to give protosolvated carbon dioxide (**176**). The gitonic superelectrophile **176** then reacts with H_2 to eventually give methane. In gas-phase studies, dicationic species such as CO_2^{2+} and OCS^{2+} have been generated using mass spectrometry techniques.[78] Both of these dicationic species have been shown to react with molecular hydrogen (deuterium), for example, producing DCO^+, DCO_2^+, and OD^+ ion products (reaction of CO_2^{2+} and D_2).

5.2.2.4 Carbon-Halogen Vicinal-Dications

Trihalomethyl cations are shown to have enhanced reactivities in superacid solution, while polyhalomethanes in the presence of excess $AlBr_3$ or $AlCl_3$ exhibit the properties of aprotic superacids.[79] The trihalomethyl cations CX_3^+ (**178**, X=Cl, Br, I) have been characterized by NMR and IR spectroscopy. The stability of these species is attributed to substantial resonance-stabilization by back-donation from the nonbonded electron pairs of the halogen atoms.[22] Trihalomethyl cations are capable of hydride abstraction from alkanes and alkyl groups when the reactions are carried out in the presence of Bronsted or Lewis superacids (eq 46–48).[80]

$$(46)$$

$$(47)$$

$$(48)$$

Sommer *et al.* has noted that the reactivities of the chloromethyl cations decrease in the order $CCl_3^+ > CHCl_2^+ \gg CH_2Cl^+$.[81] Since increasing chloro-substitution should give less reactive electrophiles (due to resonance stabilization), this trend has been interpreted to be a consequence of protosolvation/electrophilic solvation involving the halogen nonbonded electron pairs (**178** and **179**, (Scheme 17). Complexes such as CBr_4 •$2AlBr_3$, CCl_4 •$2AlCl_3$, and others have been shown to efficiently catalyze the cracking, isomerization, and oligomerization of alkanes and

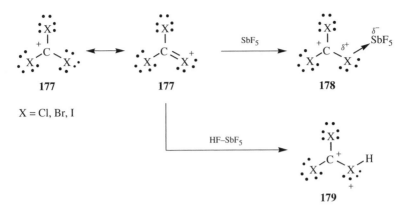

Scheme 17. Superelectrophilic activation of trihalomethyl cations.

cycloalkanes.[82] Electrophilic solvation similar to **178** can also be envisaged for these systems.[17]

Theoretical studies have examined the structures and energies of the *vicinal*-carbon-halogen dications, such as **178** and **179**.[83] An *ab initio* computational study described the structures, energies, and reactivities of protonated halomethyl cations. In the case of the trichloromethyl cation (**181**), calculations at the MP2/6-31G* level reveal a shortening of the C–Cl bond length when compared with carbon tetrachloride (Scheme 18). This is consistent with delocalization of the positive charge among the three chlorine atoms. Upon protonation, the *vicinal*-dication **182** is formed showing further shortening of the C-Cl bonds (indicating a stronger p-p interaction) and a lengthening of the C-ClH bond due to charge-charge repulsive effects. Calculations estimate a gas-phase energy barrier to proton loss of 69.5 kcal/mol, suggesting that **182** should be a viable species in mass-spectrometric studies. Diprotonated ($CCl_3H_2^{3+}$, **183**) and triprotonated ($CCl_3H_3^{4+}$, **184**) ions were also found at energy minima. In the case

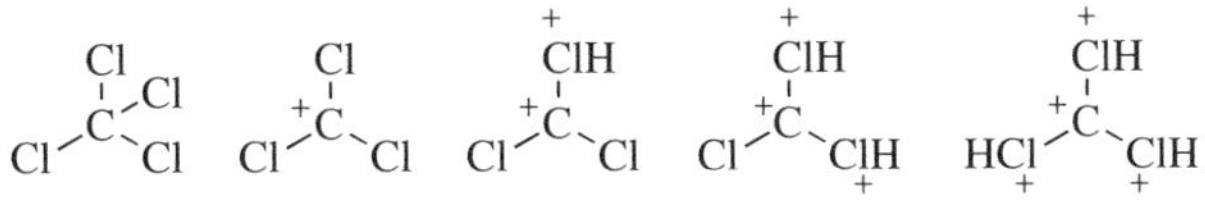

	180	**181**	**182**	**183**	**184**
C-Cl Bond Length(s), Å:	1.766	1.648	1.839 1.606 1.603	1.779 1.575	1.732
Energy Barrier to H+ Loss, kcal/mol:	N/A	N/A	69.5	27.4	4.3

Scheme 18. Calculated structural and energy parameters of prototrichloromethyl cations.

of **184**, however, deprotonation is highly exothermic (-261.7 kcal/mol) and having a barrier to proton loss of only 4.3 kcal/mol. In the same study, ^{13}C NMR chemical shifts were calculated for the species (**181–184**).[18]

To evaluate the impact of superelectrophilic activation of halogenated methyl cations, a series of isodesmic reactions were studied by *ab initio* calculations (Table 6).[83b] For the trichloromethyl cation (**181**), protonation leads to dramatically more reactive electrophiles. This is evident from the calculated reaction enthalpies in hydride transfer from propane to give the 2-propyl cation. Indeed, protosolvation of **181** to give the superelectrophile **182** leads to an increasingly exothermic reaction, more favorable by more than 140 kcal/mol (gas-phase). Another significant trend is seen in the calculated reaction enthalpies for the trichloromethyl cation (**181**), dichloromethyl cation (**185**), and the chloromethyl cation (**186**, entries 1–3). In the series **181** → **185** → **186**, the reaction enthalpy is progressively more exothermic, -3.9 → -11.9 → -26.6 kcal/mol. This is consistent with the observation that halogen substitution on a carbenium carbon leads to increasing stabilization, and the trichloromethyl cation is the most stable cation in the series. However, reactions of halogenated methanes in superacid show that the tetrahalogenated methanes produce the most reactive electrophilic systems. This has been interpreted as evidence for activation involving superelectrophilic species like **182** and **183**.

Another computational study examined the structures and energies of the trichloromethyl cation-AlCl$_3$ complexes by semi-empirical and *ab*

Table 6. Calculated isodesmic reactions between chloromethyl cations (181–186) and propane.

Entry	Reaction	ΔH (kcal/mol)
(1)	$\overset{+}{H_2CCl} + CH_3CH_2CH_3 \longrightarrow CH_2Cl_2 + CH_3\overset{+}{C}HCH_3$ **186**	-26.6
(2)	$\overset{+}{HCCl_2} + CH_3CH_2CH_3 \longrightarrow CH_2Cl_2 + CH_3\overset{+}{C}HCH_3$ **185**	-11.9
(3)	$\overset{+}{CCl_3} + CH_3CH_2CH_3 \longrightarrow HHCl_3 + CH_3\overset{+}{C}HCH_3$ **181**	-3.9
(4)	$\overset{2+}{CCl_3H} + CH_3CH_2CH_3 \longrightarrow HCCl_3\ H+ + CH_3\overset{+}{C}HCH_3$ **182**	-145.5
(5)	$\overset{3+}{CCl_3H_2} + CH_3CH_2CH_3 \longrightarrow \overset{2+}{HCCl_3H_2} + CH_3\overset{+}{C}HCH_3$ **183**	-269.4
(6)	$\overset{4+}{CCl_3H_3} + CH_3CH_2CH_3 \longrightarrow \overset{3+}{HCCl_3H_3} + CH_3\overset{+}{C}HCH_3$ **184**	-423.8

initio calculations.[84] In contrast to the study of protosolvated superelectrophiles (**182**), this study could find no evidence for the formation of the $CCl_3^+ \rightarrow AlCl_3$ complex(es), or donor-acceptor complexes analogous to **178**. The superelectrophilic character of the CCl_4-$2AlCl_3$ system was attributed to formation of a bidentate complex with substantial chlorenium ion character, $AlCl_4^- \; ClCCl^{2+} \; AlCl_4^-$, or $CCl_3^+ \; AlCl_4^-$ involving C-Cl $\rightarrow$ Cl-Al interactions. As described in Chapter 4, protonation of the dimethylbromonium cation (**187**) has been studied by *ab initio* calculations and two stable minima were found, i.e., the carbon and bromine protonated forms **188** and **189** (Figure 6).[85] Dication **188** is estimated to be more stable than **189** by about 21 kcal/mol. Monoprotonation of methyl halides occurs on the halogen atom to give the protohalonium ion. Diprotonation is shown by calculations to give the *vicinal*-dications (**190**) involving also protonation at a C-H bond, forming a 2e-3c bond. Calculations also show that the halogen atoms carry significantly more charge in the dications (**190**, CH_4XH^{2+}) than in the respective monocations (CH_3XH^+). In the case of bromomethane, the NBO charge on the bromine atom increases from 0.46 to 0.70 with formation of **190** from the monoprotonated species. Thermodynamically, diprotonation of fluoromethane is determined to be the least favorable while diprotonation of iodomethane is found to be the most favorable.

Other types of *vicinal* carbon-halogen dications have been studied by gas-phase and theoretical methods. Their full discussion is beyond the scope of this chapter and many can be found in published review articles.[86] Charge-stripping experiments have been shown to provide CH_2XH^{2+}(X=F, Cl, Br, I) dications in the gas-phase and their structures have been studied by computational methods.[87] The CX_3^{2+} (X=F, Cl) dications can also be generated in the gas-phase and have also been studied by theoretical investigations. The chemistry of CF_n^{2+} is considered important in plasma etching processes and these ions may also be important in ozone depletion chemistry.[88] In the case of CF_3^{2+}, it can be generated in the gas-phase by electron impact on CF_4. The CF_3^{2+} ion is shown to react with neutral collision partners and yield products from bond forming reactions,

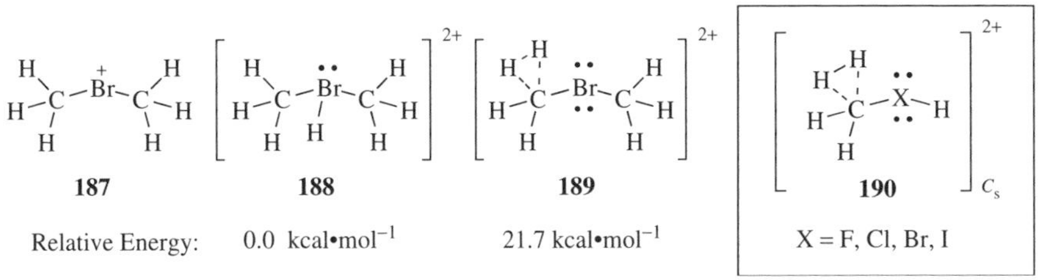

187	**188**	**189**	**190**
Relative Energy:	0.0 kcal•mol^{-1}	21.7 kcal•mol^{-1}	X = F, Cl, Br, I

Figure 6.

such as DCF_2^+ and OCF_2^+, by respective reactions with D_2 and O_2.[79] High-level quantum chemical calculations have been carried out on ions such as CF_3^{2+}.[88] As noted in the previous chapter, the CX_2^{2+} (X = F, Cl) ions have also been the subject of gas-phase and theoretical studies.[89] Likewise, CX^{2+} dications have been examined.[47] While these novel CX_n^{2+} dications may not necessarily be described as *vicinal*-type gitonic superelectrophiles, their demonstrated gas-phase reactivities are remarkably similar to the high electrophilic reactivities of condensed-phase superelectrophiles.

5.2.3   Nitrogen-Centered *Vicinal*-Dications

There are a number of well-characterized nitrogen-centered *vicinal*-dications that can be considered gitonic superelectrophiles. For example, Olah and co-workers showed that tetramethylammonium ions undergo slow hydrogen-deuterium exchange, when reacted with $DF-SbF_5$ at 20°C (Scheme 19).[90] Theoretical calculations suggest that C-H protonation (deuteration) (**192**) is the most likely mechanism of exchange ruling out the intermediacy of **191**. Structure **192** was characterized as a stable minimum (C_s symmetry) at the MP2(fu)/6-31G** level of theory. Tetra-methylammonium ion did not undergo methyl exchange with CD_3F-SbF_5, indicating that there is no involvement of a penta-coordinate nitrogen (i.e., $N(CD_3)(CH_3)_4^{2+}$).[19]

Hydrazinium dications and their related systems can be readily formed and show remarkable stability. These *vicinal*-dications show enhanced electrophilic reactivities compared to their monocationic precursors. Their chemistry has been recently reviewed,[91] therefore only a brief overview is warranted here. Hydrazine itself can be doubly protonated with HCl. When

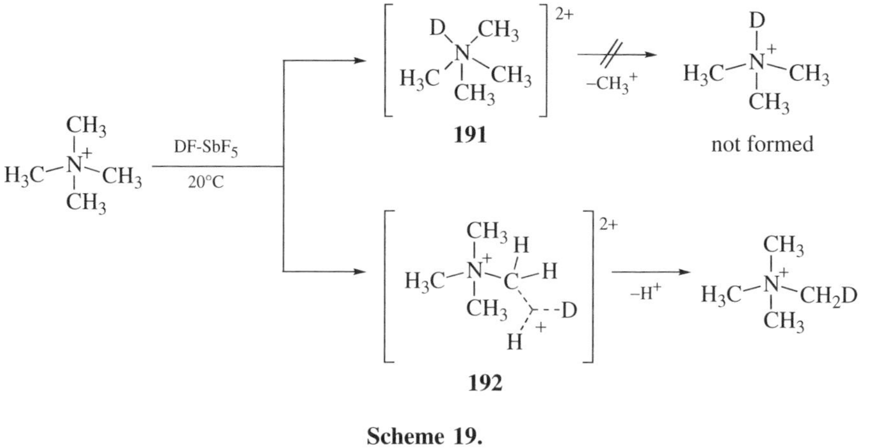

Scheme 19.

tetramethyl hydrazine was reacted with strong methylating agents, however, only monomethylation was achieved (eq 49).[92] If the monomethylated product (**193**) is then reacted in 70% H_2SO_4, the hydrazinium dication is formed to an extent of about 50% (eq 49).

$$(49)$$

Several bicyclic hydrazinium ions have been prepared and two synthetic strategies have emerged as generally useful, i.e., electrochemical oxidation and Lewis or Brønsted acid-promoted cyclization.[93] Oxidation of 1,6-diazabicyclo[4.4.4]tetradecane, for example, (**194**) gives the intermediate radical cation (**195**) and the bicylic dication (**196**, eq 50).[93a]

$$(50)$$

N–N Distance, Å: 2.81   2.30   1.53

X-ray crystal structures have been obtained for them and the nitrogen-nitrogen distance decreases with successive oxidation. When 1,5-diazabicyclo[3.3.0]octane is reacted with 3-bromo-1-propanol, followed by reaction with 40% HBF_4, the hydrazinium dication (**197**) is obtained in 80% yield (eq 51).[93b] The ^{13}C labeled hydrazinium dication (**198**) was prepared through double alkylation of tetramethyl hydrazine, the second alkylation being assisted by the silver (I) salt (eq 52).[93c]

$$(51)$$

197

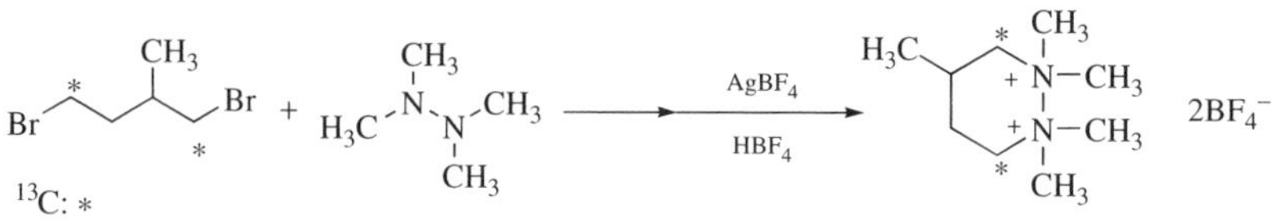

$$
\text{(52)}
$$

Although acyclic and monocyclic hydrazinium dications are readily obtainable, the propellane-type structures (i.e., **196** and **197**) have been shown to be the most easily prepared hydrazinium dications. There have been no examples reported of direct nucleophilic attack at the nitrogens of the hydrazinium ions. Nucleophilic reactions generally occur at the α-carbon atoms or at hydrogens. Even water has been shown to be capable of deprotonating the α-carbon in some of the bicyclic hydrazinium dications. This is consistent with the delocalization of positive charge in gitonic superelectrophiles and significant positive charge on the neighboring hydrogen atoms.

A related category of *vicinal*-dications involves aromatic systems having adjacent nitrogen atoms. The diquaternary salts of diazenes and diazoles can be prepared using strong alkylating agents (eqs 53–54).[94] Compound **199** is prepared in 52% yield by alkylation with trimethyloxonium salt, with some of the positive charge being delocalized onto the sulfur (eq 54).

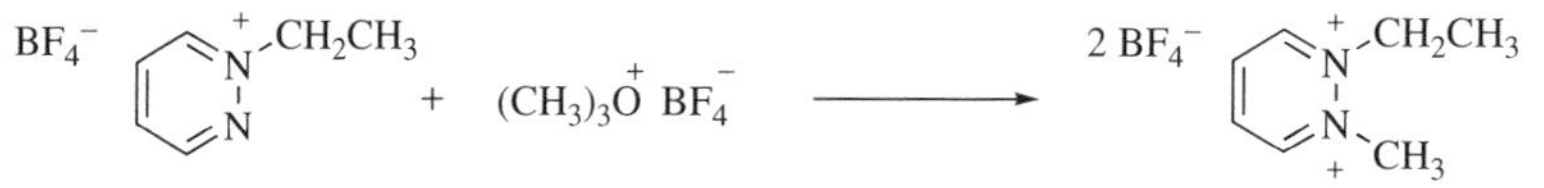

$$
\text{(53)}
$$

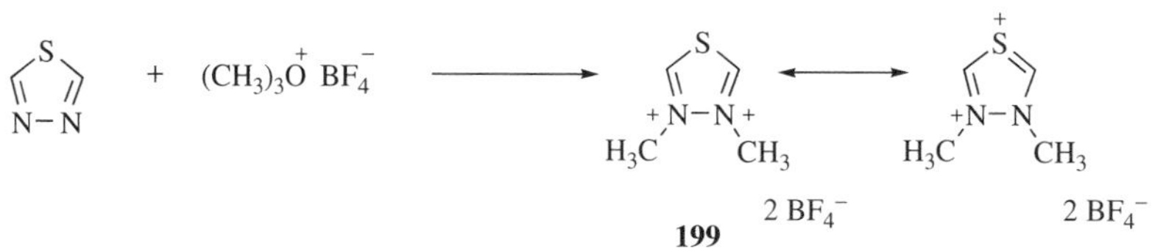

$$
\text{(54)}
$$

No systematic study of the reactivities of these electrophiles has been done, but they are known to be easily reduced to their radical cations. The *N,N*-bipyridiniums and related systems have not yet been studied in detail, but some have been prepared, such the 1-pyridinopyridinium dication **200**.[95] Little is known about the electrophilic reactivities of these compounds, but **200** has been shown to react with water at a ring carbon (eq 55). An unusual gitonic superelectrophile has been proposed involving

$$(55)$$

$$(56)$$

a tricationic intermediate **201** (eq 56).[96] Deuterium labeling suggests triple protonation of the pyrazine ring with delocalization of positive charge into the propenyl group.

Diazenium dications may also be considered gitonic superelectrophiles, but only a limited number of them have been prepared and studied.[97] In principle, diazenium dications could be prepared by double alkylation or protonation of various azo-compounds. To date, only monalkylation has been achieved with strong alkylating agents (eq 57).[97a] However, Olah *et al.* showed that diazenium dications can be produced by double protonation of azobenzenes in FSO_3H-SbF_5 at low temperature.[97b] It has also been demonstrated that the diazenium dications are prepared from electrochemical and chemical oxidations (eqs 58–59).[97c,d]

$$(57)$$

$$(58)$$

$$(59)$$

In the oxidation of **202**, the tetracyclodiazenium ion **203** is produced in 81% yield. Like the hydrazinium dications, the diazenium dications exhibit relatively high acidities of their α-hydrogen atoms.

Diazonium dications have been studied extensively in both experimental and theoretical work. The parent system for diazonium dications is doubly protonated dinitrogen (**204**). Calculations at the QCISD/6-311++G**//QCISD/6-311++G** level show that dication **204** ($D_{\infty V}$ symmetry) is a kinetically stable species in the gas-phase, having a barrier of about 30 kcal/mol for dissociative proton loss.[98] Although monoprotonated dinitrogen has been observed directly in gas-phase studies (and it is generated from protonated diazomethane, *vide infra*),[99] dication **204** has not yet been detected. Diazomethane is observed to be diprotonated under conditions of kinetic control in extremely acidic $FSO_3H–SbF_5$.[100] The experimental studies suggest protonation at both termini of diazomethane (**208**, eq 60).

| 204 | 205 | 206 | 207 |

$$\text{(60)}$$

| 208 |

Calculations at the MP2(FU)/6-31G**//MP2(FU)/6-31G** level have found that the global minimum is the *N,N*-diprotonated species **205** (gas-phase structure) while the *C,N*-diprotonated species (**208**) is 11 kcal/mol higher in energy.[98] From kinetic studies it is known that the displacement of nitrogen from the methyldiazonium ion ($CH_3N_2^+$) is a nucleophilic substitution type reaction. Weakly nucleophilic SO_2ClF is also known to displace nitrogen from methyldiazonium ion with a reaction rate that increases with the acidity of the solution. This is in accord with the formation of the protosolvated, gitonic superelectrophile **208** leading to nucleophilic attack by SO_2ClF (eq 60). Similarly, Olah and co-workers demonstrated the use of azide salts as effective aromatic aminating agents with excess $HCl-AlCl_3$.[101] Diprotonated hydrazoic acid (**206**) has been proposed as an intermediate in the conversion. This gitonic superelectrophile (**206**) was also studied by theory at the MP2(FU)/6-31G**//MP2(FU)/6-31G** level and found to be a global minimum, at least 20 kcal/mol more stable than structure **207**.[98]

Aryldiazonium dications are thought to be intermediates in the Wallach rearrangement.[97b] Azoxybenzene is shown to form the monoprotonated

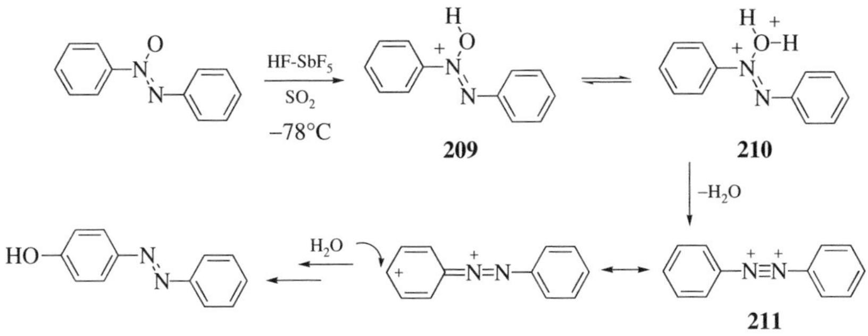

Scheme 20.

ion (**209**) in FSO$_3$H at low temperature, while the dicationic species (**210** and **211**) are directly observable by NMR in HF-SbF$_5$ at low temperature (Scheme 20).[97b] In the Wallach rearrangement, delocalization of the positive charges then leads to nucleophilic attack by water at a ring carbon. The pK$_a$ of ion **209** is estimated to be -9, indicating that the azoxybenzene is completely monoprotonated under the conditions required for the Wallach rearrangement, being likely in equilibrium with the diprotonated intermediate **210**. Water elimination then gives the diazonium dication **211**. There has been a report showing that superelectrophilic intermediates like **211** can react with relatively weak nucleophiles, such as benzene.[102]

An important class of nitrogen-based gitonic superelectrophiles are the protonitronium dication (**214**) and related systems (Scheme 21). These gitonic superelectrophiles are highly efficient for nitration of very weak nucleophiles (*vide infra*). In 1975, the observation of the enhanced electrophilic reactivity of nitronium salts in superacids was reported by Olah et al. and protosolvation of the nitronium ion (**212**) was suggested (Scheme 27).[62] While the often used nitronium tetrafluoroborate reagent reacts with a large variety of arenes, it does not react with highly deactivated ones (such as *meta*-dinitrobenzene) under aprotic conditions. In contrast the reaction in superacid gives 1,3,5-trinitrobenzene in 70% yield.[103] This suggests the formation of an activated protosolvated species or even the *de facto* protonitronium dication NO$_2$H^{2+} (**214**). A convenient system of nitration for deactivated aromatics is a mixture of nitric acid and triflatoboric acid (2 CF$_3$SO$_3$H$-$(CF$_3$SO$_3$)$_3$B; H_0 -20.5).[104] This system readily nitrates deactivated arenes such as *inter alia* pentafluorobenzene, nitrobenzene, and methyl phenyl sulfone in high yields (78–99%). Nitronium salts are reported to nitrate the trityl cation (**215**) in superacid to give **216** (eq 61).[105]

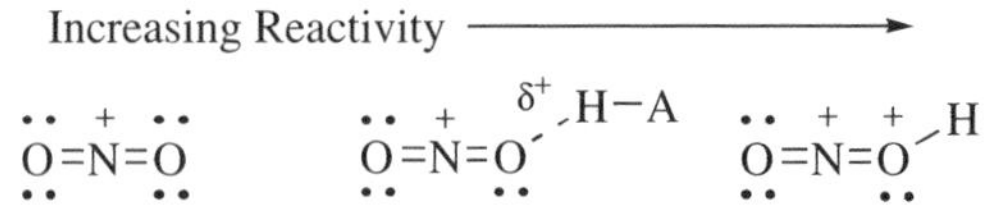

Increasing Reactivity ⟶

$$
\overset{\cdot\cdot\quad+\quad\cdot\cdot}{\underset{\cdot\cdot\quad\quad\cdot\cdot}{O=N=O}} \qquad \overset{\cdot\cdot\quad+\quad\cdot\cdot}{\underset{\cdot\cdot\quad\quad\cdot\cdot}{O=N=O}}\overset{\delta^+}{\diagdown}{}^{H-A} \qquad \overset{\cdot\cdot\quad+\quad+}{\underset{\cdot\cdot\quad\quad\cdot\cdot}{O=N=O}}\diagdown{}^{H}
$$

212   **213**   **214**

Scheme 21. The nitronium ion (**212**) and the protonitronium dication.

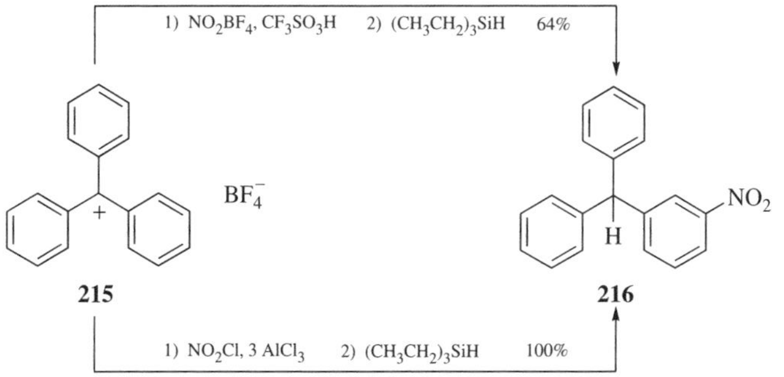

$$(61)$$

When nitronium tetrafluoroborate was attempted to react with the trityl cation in CH_2Cl_2 or sulfolane, no nitration occurred due to the deactivating effects of the carbenium ion center in **215**. Nitration of deactivated substrates is also readily accomplished by reaction with NO_2Cl with three mole excess $AlCl_3$ suggesting Lewis acidic electrophilic solvation of the nitronium cation (**217**, eq 62).[105]

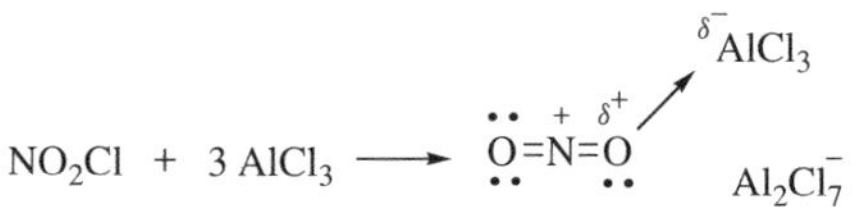

$$
NO_2Cl \ + \ 3\ AlCl_3 \ \longrightarrow \ \overset{\cdot\cdot\ +\ \delta^+}{\underset{\cdot\cdot\quad\cdot\cdot}{O=N=O}}\overset{\overset{\delta^-}{AlCl_3}}{\diagup}_{Al_2Cl_7^-} \tag{62}
$$

217

Remarkably, nitronium salts have been shown to react even with methane under superacidic conditions.[106] Nitronium hexafluorophosphate (NO_2PF_6) reacts with methane, albeit in low yields, when the reaction is carried out in the presence of boron tris(triflate) in triflic acid (eq 63).

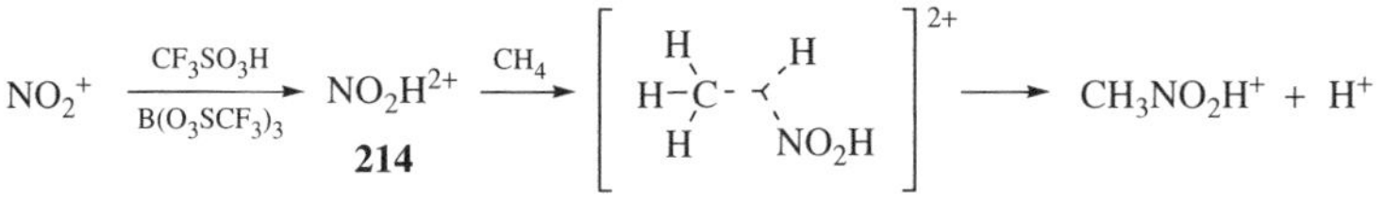

$$
NO_2^+ \xrightarrow[\text{B(O}_3\text{SCF}_3)_3]{\text{CF}_3\text{SO}_3\text{H}} NO_2H^{2+} \xrightarrow{CH_4} \left[\overset{H}{\underset{H}{H-C}}\overset{\diagdown}{\diagup}\overset{H}{\underset{NO_2H}{}} \right]^{2+} \longrightarrow CH_3NO_2H^+ \ + \ H^+ \tag{63}
$$

214   **218**

The reaction is suggested to proceed via a two-electron three-center bound carbocationic transition state (**218**) formed by insertion of dication **214** into the C-H σ-bond of methane.

The activation of the nitronium ion (**212**, NO_2^+) can be understood as a consequence of the interaction of the oxygen lone pair electrons with superacidic Brønsted or Lewis acids. Because the linear nitronium cation has no vacant orbital on the nitrogen atom nor low-lying unoccupied molecular orbital, extremely weak nucleophiles such as methane or deactivated arenes are unable to initiate the necessary polarization of the nitrogen-oxygen π-bond. Protonation (or coordination by Lewis acid) weakens the N-O π-bond character and results in the bending of the linear nitronium ion and rehybridization of the nitrogen from sp to sp^2. The rehybridization creates a developing p orbital on the nitrogen which is capable of interacting with weak π-donor or even σ-donor nucleophiles. High-level theoretical calculations have shown the protonitronium ion (**214**) to be a kinetically stable species with an estimated (gas-phase) barrier to deprotonation of 17 kcal/mol.[107] These calculations also found the optimized structure to be bent with a O–N–O bond angle of about 172°. The protonitronium ion (**214**) was experimentally observed by Schwarz in the gas phase by electron impact mass spectrometry of nitric acid.[108] Attempts to directly observe **214** under superacidic conditions by ^{15}N NMR and FT-IR (Raman) spectroscopy as a stable species have been inconclusive but the ^{17}O NMR signal is deshielded from that of the nitronium ion NO_2^+.[107b] This suggests that the protonitronium ion (**214**) is formed only in low-equilibrium concentrations even in the strongest superacids.

5.2.4   Oxygen- and Sulfur-Centered *Vicinal*-Dications

A significant number of *vicinal*-dicationic systems containing oxygen, such as superelectrophilic carboxonium ions, acylium ions, and nitronium ions, have been previously discussed. Few systems have been reported involving only oxygen cationic centers. Other than doubly ionized O_2 (known in gas-phase studies), scarce indications for such oxygen-based *vicinal*-dications have been reported. Hydrogen peroxide has been studied by low temperature NMR in HF-SbF_5 solution but no direct evidence for diprotonated species was reported and only the monoprotonated species (**219**) was observed (eq 64).[109] Even at very high acidity (ca. $H_0 -20$), proton exchange is fast compared with the NMR time scale, as both oxygen atoms are found to be equivalent by ^{17}O NMR. In a similar respect, dimethylperoxide forms trimethylperoxonium ion (**220**) under superacidic methylating conditions (eq 65).[110]

$$H_2O_2 \xrightarrow[-25°C]{HF\text{-}SbF_5} \quad \underset{\textbf{219} \quad X^-}{\overset{\overset{\displaystyle H}{|}}{H-O-O-H}} \quad \xmapsto{\ \ \ //\ \ \ } \quad \underset{2\,X^-}{\overset{\overset{\displaystyle H}{|}}{H-O-O-H}} \qquad (64)$$

$$\text{(65)}$$

220   **221**

No NMR evidence (^{1}H and ^{13}C) could be obtained for the *vicinal*-dication
(**221**). Calculations at the MP2/6-31G* level however show dication **221**
to be a stable minimum. Its structure has D_2 symmetry with the two COC
units in an almost perpendicular arrangement (dihedral angle between
two COC units is 83.7°). Protonated ozone, O_3H^+, was studied by Olah
et al. and by Italian researchers.[111] In the superacid catalyzed reactions of
ozone, diprotonated ozone is however possibly involved as the *de facto*
intermediate.

In contrast to the *vicinal*-oxonium dications, *vicinal*-sulfonium dications are well known dicationic systems. Many examples have been prepared and studied. As discussed in review articles,[91] the disulfonium
dications can be readily prepared by electrochemical or chemical oxidation of appropriate disulfides (eq 66), by reactions of mono-S-oxides
with acidic reagents (eq 67),

$$\text{(66)}$$

$$\text{(67)}$$

and other methods.[112] The reactions of mono-S-oxides with acidic reagents
have been the subject of mechanistic studies. Some of these studies suggest a complex mechanism in the formation of the *vicinal*-disulfonium
dications. On the other hand, experimental and theoretical studies of the
protonated and alkylated dications of dimethylsulfoxide (*vide infra*) have
suggested the possibility of *O,O*-diprotonation of the sulfoxide group.[113]
Thus, formation of the *vicinal*-disulfonium dication could involve diprotonation of the sulfoxide bond of **222** forming the gitonic superelectrophile **224**, which upon intramolecular nucleophilic attack gives the
vicinal-disulfonium dication (**225**, Scheme 22). It has been shown that
dications such as **225** possess moderately high electrophilic reactivities.
For example, **225** (X=CF$_3$SO$_3^-$) has been shown to react with activated

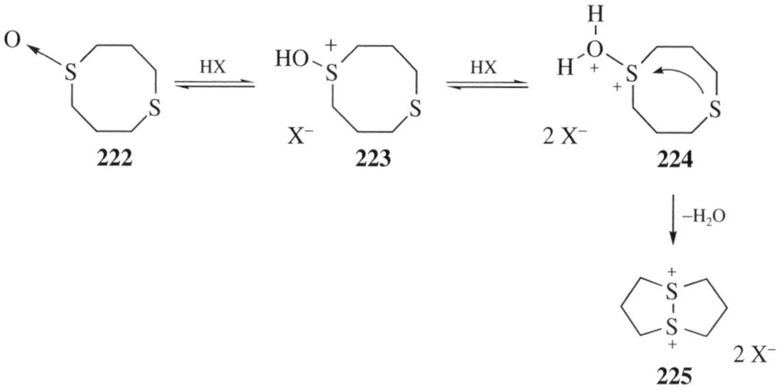

Scheme 22.

arenes (phenols and anilines) at the sulfonium center (eq 68).[114]

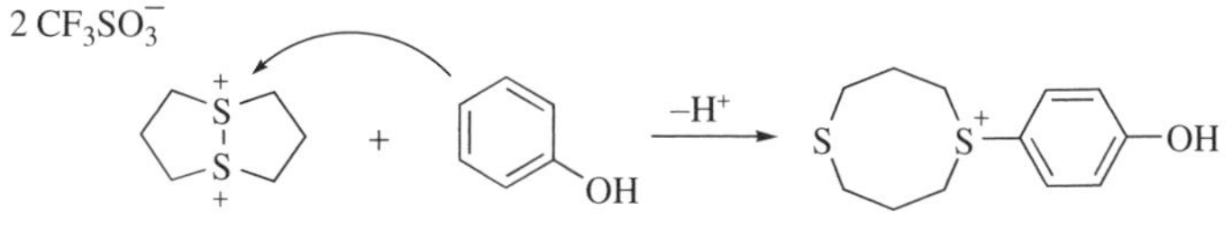

(68)

Similar chemistry has been described for the *vicinal*-dications of selenium, tellurium, and mixed chalcogen systems, both for the preparation and reactions of these *vicinal*-dications.[91]

Vicinal-sulfoxonium-oxonium dications have been studied by experimental and theoretical methods, specifically the dicationic species from dimethylsulfoxide (DMSO).[113] The structures and energies for the protonation products of DMSO were calculated at the B3LYP/6-311+G**//B3LYP/6-311+G** level (Figure 7). For the monoprotonated products, the *O*-protonated structure (**226**) is found to be 37 kcal/mol more stable than the *S*-protonated structure (**227**). Similar results are obtained from methylation of DMSO. The second protonation of *O*-protonated DMSO can take place on either oxygen or sulfur. Both the *O,O*-diprotonated species (**228**) and the *O,S*-diprotonated species (**229**) are found to be minima on the potential energy surface. However, the *O,O*-diprotonated species is found to be 20.87 kcal/mol more stable than the isomeric *O,S*-diprotonated species (**229**). Not surprisingly, there is a significant lengthening of the O–S bond with the second protonation (**226** → **228**, 1.641 Å → 1.817 Å) due to electrostatic repulsive effects. In addition to structures and energies, ^{13}C, ^{17}O, and ^{32}S NMR chemical shifts were calculated using the correlated GIAO-MP2 method with the DFT optimized geometries. Comparison of the calculated chemical shift values with the

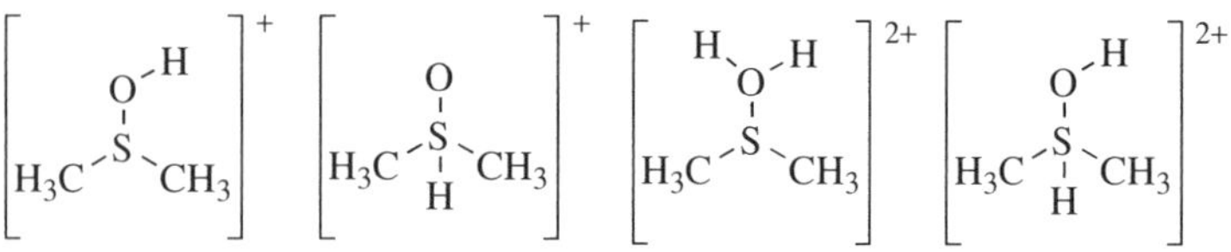

	226	**227**	**228**	**229**
Calculated ^{13}C				
Chemical Shift:	40.0 ppm	43.4 ppm	56.0 ppm	47.8 ppm

Figure 7. Protonated dimethylsulfoxide (DMSO) and calculated ^{13}C NMR chemical shift values.

experimentally determined data in $FSO_3H-SbF_5-SO_2ClF$ media indicate predominant formation of the monoprotonated species (**226**). For example, the experimentally determined value for the ^{13}C chemical shift of **226** is found at $\delta^{13}C$ 34.3, while GIAO-MP2 value is estimated to be $\delta^{13}C$ 40 ($\delta^{13}C$ 35 at the GIAO-SCF level). The diprotonated species is estimated to be further deshielded to $\delta^{13}C$ 56.0.

5.2.5   Halogen-Centered *Vicinal*-Dications

Despite the importance of halogens in synthetic and mechanistic chemistry, there has been very little work on the superelectrophililc activation of halogens, of *vicinal*-halogen dications, and related systems. Gas-phase and theoretical studies have demonstrated the possibility of multiply-ionization of diatomic halogens, such as producing I_2^{3+}, Br_2^{3+}, and Cl_2^{3+}.[115] In the condensed phase, acid-catalyzed halogenation of weak nucleophiles is thought to occur by interaction of halogen lone pair electrons with the acid (eq 69).

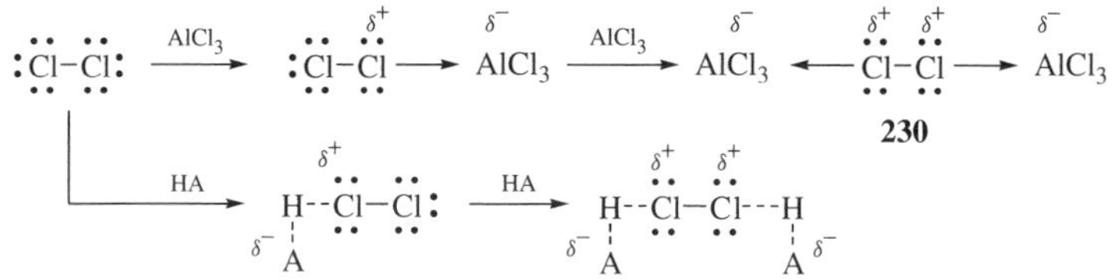

(69)

Complexation with a second equivalent of Lewis acid could produce a superelectrophilic species (**230**) having the *vicinal*-dication structure, or alternatively, double protonation could conceivably generate superelectrophile **231**. Olah and Rasul recently reported a calculational study of both protonated and methylated vicinal dihalogen dications ($HXXH^{2+}$, $CH_3XXCH_3^{2+}$, etc.).[116] Superelectrophilic activation can also be envisaged for hypohalogen systems and other oxygenated halogens. Diprotonation of iodosyl systems or hypochlorous acid could potentially form the *vicinal*-dications (**232–234**) (eqs 70–71).

$$X{-}I{=}O \xrightarrow{H^+} X{-}\overset{+}{I}{=}O{-}H \xrightarrow{H^+} \overset{+}{X}{-}I{-}\overset{+}{O}\overset{H}{\underset{H}{\diagdown}} \text{ or } \overset{+}{X}{-}\overset{+}{I}{=}\underset{H}{\overset{H}{\underset{|}{O}}}{-}H \qquad (70)$$

$$\hspace{6cm} \mathbf{232} \hspace{2cm} \mathbf{233}$$

$$H{-}O{-}Cl \xrightarrow{H^+} \underset{H}{\overset{H}{\diagdown}}\overset{+}{O}{-}Cl \xrightarrow{H^+} \underset{H}{\overset{H}{\diagdown}}\overset{+}{O}{-}\overset{+}{Cl}{-}H \qquad (71)$$

$$\hspace{5.5cm} \mathbf{234}$$

Protonation of perchloric acid could form the corresponding dication **235** and trication **236** (eq 72).

$$\underset{\overset{||}{O}}{\overset{\overset{O}{||}}{HO{-}Cl{=}O}} \xrightarrow{H^+} \underset{\overset{||}{O}}{\overset{\overset{O}{||}+}{HO{-}Cl{-}OH}} \xrightarrow{H^+} \underset{\overset{+}{OH}}{\overset{\overset{O}{||}+}{HO{-}Cl{-}OH}} \xrightarrow{H^+} \underset{\overset{+}{OH}}{\overset{\overset{+OH}{||}+}{HO{-}Cl{-}OH}} \qquad (72)$$

$$\hspace{4.5cm} \mathbf{235} \hspace{2cm} \mathbf{236}$$

These systems await future studies.

5.2.6 Noble Gas-Centered *Vicinal*-Dications

As noted in Chapter 1, the helium dimer dication, He_2^{2+}, was predicted by Pauling to be a stable species.[117] This unusual dication indeed can be generated in the gas-phase and it has been studied by several theoretical investigations.[118] Many other noble gas diatomic species have been observed by gas-phase studies, including: Ne_2^{2+}, $ArNe^{2+}$, $XeNe^{2+}$, and $KrHe^{2+}$.[86c] Dicationic species have also been observed with noble gas bound to other atoms. For example, noble gas chloride dications ($NeCl^{2+}$, $KrCl^{2+}$, and $XeCl^{2+}$) are reported in mass spectrometric studies.[86c] Bonding to metal and nonmetal centers have also been seen in dicationic and tricationic species, for example, in $GeNe^{2+}$, $PtHe^{2+}$, VHe^{3+}, $SiNe^{2+}$, and CAr^{2+}.[86b,c] Other related dicationic and tricationic species have been studied by theoretical methods. Not surprisingly, the *vicinal*-dications (and trications) of noble gas clusters are limited to the gas-phase, and no analogous condensed phase species have been reported. It may be possible to form a doubly electron deficient species (**237**, eq 73)

$$F{-}Xe{-}F \xrightarrow{SbF_5} \overset{\delta^+}{F}{-}Xe{-}\overset{\delta^-}{F}{-}{-}SbF_5 \xrightarrow{SbF_5} SbF_5{-}{-}{-}\overset{\delta^-}{F}{-}\overset{\delta^{++}}{Xe}{-}\overset{\delta^-}{F}{-}{-}{-}SbF_5 \qquad (73)$$

$$\hspace{8cm} \mathbf{237}$$

by complexation of xenon fluorides by Lewis acids, although it is uncertain if such a species could be persistent in the condensed phase.

Olah and colleagues have studied the heliomethonium dication, CH_4He^{2+} (**238**)

$$\left[\begin{array}{c} H\text{-}\;\;\text{-}H \\ \backslash\;/ \\ H-C\diagdown \\ /\quad He \\ H \end{array} \right]^{2+}$$

238

by *ab initio* calculations at the MP2/6-31G** and MP2/6-311+G(2 d,p) levels.[119] This gitonic superelectrophile **238** is a stable minimum (C_s symmetry) on its potential energy surface and its structure is isoelectronic with CH_5^+. The transition state structure for proton loss was calculated and dication **238** is protected from dissociation by a significant energy barrier (21.3 kcal/mol). Based on calculations, it is suggested that dication **238** may be produced by the reaction of CH_4^{2+} and He.

REFERENCES

(1) K. Lammertsma; M. Barzaghi; G. A. Olah; J. A. Pople; A. J. Kos; P. v. R. Schleyer *J. Am. Chem. Soc.* **1983**, *105*, 5252.

(2) G. A. Olah; N. Hartz; G. Rasul; G. K. S. Prakash *J. Am. Chem. Soc.* **1993**, *115*, 6985.

(3) G. A. Olah; J. L. Grant; R. J. Spear; J. M. Bollinger; A. Serianz; G. Sipos *J. Am. Chem. Soc.* **1976**, *98*, 2501.

(4) (a) S. Saito; Y. Sato; T. Ohwada; K. Shudo *J. Am. Chem. Soc.* **1994**, *116*, 2312. (b) Y. Sato; M. Yato; T. Ohwada; S. Saito; K. Shudo *J. Am. Chem. Soc.* **1995**, *117*, 3037.

(5) K. Y. Koltunov; S. Walspurger; J. Sommer *Tetrahedron Lett.* **2004**, *45*, 3547.

(6) S. Saito; T. Ohwada; K. Shudo, *J. Am. Chem. Soc* **1995**, *117*, 11081.

(7) G. A. Olah; G. Rasul; C. T. York; G. K. S. Prakash *J. Am. Chem. Soc.* **1995**, *117*, 11211.

(8) T. Drewello; W. Koch; C. B. Lebrilla; D. Stahl; H. Schwarz *J. Am. Chem. Soc.* **1987**, *109*, 2922.

(9) G. A. Olah in *Borane, Carborane, Carbocation Continuum*, J. Casanova, Ed.; Wiley: New York, **1998**, pp 131–145.

(10) E. Muetterties *J. Am. Chem. Soc.* **1959**, *81*, 2597.

(11) G. A. Olah *Angew. Chem. Int. Ed. Engl.* **1993**, *32*, 767.

(12) Reference 9, p. 135.

(13) K. Lammertsma; P. v. R. Schleyer; H. Schwarz *Angew. Chem. Int. Ed. Engl.* **1989**, *28*, 1321.

(14) G. A. Olah; G. K. S. Prakash; G. Rasul *ARKIVOC* **2002**, *2*, 7.

(15) W. E. Piers; S. C. Bourke; K. D. Conroy *Angew. Chem. Int. Ed.* **2005**, *44*, 5016 and references cited therein.

(16) D. Stahl; F. Maquin; T. Gaumann; H. Schwarz; P.-A. Carrupt; P. Vogel *J. Am. Chem. Soc.* **1985**, *107*, 5049.

(17) (a) G. Rasul; G. A. Olah *Inorg. Chem.* **1997**, *36*, 1278. (b) G. A. Olah; G. Rasul *J. Am. Chem. Soc.* **1996**, *118*, 12922.

(18) G. A. Olah, unpublished results.

(19) G. A. Olah; A. Burrichter; G. Rasul; R. Gnann; K. O. Christe; G. K. S. Prakash *J. Am. Chem. Soc.* **1997**, *119*, 8035.

(20) M. Attina; F. Cacace; F. Grandinetti; G. Occhiucci; A. Ricci *Int. J. Mass Spectrom.* **1992**, *117*, 47.

(21) C. H. DePuy; R. Gareyev; J. Hankin; G. E. Davico; R. Damrauer *J. Am. Chem. Soc.* **1997**, *119*, 427.

(22) G. A. Olah; K. K. Laali; Q. Wang; G. K. S. Prakash *Onium Ions*, Wiley, New York, **1998**.

(23) (a) B. S. Jursic *Theochem* **2000**, *498*, 149. (b) K. Lammerstma; G. A. Olah; M. Barzaghi; M. Simonetta *J. Am. Chem Soc.* **1982**, *104*, 6851.

(24) (a) G. Frenking *J. Am. Chem. Soc.* **1991**, *113*, 2476. (b) Reference 13.

(25) (a) N. C. Baenzinger; R. E. Buckles; T. D. Simpson *J. Am. Chem. Soc.* **1967**, *89*, 3405. (b) Reference 3.

(26) R. Rathore; S. V. Lindeman; A. S. Kumar; J. K. Kochi *J. Am. Chem. Soc.* **1998**, *120*, 6931.

(27) T. Mori; Y. Inoue *J. Phys. Chem. A* **2005**, *109*, 2728.

(28) (a) G. A. LePage; R. M. Elofson; K. F. Schulz; J. Laidler; K. P. Kowalewski; A. S. Hay; R. J. Crawford; D. Tanner; R. B. Sandin *J. Med. Chem.* **1983**, *26*, 1645. (b) R. M. Elofson; D. H. Anderson; H. S. Gutowsky; R. B. Sandin; K. F. Shulz *J. Am. Chem. Soc.* **1963**, *85*, 2622.

(29) (a) H. Hart; T. Sulzberg; R. R. Rafos *J. Am. Chem. Soc.* **1963**, *85*, 1800. (b) Reference 3.

(30) D. A. Klumpp; D. N. Baek; G. K. S. Prakash; G. A. Olah *J. Org. Chem.* **1997**, *62*, 6666.

(31) T. Ohwada; K. Shudo *J. Am. Chem. Soc.* **1988**, *110*, 1862.

(32) (a) T. Ohwada; T. Suzuki; K. Shudo *J. Am. Chem. Soc.* **1998**, *120*, 4629. (b) T. Ohwada; K. Shudo *J. Am. Chem. Soc.* **1988**, *110*, 34.

(33) G. A. Olah; A. M. White *J. Am. Chem. Soc.* **1967**, *89*, 4752.

(34) G. A. Olah; A. T. Ku; J. Sommer *J. Org. Chem.* **1970**, *35*, 2159.

(35) T. Yamazaki; S.-i. Saito; T. Ohwada; K. Shudo *Tetrahedron Letters* **1995**, *36*, 5749.

(36) D. A. Klumpp; M. Garza; S. Lau; B. Shick; K. Kantardjieff *J. Org. Chem.* **1999**, *64*, 7635.

(37) (a) D. A. Klumpp; D. Do; M. Garza; M. C. Klumpp; K. Kantardjieff; G. K. S. Prakash; G A. Olah *Abstract of Papers*, 219th National Meeting of the American Chemical Society, San Francisco, CA, March 2000; American Chemical Society: Washington, DC, **2000**; Abstract ORGN 321. (b) M. G. Zolotukhin; L. Fomina; R. Salcedo; L. E. Sansores; H. M. Colquhoun; L. M. Khalilov *Macromolecules* **2004**, *37*, 5140. (c) M. G. Zolotukhin; S. Fomine; L. M. Lazo; R. Salcedo; L. E. Sansores; G. G. Cedillo; H. M. Colquhoun; J. M. Fernandez-G.; L. M. Khalilov *Macromolecules* **2005**, *38*, 6005.

(38) D. A. Klumpp, unpublished data.

(39) G. A. Olah; J. Bausch; G. Rasul; H. George; G. K. S. Prakash *J. Am. Chem. Soc.* **1993**, *115*, 8060.

(40) (a) D. A. Klumpp; K. Y. Yeung; G. K. S. Prakash; G. A. Olah *J. Org. Chem.* **1998**, *63*, 4481. (b) M. Smet; E. H. Schacht; W. Dehaen *Angew. Chem., Int. Ed.* **2002**, *41*, 4547. (c) Y. Fu; A. Vandendriessche; W. Dehaen; M. Smet *Macromolecules* **2006**, *39*, 5183. (d) M. G. Zolotukhin; M. d. C. G. Hernandez; A. M. Lopez; L. Fomina; G. Cedillo; A. Nogales; T. Ezquerra; D. Rueda; H. M. Colquhoun; K. M. Fromm; A. Ruiz-Trevino; M. Ree *Macromolecules* **2006**, *39*, 4696. (e) D. A. Neel; M. L. Brown; P. A. Lander; T. A. Grese; J. M. Defauw; R. A. Doti; T. Fields; S. A. Kelly; S. Smith; K. M. Zimmerman; M. I. Steinberg; P. K. Jadhav *Bioorg. Med. Chem. Lett.* **2006**, *15*, 2553.

(41) T. Ohwada; T. Yamazaki; T. Suzuki; S. Saito; K. Shudo *J. Am. Chem. Soc.* **1996**, *118*, 6220.

(42) H. Bock; K. Ruppert; K. Merzweiler; D. Fenske; H. Geosmann *Angew. Chem. Int. Ed. Engl.* **1989**, *28*, 1684.

(43) (a) G. A. Olah; N. Hartz; G. Rasul; G. K. S. Prakash; M. Burkhart; K. Lammertsma *J. Am. Chem. Soc.* **1994**, *116*, 3187. (b) Reference 2.

(44) Y. Sato; M. Yato; T. Ohwada; S. Saito; K. Shudo *J. Am. Chem. Soc.* **1995**, *117*, 3037.

(45) W. Koch; N. Heinrich; H. Schwarz *J. Am. Chem. Soc.* **1986**, *108*, 5400.

(46) G. Rasul; G. K. S. Prakash; G. A. Olah *J. Mol. Struct. (Theochem)* **1999**, *466*, 245.

(47) F. Maquin; D. Stahl; A. Sawaryn; P. v. R. Schleyer; W. Koch; G. Frenking; H. Schwarz *J. Chem. Soc., Chem. Commun.* **1984**, 504.

(48) A. Yokoyama; T. Ohwada; K. Shudo *J. Org. Chem.* **1999**, *64*, 611.

(49) G. Rasul; G. K. S. Prakash; G. A. Olah *Inorg. Chem.* **2002**, *41*, 5589.

(50) A. Berkessel; R. K. Thauer *Angew. Chem. Int. Ed. Engl.* **1995**, *34*, 2247.

(51) M. K. Denk; J. M. Rodezno; S. Gupta; A. J. Lough *J. Organomet. Chem.* **2001**, *617–618*, 242.

(52) J. Cioslowski; G. Boche *Angew. Chem. Int. Ed. Engl.* **1997**, *36*, 107.

(53) C. Widauer; G. S. Chen; H. Grutzmacher *Chem. Eur. J.* **1998**, *4*, 1154.

(54) (a) N. Hartz; G. Rasul; G. A. Olah *J. Am. Chem. Soc.* **1993**, *115*, 1277. (b) A. J. Illies; M. F. Jarrold; M. T. Bowers *J. Chem. Phys.* **1982**, *77*, 5847. (c) Reference 11.

(55) P. J. F. de Rege; J. A. Gladysz; I. T. Horvath *Science*, **1997**, *276*, 776.

(56) G. A. Olah; G. K. S. Prakash; T. Mathew; E. R. Marinez *Angew. Chem., Int. Ed.* **2000**, *39*, 2547.

(57) G. A. Olah; F. Pelizza; S. Kobayashi; J. A. Olah *J. Am. Chem. Soc.* **1976**, *98*, 296.

(58) (a) G. A. Olah; A. Burrichter; T. Mathew; Y. D. Vankar; G. Rasul; G. K. S. Prakash *Angew. Chem. Int. Ed. Engl.* **1997**, *36*, 1875. (b) Reference 57.

(59) D. Stahl; F. Maquin *Chem. Phys. Lett.* **1984**, *106*, 531.

(60) M. W. Wong; B. F. Yates; R. H. Nobes; L. Radom *J. Am. Chem. Soc.* **1987**, *109*, 3181.

(61) G. A. Olah; D. A. Klumpp *Acc. Chem. Res.* **2004**, *37*, 211.

(62) G. A. Olah; A. Germain; H. C. Lin; D. Forsyth *J. Am. Chem. Soc.* **1975**, *97*, 2928.

(63) M. Vol'pin; I. Akhrem; A. Orlinkov *New J. Chem.* **1989**, *13*, 771.

(64) G. A. Olah; A. Burrichter; G. Rasul; G. K. S. Prakash; M. Hachoumy; J. Sommer *J. Am. Chem. Soc.* **1996**, *118*, 10423.

(65) See also: J. P. Hwang; G. K. S. Prakash; G. A. Olah *Tetrahedron* **2000**, *56*, 7199.

(66) W. Koch; G. Frenking; H. Schwarz; F. Maquin; D. Stahl *Int. J. Mass Spec. Ion Proc.* **1985**, *63*, 59.

(67) (a) B. F. Yates; W. J. Bouma; L. Radom *J. Am. Chem. Soc.* **1986**, *108*, 6545. (b) Reference 66.

(68) J. Bausch; G. K. S. Prakash; G. A. Olah *J. Am. Chem. Soc.* **1991**, *113*, 3205.

(69) J. Bukala; J. C. Culmann; J. Sommer *J. Chem. Soc., Chem. Commun.* **1992**, 482.

(70) D. M. Brouwer; A. A. Kiffen *Recl. Trav Chim. Pays-Bas* **1973**, *92*, 689.

(71) G. A. Olah; A.-H. Wu *Synlett* **1990**, 599.

(72) (a) W. J. Bouma; L. Radom *J. Am. Chem. Soc.* **1983**, *105*, 5484. (b) Reference 47.

(73) T. Suzuki; T. Ohwada; K. Shudo *J. Am. Chem. Soc.* **1997**, *119*, 6774.

(74) G. A. Olah; T. Mathew; E. R. Marinez; P. M. Esteves; M. Etzkorn; G. Rasul; G. K. S. Prakash; *J. Am. Chem. Soc.* **2001**, *123*, 11556.

(75) D. A. Klumpp; M. Garza; S. Lau; B. Shick; K. Kantardjieff *J. Org. Chem.* **1999**, *64*, 7635.

(76) C. Blackburn; R. F. Childs *J. Chem. Soc., Chem. Commun.* **1984**, 812.

(77) (a) G. A. Olah; A. M. White *J. Am. Chem. Soc.* **1967**, *90*, 1884. (b) G. A. Olah; K. K. Laali; Q. Wang; G. K. S. Prakash *Onium Ions*, Wiley, New York, **1998**, pp 450–451.

(78) S. D. Price; M. Manning; S. R. Leone *J. Am. Chem. Soc.* **1994**, *116*, 8673.

(79) R. D. Bach; R. C. Badger *Synthesis* **1979**, 529.

(80) (a) H. Vancik; K. Percac; D. E. Sunko *J. Am. Chem. Soc.* **1990**, *112*, 7418. (b) A. Martin; M.-P. Jouannetaud; J.-C. Jacquesy *Tetrahedron Lett.* **1996**, *37*, 2967. (c) Reference 11.

(81) J. Sommer; J. Bukala *Acc. Chem. Res.* **1993**, *26*, 370.

(82) I. S. Akhrem; I. M. Churilova; S. V. Vitt *Russ. Chem. Bull., Int. Ed.* **2001**, *50*, 81.

(83) (a) G. A. Olah; G. Rasul; A. K. Yudin; A. Burrichter; G. K. S. Prakash; A. L. Chistyakov; I. V. Stankevich; I. S. Akhrem; N. P. Gambaryan; M. E. Vol'pin *J. Am. Chem. Soc.* **1996**, *118*, 1446. (b) Reference 63.

(84) I. S. Akhrem; A. L. Chistyakov; N. P. Gambaryan; I. V. Stankevich; M. E. Vol'pin *J. Organomet. Chem.* **1997**, *536–537*, 489.

(85) G. A. Olah; G. Rasul; M. Hachoumy; A. Burrichter; G. K. S. Prakash *J. Am. Chem. Soc.* **2000**, *122*, 2737.

(86) (a) K. Lammertsma; P. v. R. Schleyer; H. Schwarz *Angew. Chem. Int. Ed. Engl.* **1989**, *28*, 1321. (b) W. Koch; F. Maquin; D. Stahl; H. Schwarz *Chimia* **1985**, *39*, 376. (c) D. Schröder; H. Schwarz *J. Phys. Chem. A.* **1999**, *103*, 7385.

(87) (a) G. Rasul; G. K. S. Prakash; G. A. Olah *J. Mol. Struct. (Theochem.)* **1999**, *466*, 245. (b) Reference 47.

(88) J. Hrusak; N. Sandig; W. Koch *Int. J. Mass Spectrom.* **1999**, *185–187*, 701.

(89) (a) K. Leiter; K. Stephean; E. Mark; T. D. Mark *Plasma Chem. Plasma Processes* **1984**, *4*, 235. (b) C. J. Porter; C. J. Proctor; T. Ast; J. H. Beynon *Croat. Chem. Acta* **1981**, *54*, 407. (c) Y.-Y. Lee; S. R. Leone *J. Phys. Chem. A* **1995**, *99*, 15438. (d) Z. Herman; J. Zabka; Z. Dolejsek; M. Farnik *J. Mass Spectrom.* **1999**, *192*, 191.

(90) G. A. Olah; A. Burrichter; G. Rasul; G. K. S. Prakash *J. Am. Chem. Soc.* **1997**, *119*, 4594.

(91) V. G. Nenajdenko; N. E. Shevchenko; E. S. Balenkova; I. V. Alabugin *Chem. Rev.* **2003**, *103*, 229.

(92) M. G. Ahmed; R. W. Alder; G. H. James; M. L. Sinnott; M. C. Whiting *J. Chem. Soc., Chem. Commun.* **1968**, 1533.

(93) (a) R. W. Alder; A. G. Orpen; J. M. White *J. Chem. Soc., Chem. Commun.* **1985**, 494. (b) R. W. Alder; R. B. Sessions; J. M. Mellor; M. F. Rawlins *J. Chem. Soc., Chem. Commun.* **1977**, 747. (c) F. M. Menger; L. L. D'Angelo *J. Org. Chem.* **1991**, *56*, 3467. (d) R. W. Alder; R. B. Sessions; J. M. Mellor; M. F. Rawlins *J. Chem. Soc., Perkin 1* **1982**, 603.

(94) (a) T. J. Curphey; K. S. Prasad *J. Org. Chem.* **1978**, *37*, 2259. (b) M. Schmittel; M. Lal; K. Graf; G. Jeschke; I. Suske; J. Salbeck *Chem. Commun.* **2005**, *45*, 5650.

(95) A. R. Katritzky; M. P. Sammes *J. Chem. Soc., Chem. Commun.* **1975**, 247.

(96) Y. Zhang; J. Briski; Y. Zhang; R. Rendy; D. A. Klumpp *Org. Lett.* **2005**, *7*, 2505.

(97) (a) A. N. Ferguson *Tetrahedron Lett.* **1973**, *30*, 2889. (b) G. A. Olah; K. Dunne; D. P. Kelly; Y. K. Mo *J. Am. Chem. Soc.* **1972**, *94*, 7438. (c) S. F. Nelson; S. C. Blackstock; K. J. Haller *Tetrahedron* **1986**, *42*, 6101. (d) S. F. Nelson; W. C. Hollinsed; C. R. Kessel; J. C. Calabrese *J. Am. Chem. Soc.* **1978**, *100*, 7876.

(98) G. Rasul; G. K. S. Prakash; G. A. Olah *J. Am. Chem. Soc.* **1994**, *116*, 8985.

(99) (a) H. Kenso *Bull. Chem. Soc. Jpn.* **1979**, 1578. (b) Y.-H. Li; A. G. Harrison *Int. J. Mass Spectrom. Ion Phys.* **1978**, *28*, 289.

(100) J. F. McGarrity; D. P. Cox *J. Am. Chem. Soc.* **1983**, *105*, 3961.

(101) A. Mertens; K. Lammertsma; M. Arvanaghi; G. A. Olah *J. Am. Chem. Soc.* **1983**, *105*, 5657.

(102) K. Shudo; T. Ohta; T. Okamoto *J. Am. Chem. Soc.* **1981**, *103*, 645.

(103) G. A. Olah; H. C. Lin *Synthesis* **1974**, 444.

(104) G. A. Olah; A. Orlinkov; A. B. Oxyzoglou; G. K. S. Prakash *J. Org. Chem.* **1995**, *60*, 7348.

(105) G. A. Olah; Q. Wang; A. Orlinkov; P. Ramaiah *J. Org. Chem.* **1993**, *58*, 5017.

(106) G. A. Olah; P. Ramaiah; G. K. S. Prakash *Proc. Natl. Acad. Sci. USA.* **1997**, *94*, 11783.

(107) (a) G. A. Olah; G. Rasul; R. Aniszfeld; G. K. S. Prakash *J. Am. Chem. Soc.* **1992**, *114*, 5608. (b) G. K. S. Prakash; G. Rasul; A. Burrichter; G. A. Olah in *Nitration—Recent Laboratory and Industrial Developments*, L. F. Albright; R. V. C. Carr; R. J. Schmitt (Eds.), ACS Symposium Series 623, American Chemical Society, Washington D.C., **1996**, p. 10.

(108) T. Weiske; W. Koch; H. Schwarz *J. Am. Chem. Soc.* **1993**, *115*, 6312.

(109) (a) K. O. Christe; W. W. Wilson; E. C. Curtis *Inorg. Chem.* **1979**, *18*, 2578. (b) G. A. Olah; A. L. Berrier; G. K. S. Prakash *J. Am. Chem. Soc.* **1982**, *104*, 2373.

(110) G. A. Olah; G. Rasul; A. Burrichter; M. Hachoumy; G. K. S. Prakash; R. I. Wagner; K. O. Christe *J. Am. Chem. Soc.* **1997**, *119*, 9572.

(111) (a) G. A. Olah; N. Yoneda; D. G. Parker *J. Am. Chem. Soc.* **1976**, *98*, 5261. (b) M. Ceotto; F. A. Gianturco; D. M. Hirst *J. Phys. Chem. A* **1999**, *103*, 9984. (c) M. Aschi; A. Largo *Int. J. Mass Spectrom.* **2003**, *228*, 613.

(112) (a) V. G. Nenajdenko; P. V. Vertelezkij; A. B. Koldobskij; I. V. Alabugin; E. S. Balenkova *J. Org. Chem.* **1997**, *62*, 2483. (b) K. Ohkata; K. Okada; K. Akida *Heteroat. Chem.* **1995**, *6*, 145.

(113) G. Rasul; G. K. S. Prakash; G. A. Olah *J. Org. Chem.* **2000**, *65*, 8786.

(114) H. Fujihara; R. Akaishi; N. Furukawa *Chem. Lett.* **1988**, 709.

(115) H. Sakai; H. Stapelfeldt; E. Constant; M. Y. Ivanov; D. R. Matusek; J. S. Wright; P. B. Corkum *Phys. Rev. Lett.* **1998**, *81*, 2217.

(116) G. Rasul; G. A. Olah, manuscript submitted.

(117) L. Pauling *J. Chem. Phys.* **1933**, *1*, 56.

(118) J. D. Dunitz; T. K. Ha *J. Chem. Soc. Chem. Commun.* **1972**, 568.

(119) G. A. Olah; G. K. S. Prakash; G. Rasul *J. Mol. Struct. (Theochem.)* **1999**, *489*, 209.

6

GITONIC 1,3-SUPERELECTROPHILES

6.1 STRUCTURAL CONSIDERATIONS

Multiply charged onium cations have been actively studied over the years.[1] With the development of the concept of superelectrophiles and the studies of varied related onium dications and trications, it has become clear that electrophilic reactivities drop off rapidly with an increasing distance between the charge centers. A large enough separation of the charge centers in onium dications leads to electrophilic reactivities similar to monocationic electrophiles. As discussed in the previous chapters, geminal and vicinal-type dications are very reactive gitonic superelectrophiles. A number of studies have also demonstrated that 1,3-dicationic systems can exhibit superelectrophilic activity, despite the increased charge-charge separation. These 1,3-dicationic superelectrophiles are discussed in this chapter. First superelectrophiles composed of 1,3-carbodications, specifically containing carbenium and carbonium ion centers, are considered. This is followed by those containing oxonium and carboxonium ions, acyl dications, and azacarbodications.

6.2 1,3-DICATIONIC SYSTEMS

6.2.1 1,3-Carbodications

The destabilization and connected superelectrophilic character of some 1,3-carbodications can be seen in the unsuccessful attempts to prepare

Superelectrophiles and Their Chemistry, by George A. Olah and Douglas A. Klumpp
Copyright © 2008 John Wiley & Sons, Inc.

certain persistent 1,3-carbodications. For example, the *tert*-butylcation
(**1**) is known to be a remarkably stable species in superacidic solution
(eq 1).[2] In contrast, ionization of 2,3,3,4-tetramethyl-2,4-pentanediol (**2**)
does not produce the expected 1,3-carbodication (**3**), but instead disproportionation products 2,3-dimethyl-2-butyl cation (**4**) and protonated acetone
(**5**) are formed (eq 2).[3]

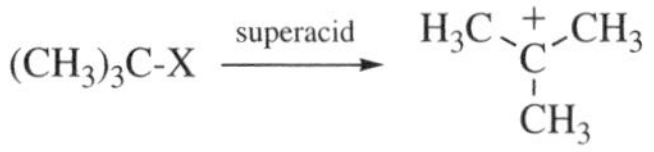

$$(CH_3)_3C\text{-}X \xrightarrow{\text{superacid}} \underset{\underset{CH_3}{|}}{H_3C\overset{+}{\underset{}{C}}CH_3} \tag{1}$$

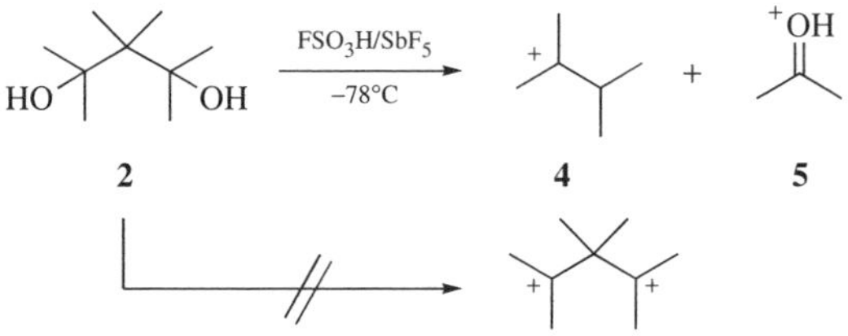

1

(2)

In a similar respect, ionization of 2,4-dichloro-2,4-dimethylpentane (**6**)
does not give the 1,3-carbodication (**7**, eq 3).[3] Despite the superacidic
conditions, deprotonation occurs to give the allylic cation (**8**). Even substitution by phenyl groups is not enough to stabilize the 1,3-dication.
For example 1,1,3,3-tetraphenyl-1,3-propanediol (**9**) also undergoes the
deprotonaton or disproportionation reactions (eq 4).[3]

In an attempt to prepare an adamanta-1,3-diyl dication **12**, only the
monocationic, donor-acceptor complex (**11**) could be observed experimentally in the superacid-promoted reaction of the difluoride (**10**, eq 5).[4]

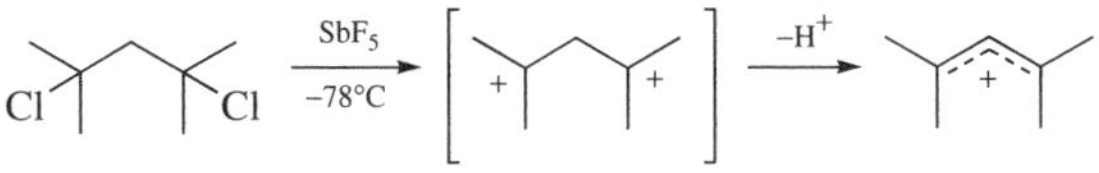

(3)

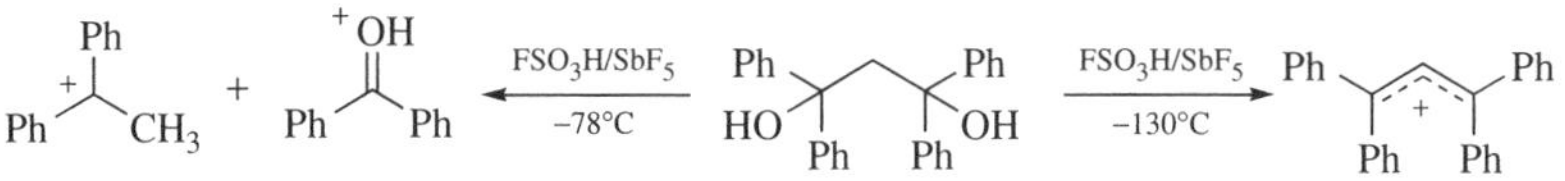

(4)

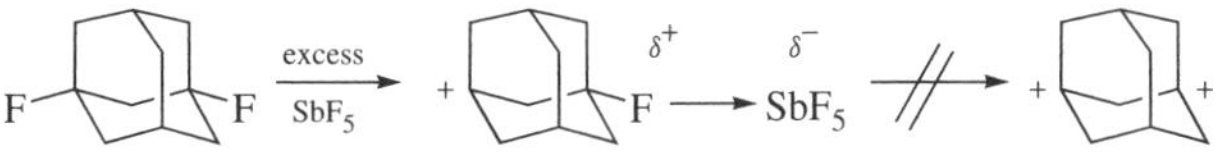

(5)

Although the fully formed dicationic structure (**12**) is not formed, the donor-acceptor complex **11** may have partial superelectrophilic character by interaction with SbF$_5$. The adamanta-1,3-diyl dication **12** has been found to be the global minimum structure on the C$_{10}$H$_{14}^{2+}$ potential energy surface.[5] Theoretical studies at the B3LYP/6-31G** level have shown **12** to be 0.4 kcal/mol more stable than the isomeric 1,4-dication **13**. The loss of stabilization on going from the 3° carbocationic center to the 2° carbocationic center is compensated by an increasing distance between charge centers. The next most stable structure is the 1,5-dication **14**, which is found to be 3.3 kcal/mol less stable than **12**. The structures and energies for the *protio*-adamantyl dications (C$_{10}$H$_{16}^{2+}$) were also studied by theoretical calculations.[5] At the B3LYP/6-31G** level, five energy minima were located. The global energy minimum for the *protio*-adamantyl dication was found for structure **15**, which can be considered as the C(3)–C(9) protonated 1-adamantyl cation. The structure contains a two electron–three center (2e–3c) bond involving two carbons and a carbenium center separated by a methylene group. The other C$_{10}$H$_{16}^{2+}$ structures are energy minima with considerably higher energy (> 18 kcal/mol) than dication **15**.

	12	**13**	**14**
Rel. Energy: (kcal/mol)	0.0	0.4	3.3

As in the case of vicinal dications, stabilization of the carbocationic centers has provided several examples of observable 1,3-carbodications. For example, Olah and co-workers were able to generate the cyclopropyl stabilized 1,3-carbodication (**17**) by ionization of the diol (**16**) in FSO$_3$H–SbF$_5$ at low temperature (eq 6).[6]

The carbocationic centers of **17** exhibit a δ^{13}C resonance of 262.8, which agrees reasonably with the calculated IGLO DZ//B3LYP/6-31G* chemical shift of $\delta^{13}C$ 282.6. The cationic centers of the carbodication (**17**) are shielded by 12 ppm, when compared with the 1,1-dicyclopropylethyl cation (**18**), showing an increased delocalization of the charge from the cationic centers into the cyclopropyl groups. The trimethylenemethane dication (**20**) is an interesting example of a prepared 1,3-dication.[7] Ionization of the diol (**19**) in $FSO_3H–SbF_5$ produces the dication (**20**), which was observed by NMR (eq 7).

$$(7)$$

Another class of gitonic superelectrophiles (based on the 1,3-carbodication structure) are the Wheland intermediates or sigma complexes derived from electrophilic aromatic substitution of carbocationic systems (eq 8).

$$(8)$$

Few examples of such intermediates have been reported, but one example involves the superelectrophilic nitration of the trityl cation.[8] Attack of the trityl cation (**21**) by the protosolvated nitronium ion (**22**, HNO_2^{2+}) leads to a sigma complex (**23**) having the 1,3-dicationic structure. It is expected that delocalization of the charges leads to significant contributions from charge separated resonance forms (eq 8). In general, these types of dicationic sigma complexes will also be formed when the reacting arene contains a potentially cationic substituent group. This would include the superacid-catalyzed reactions of nitrobenzenes, aryl ketones, aryl aldehydes, and anilines, many of which are fully protonated in superacids. The nitration of phenalenone is another example in which a dicationic species has been proposed as a key intermediate (eq 9).[9]

$$(9)$$

In strongly acidic media, 2-nitrophenalenone is produced which is consistent with formation of dication **24**. There are also examples of diprotonated phenalenones, similar to intermediate **24**, that have been directly observed by NMR spectroscopy.

A similar type of intermediate has been proposed in the superacid-catalyzed hydrogen-deuterium exchange involving 9-substituted-9-fluorenyl cations.[10] For example, 9-methyl-9-fluorenyl cation (**25**) can be generated in deuterated fluorosulfonic acid solution from ionization of the 9-methyl-9-fluorenol (Scheme 1). Deuterium exchange is observed at the C-2/7, C-4/5, and at the methyl carbon. These exchange products have been rationalized by formation of the ring-deuterated, dicationic species (**26**). Formation of the dicationic species likely increases the acidity of the methyl hydrogens (making them superelectrophilic) and deprotonation gives monocation **27**, which undergoes subsequent deuteration.

There are several other carbodications that may be represented as 1,3-dicationic systems (or nonclassical, highly delocalized systems), including Schleyer's 1,3-dehydro-5,7-adamantanediyl dication (**28**),[11] Hogeveen's pyramidal dication (**29**),[12] diallyl dications (i.e., **30**),[13] dicationic ethers (i.e., **31**),[14] diprotonated pyrones (**32**),[15] as well as various aromatic dications.

Though interesting as dicationic species, these ions are not generally considered gitonic superelectrophiles. Several recent reviews have summarized their chemistry.[16]

As described in the case of the 1-adamantyl cation, protosolvation of carbon-carbon and carbon-hydrogen sigma bonds can lead to cationic 2e−3c bonding and the formation of carbonium ion centers. The role of this interaction in the chemistry of superelectrophiles is typified by protosolvation of the *tert*-butyl cation (**1**) in superacid to provide the dicationic

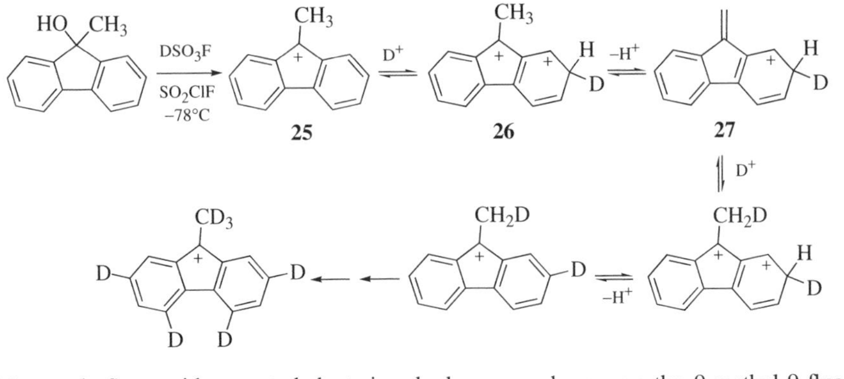

Scheme 1. Superacid-promoted deuterium-hydrogen exchange on the 9-methyl-9-fluo-renyl cation (**25**)

species (**33**, eq 10).[17] As described in the previous chapter, this leads to isotopic exchange when deuterated superacid is used. Although such protosolvation of trialkylcarbenium ions has only been reported for the *tert*-butyl cation and 2-propyl cation, similar dicationic species should also be possible for larger systems leading 1,3-dicationic superelectrophiles. For example, the 3-ethyl-3-pentyl cation (**34**) should preferentially be protosolvated at the methyl groups (**35**) instead of the methylene groups (**36**), due to the charge-charge repulsive effects (eq 11).

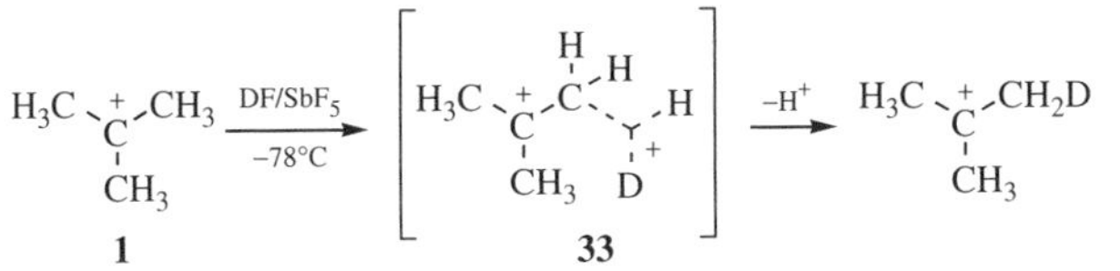

$$(10)$$

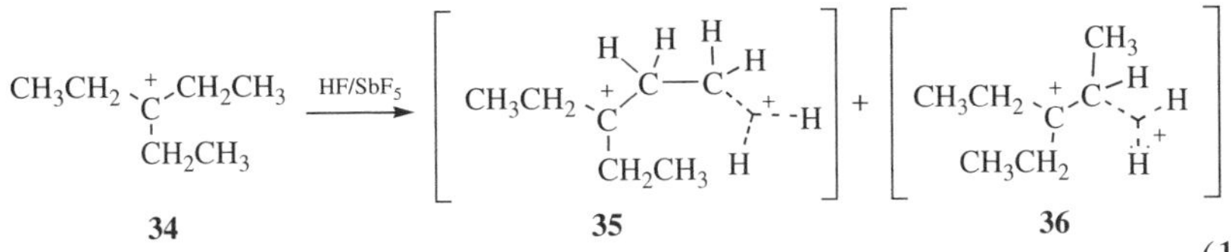

$$(11)$$

In addition to the carbenium-carbonium dications (i.e. **35**), bis-carbonium dications have been studied by theoretical methods. Examination of the potential energy surface of $C_3H_{10}^{2+}$ (diprotonated propane) finds three structures at energy minima.[18] The global minimum corresponds to the structure (**37**) in which both terminal carbons are involved in electron-deficient 2e−3c bonds. The next most stable $C_3H_{10}^{2+}$ structure (**38**) lies 6.7 kcal/mol higher in energy and it possesses two 2e−3c bonds, one involving the C–H bond and the other involving the C–C bond. In a similar respect, the global minimum on the $C_4H_{12}^{2+}$ potential energy surface is found at a structure (**39**) involving 2e–3c bonds of the terminal carbons.[18]

It might be argued that structures such as **37** and **39** should actually be described as distonic superelectrophiles, because three carbon atoms separate the protosolvated, 2e–3c bonds. However, as it is understood that some of the positive charge resides at the carbon atoms of the $-CH_4^+$ groups, the designations as gitonic superelectrophiles can be considered appropriate for **37** and **39**.

6.2.2   Carboxonium-Centered 1,3-Dications

There have been a significant number of reports of gitonic superelectrophilic systems involving carboxonium-centered dications. Many of these can be represented as 1,3-dicationic systems, although charge delocalization may increase the distance between charge centers. Some superelectrophiles of this type could be considered gitonic or distonic, depending upon which resonance form predominates in describing the structure. For example, acidic carboxonium ions can be described as either the carbenium or oxonium structures (**40a** and **40b**, respectively).

With the adjacent carbocationic center, structure **40a** would be considered a gitonic superelectrophile while **40b** would be a distonic superelectrophile. Although it is generally recognized that the oxonium-type structure (**40b**) is the more important resonance form in the description of many carboxonium ions,[1b] for demonstrating the raised point we use the hydroxycarbenium ion form for some of the dicationic structures in this section. These include carboxonium dications from 1,3-dicarbonyl groups, α,β-unsaturated carbonyl compounds, and others. The oxonium-type of representation is, however, used in the section describing gitonic superelectrophiles from diprotonated esters and carboxylic acids.

6.2.2.1 *Carboxonium-Carbenium Dications* The protonation of mesityl oxide (**41**) has been studied by several groups.[19] It has been proposed that mesityl oxide can form equilibria in superacid with the O,O-diprotonated species (**42**) and (at more elevated temperatures) with the C,O-diprotonated species (**40**, eq 12).[19a]

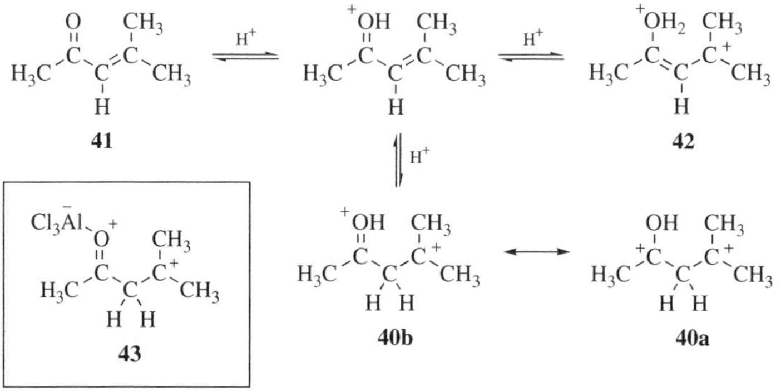

(12)

Evidence for the involvement of the diprotonated species **40** and **42** includes ^{13}C NMR data, which shows deshielding of the involved carbons as the acidity of the media increases from H_0 -8 to -26 (mesityl oxide is estimated to be fully monoprotonated in acids of about H_0 -8). Dication **42** has been shown to be capable of reacting even with very weak nucleophiles, for example, abstracting hydride from cyclohexane.[19b] An analogous species (**43**) has been proposed in interaction with excess $AlCl_3$. Protonated cyclohex-2-enone is converted to 3-methylcyclo-pent-2-enone in HF–SbF_5 solution at 50°C.[20] The reaction mechanism is thought to involve the dicationic intermediate **44** (eq 13). Likewise, aryl-substituted indenones are converted to the dications by reaction in superacid (eq 14).[21]

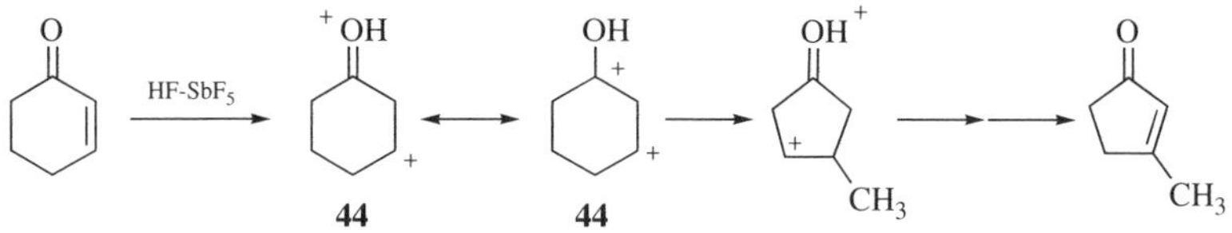

(13)

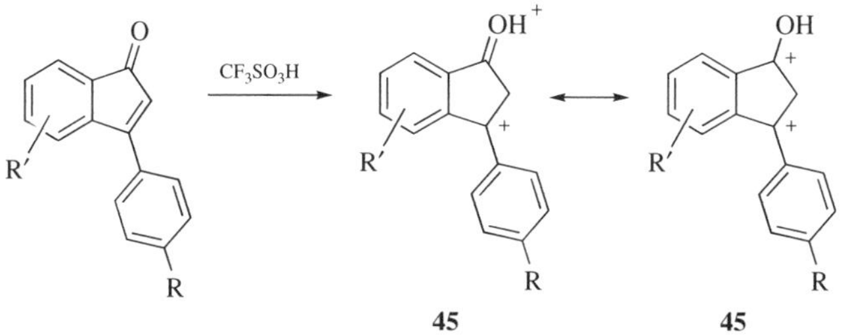

(14)

Dications (**45**) can be directly observed in some cases (R = $-CH_3$ and $-OCH_3$). Depending on the substituents, the carboxonium carbon (C-1) is observed between $\delta^{13}C$ 214–217, C-2 is observed at $\delta^{13}C$ 44–46, and C-3 is observed at about $\delta^{13}C$ 197. These data are consistent with the formation of diprotonated species **45**. For $\alpha\beta$-unsaturated ketones, both *C,O*-diprotonated and *O,O*-diprotonated species have been observed. For example, dication **46**, arising from *C,O*-diprotonation, is the product from 4-phenyl-3-buten-2-one in superacid,[22] while dication **47**, arising from *O,O*-diprotonation, is the product from chalcone in superacid.[23]

Carboxonium-carbenium dications have also been proposed in the reactions of aryl ethers, phenols, and naphthols, with superacids. When 2-naphthol is reacted with an excess of $AlCl_3$ (3 equivalents) and cyclohexane, the product of ionic hydrogenation is observed in 59% yield (eq 15).[24] This conversion is thought to occur by double protonation of the 2-naphthol ring to give the dication **48**, which is capable of abstracting hydride from cyclohexane. Similar intermediates are formed by the reactions of $HF-SbF_5$ with naphthyl ethers (eq 16).[25]

Dication **49** can be directly observed by low temperature NMR. Like other carboxonium ions, dication **49** forms an equilibrium mixture of the *E* and *Z* stereoisomers.

A variety of 1,3-dicationic carboxonium superelectrophiles have been generated from protonation of carboxylic acids and their derivatives. For example, carboxonium-carbenium dications have been proposed in the superacid promoted reaction of cinnamic acid (**50**) and related compounds.[26] When cinnamic acid is reacted with triflic acid and benzene, the indanone product **51** is formed in good yield (eq 17).

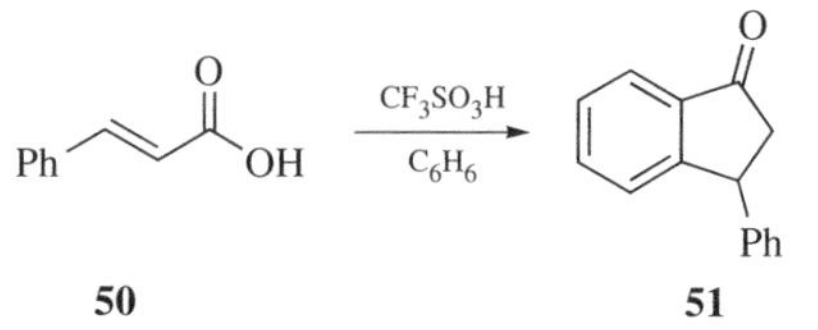

$$ \text{(17)} $$

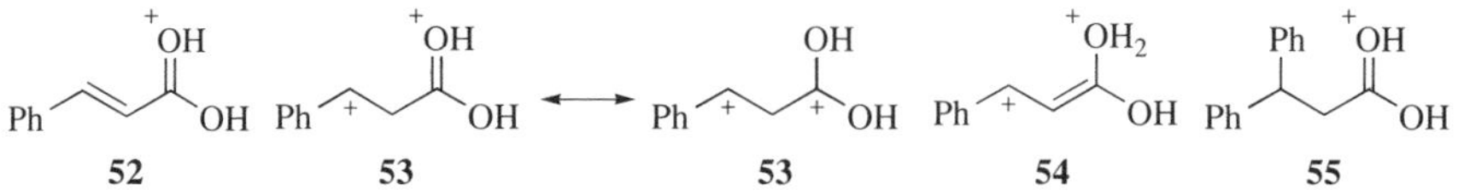

Mechanistically, it is suggested to involve protonation of the carboxylic acid group to give the monocation **52**, followed by a second protonation at the C-2 position to give the superelectrophile **53**.[26a] Cyclization of the propionic acid derivative (**55**) then gives the indanone product. *Ab initio* calculations suggest that **53** is the key intermediate, rather than dication **54** (arising from *O,O*-diprotonation).[26a] Structure **54** is found to lie 28 kcal/mol above **53** on the potential energy surface. Although superelectrophile **53** could not be directly observed using low temperature ^{13}C NMR, the dication (**57**) from β-phenylcinnamic acid (**56**) can be observed in $FSO_3H-SbF_5-SO_2ClF$ solution.[26a] The superelectrophilic character of dication **57** is seen by comparing it with a monocationic analogue, the 1,1-diphenylethyl cation **59**. While dication **57** reacts readily with benzene, and gives the indanone (**58**) in high yield from a cyclization (eq 18), the 1,1-diphenylethyl cation (**59**) is unreactive towards benzene (eq 19).

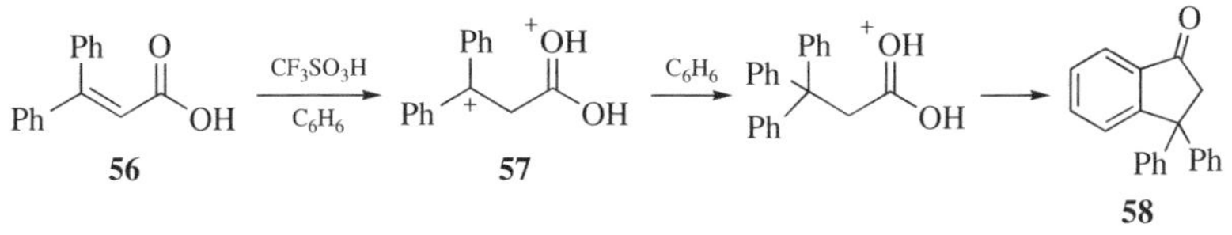

$$ \text{(18)} $$

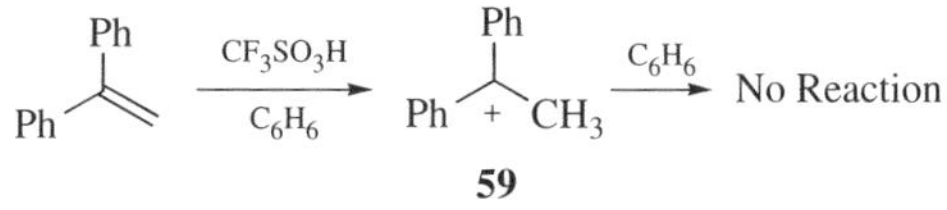

$$ \text{(19)} $$

This suggests that the protonated carboxylic acid group enhances the electrophilic reactivity of the carbocationic site (i.e., makes it superelectrophilic), despite the resonance stabilization by the two phenyl groups.

Several types of unsaturated amides have produced gitonic superelectrophiles in superacidic media. For example, reaction of the cinnamamide (**60**) with benzene in CF_3SO_3H gives the addition product in quantitative yield (eq 20).[27]

$$\text{(20)}$$

The dicationic species (**61**) is proposed as the key intermediate in the reaction. Dicationic intermediates such as **61** have also been proposed in conversions of unsaturated amides with polyphosphoric acid, sulfated zirconia, and zeolite catalysts, such as HUSY.[28] Likewise, ionic hydrogenation of 2-quinolinol (**62**) can be best understood by the involvement of the dicationic species (**63**), which is sufficiently electrophilic to react with cyclohexane (eq 21).[29]

$$\text{(21)}$$

It was also noted that dications like **63** may only be discrete intermediates in reactions in very strong liquid superacids, while protosolvated species like **64** may be more likely intermediates in reactions involving weaker superacids or zeolites.

There have been two reports involving gitonic superelectrophiles composed of carboxonium ions and vinylic carbocations in a 1,3-relationship. In the reaction of 3-phenylpropynoic acid (**65**) with benzene in superacid the novel carboxonium-vinyl dication **66** is generated, followed by reaction with benzene and then cyclization (eq 22).[26a] Likewise, the unsaturated amide (**67**) gives the cyclization product in high yields (70–97%) in very strong acids (polyphosphoric acid, CF_3SO_3H, Nafion SAC-13, or HUSY; eq 23).[30]

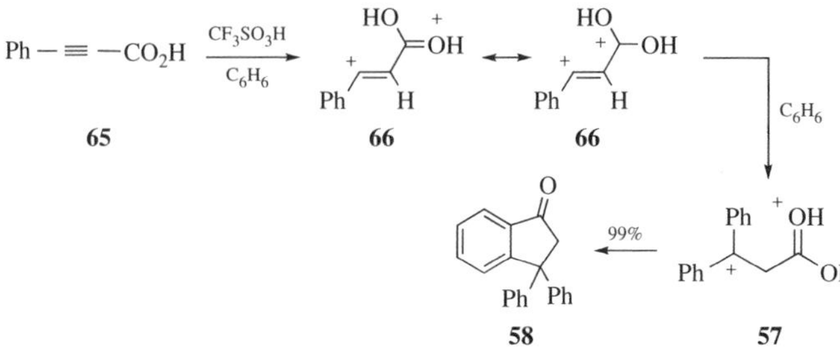

$$\text{Ph} - \equiv - \text{CO}_2\text{H} \xrightarrow[\text{C}_6\text{H}_6]{\text{CF}_3\text{SO}_3\text{H}}$$

(22)

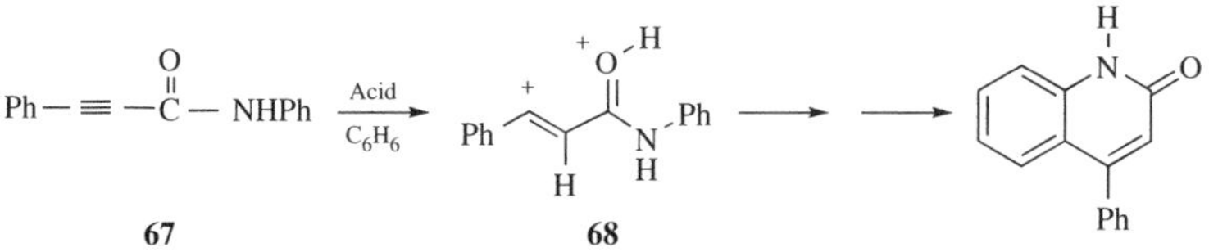

The carboxonium-vinylic dication (**68**) is considered the key intermediate leading to the cyclization product. The analogous vinylic dications (**69–71**) have also been generated in superacid.[31] Each of the species exhibits high electrophilic reactivity.

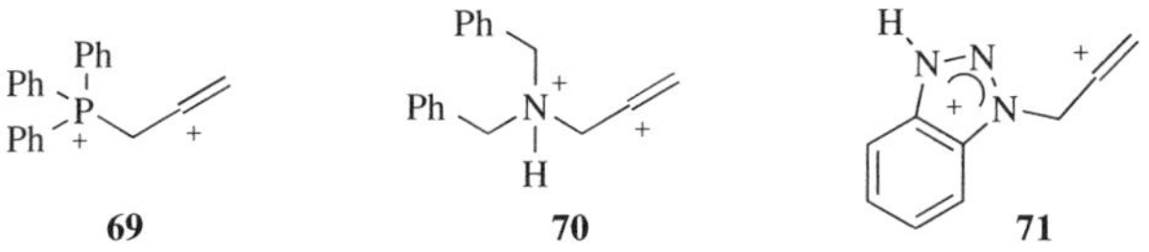

6.2.2.2 Bis-carboxonium Dications

A number of reports describe gitonic superelectrophiles arising from carboxonium ion groups separated by a carbon, oxygen, or nitrogen atom (Table 1). As noted, acidic carboxonium ions can be represented as either the oxonium-type or hydroxycarbenium-type structures. Depending on which resonance form is considered, these bis-carboxonium dications are considered either a gitonic or distonic superelectrophiles. For example, aliphatic 1,3-diketones can be diprotonated in $FSO_3H-SbF_5-SO_2$ solution at low temperature (eq 24).[32a]

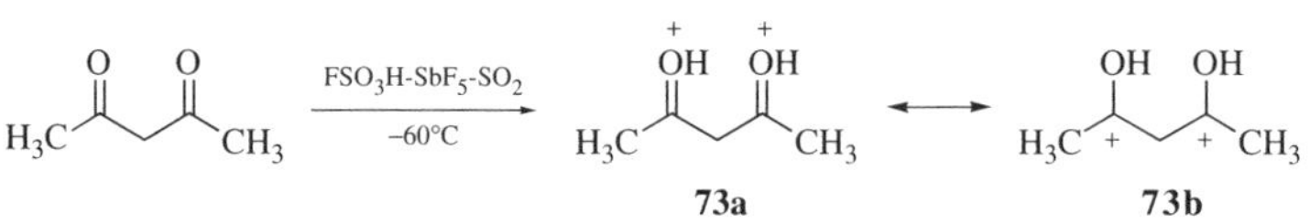

$$\xrightarrow[-60°\text{C}]{\text{FSO}_3\text{H-SbF}_5\text{-SO}_2}$$

(24)

The two important resonance structures for the dication are the bis-oxonium structure (**73a**, a distonic superelectrophile) and the bis-carbenium structure (**73b**, a gitonic superelectrophile). Although it is understood that various factors should favor structure **73a** (including charge-charge repulsion), a considerable amount of positive charge resides at the carbon. Thus, due to their (partial) 1,3-dicationic character, these types of dicationic species are considered as gitonic superelectrophiles.

Bis-carboxonium ions such as **73** can be directly observed using low-temperature NMR. In the case of **73**, 2,4 pentanedione is dissolved in $FSO_3H-SbF_5-SO_2$ solution at $-60°C$ and the ^1H NMR shows three absorptions, including the carboxonium protons.[32] In some cases (especially in weaker superacid systems), the diprotonated species form equilibrium mixtures with the monoprotonated species. When either 1,3-cyclohexane-dione or 2-methyl-1,3-cyclopentanedione is reacted in very strong superacids, only the monoprotonated species are observed.[32] This is attributed to increased stability of the enol-type cations, **90** and **91**, when compared with the acyclic systems.

However, diprotonated and triprotonated indane derivatives (**92** and **93**) have been reported.[33]

The dicationic species have also been obtained from β-ketoacids, β-ketoesters, and β-ketoamides in superacid solutions (Table 1, entries 2–4). Diprotonated acetoacetic acid (**75**) can be observed by low-temperature NMR under stable ion conditions.[34] Likewise, diprotonated methylacetoacetate (**77**) can be observed by NMR at temperatures lower than $-80°C$ in $FSO_3H-SbF_5-SO_2$ solution.[35] With ethyl acetoacetate in $HF-SbF_5$, the equilibrium constant for the dication-monocation equilibrium has been estimated to be at least 10^7, indicating virtually complete conversion to the superelectrophile.[35] The β-ketoamide (**78**) is found to give the condensation products **95** in good yield from CF_3SO_3H and the superelectrophile **79** is proposed as the key intermediate in the condensation reaction (eq 25).[27]

(25)

Table 1. Dications formed upon protonation of 1,3-dicarbonyl groups and related precursors.

Entry	Precursor	Dication	Acid
(1)	**72**	**73**	$FSO_3H\text{-}SbF_5$
(2)	**74**	**75**	$FSO_3H\text{-}SbF_5$
(3)	**76**	**77**	$FSO_3H\text{-}SbF_5$
(4)	**78**	**79**	CF_3SO_3H
(5)	**80**	**81**	$FSO_3H\text{-}SbF_5$
(6)[a]	**82**	**83**	$FSO_3H\text{-}SbF_5$
(7)	**84** X = H, Cl, Br, I	**85**	CF_3SO_3H $FSO_3H\text{-}SbF_5$ $BF_3\text{-}H_2O$
(8)	**86**	**87**	CF_3SO_3H
(9)	**88**	**89**	$HF\text{-}SbF_5$

[a]Cyclic dication **83** is in equilibrium with open chain dication, see text.

Following electrophilic attack on benzene, the carbenium-carboxonium dication **94** is generated, which then gives the product **95**. There have been several studies of the chemistry of malonic acid (**80**) and its esters in superacidic media. It has been shown that diprotonated products (i.e., **81**) are formed (Table 1, entry 5).[36]

A number of cyclic systems have produced bis-carboxonium dications (Table 1, entries 6–9). In the case of succinic anhydride (**82**), however, the product formed at $-80°C$ in $FSO_3H-SbF_5-SO_2$ solution was the acyl-carboxonium dication (**96**, eq 26).[37]

$$\text{(26)}$$

When the solution is warmed to $-40°C$, NMR suggests that the acyl-carboxonium dication **96** rearranges through the bis-carboxonium dication **83** leading to the acyl-carboxonium dication **96**.

The succinimide ring system is also thought to form diprotonated, bis-carboxonium type dications (entry 7).[38,39] When succinimide (**84**, X = H) is dissolved in $FSO_3H-SbF_5-SO_2$ solution, the dicationic species (**85**, X = H) can be directly observed by low temperature NMR.[39] Recently, Olah and co-workers demonstrated that N-halo-succinimide (**84**, X = Cl, Br, I)–BF_3-H_2O is a highly effective halogenating system for arenes.[39] A remarkable aspect of this halogenating system is its ability to even readily halogenate deactivated arenes. For example, 2-fluoronitrobenzene is converted to the halogenated products (**97**, X = Cl, Br, I) in high yields from the corresponding N-halosuccinimides and BF_3-H_2O (eq **27**).

X	Yield
Cl	81%
Br	95%
I	92%

$$\text{(27)}$$

To account for these conversions, two mechanistic proposals have been made: either the halogenating system involves a highly reactive and solvated X^+, or the reaction involves X^+ transfer from a protonated form of N-halosuccinimide. In the later case, X^+ transfer could occur from the neutral N-halosuccinimide, monoprotonated, or multiply protonated, superelectrophilic species. These mechanistic possibilities were further examined using DFT calculations. For N-chlorosuccinimide and its protosolvated intermediates, a series of calculations were done at the B3LYP/

**Table 2. Calculated reaction enthalpies for the generation of Cl⁺ from
N-chlorosuccinimide and its protonated forms.**

	ΔH, kcal/mol

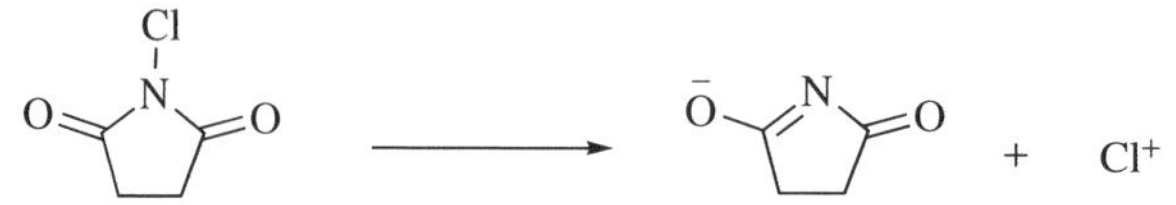

324.4

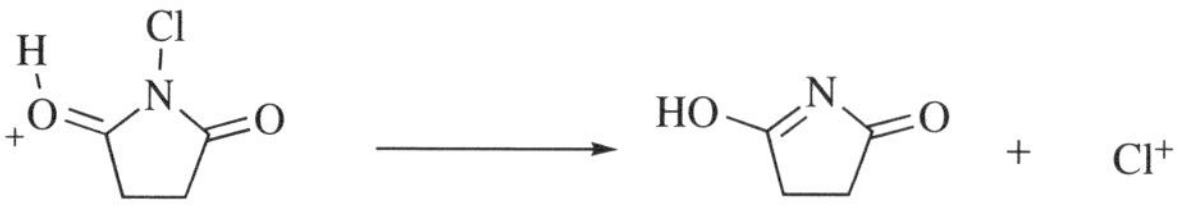

192.2

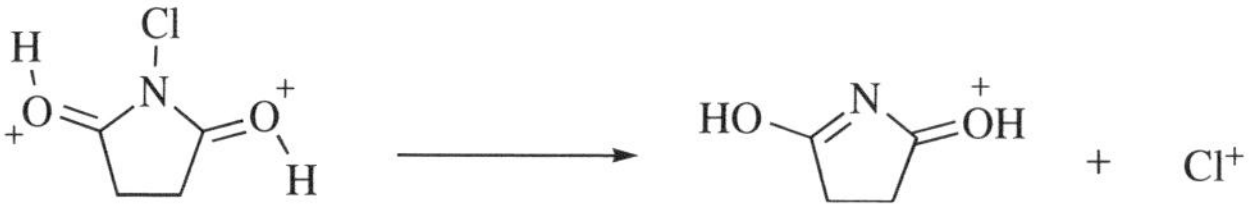

70.2

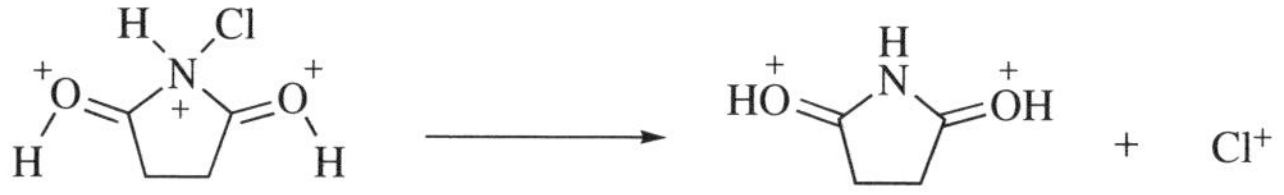

−102.0

98

6-311++G**//B3LYP/6-31G* level to estimate the reaction enthalpies
for the generation of Cl⁺ (Table 2).[39b] Optimized structures and energies
were calculated for *N*-chlorosuccinimide, its mono-, di-, and triprotonated forms, and the dehalogenated product. Not surprisingly, cleavage
of the nitrogen-chlorine bond to form Cl⁺ becomes increasingly favorable with a greater degree of protonation of *N*-chlorosuccinimide. Only
in the case of the triprotonated species (**98**), however, is the release of
Cl⁺ predicted to be exothermic. Although it is not presently clear whether
this particular halogenating system involves the release of solvated Cl⁺
or the direct transfer of Cl⁺ to the arene nucleophile, the calculations suggest that protosolvation of the *N*-chlorosuccinimide leads to ground state
destabilization. If Cl⁺ is directly transferred to the arene nucleophiles,
this estabilization of the *N*-chlorosuccinimide by superelectrophilic activation should lead to lowered activation barriers leading to the product,
compared to the uncatalyzed reactions (eqs 28–29).

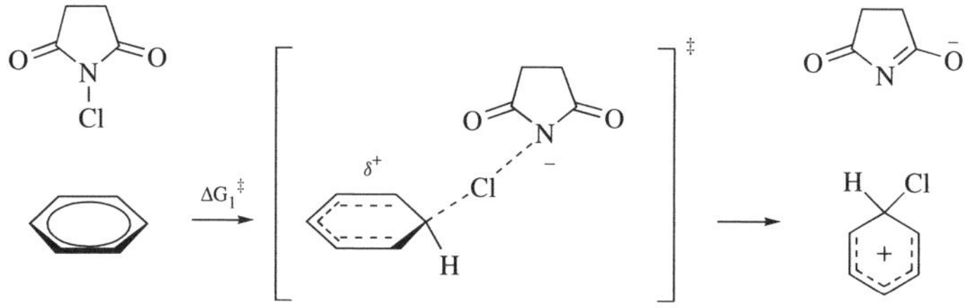

(28)

$$\Delta G_1{}^{\ddagger} > \Delta G_2{}^{\ddagger}$$

(29)

Transfer of Cl^+ to the arene provides some relief of the Coulombic repulsion in the multiply charged, superelectrophilic system. Under the reaction conditions, it is not yet known to what extent the N-halosuccinimides are protonated in BF_3-H_2O, but this acid-catalyst has an estimated acidity around H_0 -12.

Parabanic acid (**86**) has been shown to produce highly electrophilic species in superacidic CF_3SO_3H, and the resulting electrophile is capable of reacting with C_6H_6 and moderately deactivated arenes.[40] The superacid-promoted condensation reaction of parabanic acid with benzene (and other arenes) in CF_3SO_3H provides satisfactory yields of the 5,5-diarylhydantoins (eq 30),

(30)

well known for their use as anticonvulsant drugs. To explain the high electrophilic reactivity of parabanic acid in CF_3SO_3H, the superelectrophilic intermediate (**87**) is suggested as the key intermediate. Alternatively, a protosolvated ion with partial dicationic character may also be involved.

A series of protonated napthalenediols have also been studied and in some cases the bis-carboxonium dications can be observed by low temperature NMR from $HF-SbF_5$ solution.[25] In the case of 1,3-napthalenediol (**88**), the bis-carboxonium ion (**89**) is formed having the 1,3-dicationic character (Table 1, entry 9).

6.2.2.3 Carboxonium-Ammonium and Related Dications

A wide variety of species have been generated in which the 1,3-dicationic structure arises from carboxonium ion centers being adjacent (separated by one carbon) to an ammonium or related charge center. These intermediates may be described as reactive dications, yet they have been shown to exhibit electrophilic reactivities comparable to superelectrophiles.

It was demonstrated that α-aminoacids can be protonated in superacids to form well-defined dicationic species (**99**, eq 31).[41] It is not known if racemization of the α-aminoacid dications occurs from deprotonation at the α-carbon. Given the high acidity of the media, however, this seems unlikely. Triprotonated amino acids were likewise observed with amino acids lysine, methionine, aspartic acid, and glutamic acid. Interestingly, there were no examples of the α-aminoacid dications (i.e., **99**) cleaving to the corresponding acyl dications (i.e., **100**). Increasing the charge separation of the cationic centers however facilitates the dehydration to the acyl dication (**102**, eq 32).

$$(31)$$

$$(32)$$

The different chemistry of the dications **99** and **101** seems to reflect the superelectrophilic nature of the gitonic dication. It has also been shown that simple peptides may be multiply protonated in acids like $FSO_3H–SbF_5$, generally being protonated at the terminal amino group, the carboxyl group, and at the peptide bonds. In a study of the chemistry of N-tosylated phenylalanine derivatives, the diprotonated intermediate (**103**) was proposed in a reaction with superacid CF_3SO_3H (eq 33).[42]

$$(33)$$

It has also been shown that gitonic dications may be generated from superacid-promoted reactions of aminoacetals (eq 34).[43]

$$(34)$$

Reaction of acetal **104** with benzene in the presence of CF_3SO_3H leads to product **107** in high yield. This conversion involves formation of the ammonium-carboxonium dication (**105**), a reactive dication possessing some 1,3-dicationic character. Reaction with benzene and subsequent loss of methanol generates another reactive dication (**106**), which then gives the product. The superelectrophilic character of the ammonium-carboxonium dications is indicated by their reactions with moderately deactivated arenes, such as o-dichlorobenzene.

The N-heterocyclic systems have likewise been shown to produce reactive dications having 1,3-dicationic character. Piperidones and related systems are diprotonated in superacid and the resulting intermediates are capable of reacting with benzene in condensation reactions (eq 35).[44]

$$ (35) $$

Several types of nitrogen-containing heteroaromatic compounds are also capable of producing carboxonium-centered dications (Table 3).[45] Among the dications **108–113**, all have been shown to react with weak nucleophiles such as benzene, deactivated arenes, and even saturated hydrocarbons. Moreover, their reactivities greatly exceed that of comparable monocationic electrophiles. In the case of dication **111**, for example, it is shown that it will condense with benzene in a hydroxyalkylative conversion (eq 36).[45d]

$$ (36) $$

Under the same reaction conditions, however, protonated acetone does not react with benzene. Dications **112** and **113** have been shown to undergo ionic hydrogenation in the presence of cyclohexane, an exceptionally weak nucleophile.[45e,f]

The chemistry of N-acyliminium ions has great synthetic value and many of the conversions are done in highly acidic media.[46] There are several examples of N-acyliminium ion reactions that likely involve gitonic superelectrophiles. For example, reaction of the 3-chlorophthalimidine **114** with excess $AlCl_3$ produces an electrophile capable of reacting with m-dichlorobenzene (eq 37).[47]

Table 3. Dicationic species (108–113) arising from diprotonation of N-heteroaromatic compounds.

Entry	Substrate	Dication	Acid
(1)		**108**	CF₃SO₃H
(2)		**109**	CF₃SO₃H
(3)		**110**	CF₃SO₃H
(4)		**111**	CF₃SO₃H
(5)		**112**	CF₃SO₃H-SbF₅
(6)		**113** X = H or AlₙBr₃ₙ⁻	HBr-AlBr₃

(37)

As a reasonable explanation for the high electrophilic reactivity, the N-acyliminium ion (**115**) may be interacting with excess Lewis acid to produce the gitonic superelectrophile (**116**). Another superelectrophilic N-acyliminium ion is involved in the superacid-promoted Tscherniac amidomethylation reaction of aromatic compounds.[48] It was observed that N-hydroxymethylphthalimide reacts with benzene and deactivated arenes in superacidic CF_3SO_3H (Scheme 2). Given the high electrophilic reactivity, the gitonic superelectrophiles (**117** and **118**) were proposed as probable intermediates. The cyclic N-acyliminium (**120**) ion is obtained from ionization of the 5-hydroxypyrrolidone (**119**) and indirect evidence suggests the formation of the gitonic superelectrophile (**121**) in superacids (eq 38).[49] A dramatic acidity effect is seen in the reactions of **119** with arenes and CF_3SO_3H (H_0 −14) or CF_3CO_2H (H_0 −2.7). Compound **119** reacts with a moderately deactivated arene, p-dichlorobenzene, in CF_3SO_3H. In contrast, the same reaction with CF_3CO_2H gives no arylated product (**122**). However, when an activated arene is used in the reaction with **119** and CF_3CO_2H, the arylated product (**123**) is obtained (eq 39), indicating that the N-acyliminium ion (**120**) is generated in the CF_3CO_2H media. This is in accord that the superacidic CF_3SO_3H protosolvates the N-acyliminium ion (**120**) to produce the superelectrophile (**121**, eq 38). This chemistry was also studied by calculations (Figure 1) that indicate that the superelectrophilic N-acyliminium ion (**121**) is far more reactive towards benzene than the monocationic N-acyliminium ion (**120**). Calculations using gas-phase modeling indicate that formation of the σ-complex from the monocationic N-acyliminium ion (**120**) and benzene is significantly endothermic (at the MP2/6-31+G* level no minimum

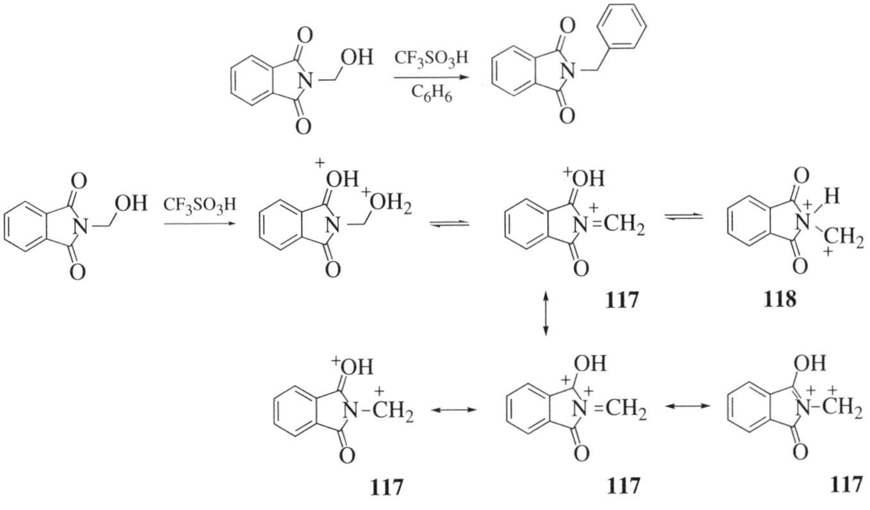

Scheme 2.

could be found for the σ-complex, and without a fixed ring-ring bond, the structure reverts to benzene and **120**). However, formation of the σ-complex from the superelectrophilic *N*-acyliminium ion (**121**) and benzene is found to highly exothermic. The more favorable thermodynamics is clearly the result of effective charge dispersal in the formation of the σ-complex from the superelectrophile. Given the highly acidic conditions and high temperatures often used in the reactions of *N*-acyliminium ions with arenes, it is likely that other examples of superelectrophilic intermediates arising from protosolvation, or interaction with excess Lewis acids are to be found.

could be found for the σ-complex, and without a fixed ring-ring bond,

$$
\begin{array}{cccccc}
\textbf{119} & & \textbf{120} & & \textbf{121} & & \textbf{122}
\end{array} \tag{38}
$$

$$
\begin{array}{ccccc}
\textbf{119} & & \textbf{120} & & \textbf{123}
\end{array} \tag{39}
$$

A series of phosphonium-carboxonium dications have also been studied in superacid catalyzed reactions.[31a] When the dicationic electrophiles are compared with similar monocationic species, it is clear that the phosphonium group enhances the electrophilic character of the carboxonium center. For example, protonated acetone is incapable of reacting with benzene in condensation reactions, however, the phosphonium-substituted carboxonium ion (**124**) reacts in high yield (eq 40).

$$
\textbf{124} \tag{40}
$$

Phosphonium groups are well known for their ability to stabilize adjacent anionic sites (i.e., Wittig reagents), but the results with the dicationic species indicate that phosphonium groups can also destabilize adjacent cationic groups producing their superelectrophilic reactivities.

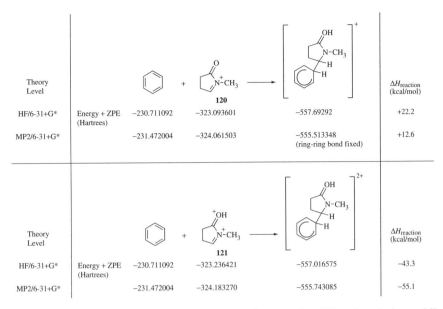

Figure 1. Calculated energetics for the reactions of monocation **120** and superelectrophile **121** with benzene to form the initial σ-complexes.

6.2.2.4 Diprotonated Esters and Carboxylic Acids A significant class of 1,3-dicationic gitonic superelectrophiles are the diprotonated esters and carboxylic acids. Acid-catalyzed ester cleavage has been extensively studied in organic chemistry and evidence has emerged to suggest that superelectrophiles can play a role in this chemistry. The acid catalyzed $A_{ac}1$ cleavage mechanism involves initial protonation at the acyl oxygen, followed by proton transfer to the ether oxygen, followed by direct cleavage to the acyl cation and alcohol (eq 41). In low-temperature NMR studies, methyl acetate is found to be completely protonated at the acyl oxygen (**125**) in superacidic $FSO_3H/SbF_5/SO_2$ solution.[50] There is no indication of an equilibrium with the neutral ester or the ether protonated isomer (**126**). However, even at $-78°C$, protonated methyl acetate undergoes slow acyl oxygen cleavage to acetyl cation and methyloxonium ion (eq 42).

$$(41)$$

$$H_3C \overset{O}{\underset{}{\underset{\|}{C}}} O\text{-}CH_3 \xrightarrow[\substack{SO_2 \\ -78°C}]{FSO_3H:SbF_5} H_3C \overset{\overset{+}{O}H}{\underset{}{\underset{\|}{C}}} O\text{-}CH_3 \underset{}{\overset{H^+}{\rightleftharpoons}} H_3C \overset{\overset{+}{O}H}{\underset{}{\underset{\|}{C}}} \overset{+}{\underset{\underset{H}{|}}{O}}\text{-}CH_3 \longrightarrow H_3C \overset{O}{\underset{}{\underset{\|}{C}}}{}^+ + \overset{+}{C}H_3OH_2$$

$$\qquad\qquad\qquad\qquad\qquad\qquad\quad \mathbf{125} \qquad\qquad\qquad \mathbf{127}$$

$$(42)$$

This is indicative of further protolytic activation *via* the superelectrophilic *gitonic* carboxonium dication (**127**). The protonated species arising from methyl acetate were also studied by calculational methods.[50] Consistent with the experimental results, the most stable monoprotonated species is found to be the acyl-oxygen protonated monocation (**125**). Several diprotonated structures were found as minima on the potential energy surface (MP4(SDTQ)/6-31G*//MP2/6-31G* level) and a stereoisomer of **127** is found to be the most stable. Structurally, the second protonation on **125** leads to a lengthening of the oxonium centers oxygen-carbon bond and a shortening of the oxygen-carbon bond of the acyl center. Another minimum on the potential energy surface is the dication arising from double protonation at the acyl oxygen. It is found just 3.0 kcal/mol higher in energy than the global minimum (i.e., **127**).

Further evidence for dicationic intermediates comes from a study of the 1,1-di-methoxyethyl cation (**128**) and its reaction with $CD_3F\text{-}SbF_5\text{-}SO_2$ (eq 43).[50] At $-30°C$, slow methyl exchange occurs at the methoxy oxygen. Since demethylation (to form the neutral ester) is unlikely in such highly acidic solution, the methyl exchange is best interpreted by electrophilic solvation of **128** by the $CD_3F{:}SbF_5$ and methylation to give the dication **129**. Subsequent demethylation then gives the exchange product **130**. The superacid-promoted ring opening of the β-lactone **131** has also been shown to produce the distonic acyl-oxonium dication **133**.[51]

$$H_3C \overset{+}{\underset{}{\underset{O\text{-}CH_3}{\overset{O\text{-}CH_3}{\diagup}}}}{} \xrightarrow[-30°C]{CD_3F{:}SbF_5/SO_2} H_3C \overset{+}{\underset{}{\underset{\underset{D_3C}{+}}{\overset{O\text{-}CH_3}{\diagup}}}}{O\text{-}CH_3} \underset{}{\overset{Nu \quad (NuCH_3)^+}{\rightleftharpoons}} H_3C \overset{+}{\underset{}{\underset{O\text{-}CD_3}{\overset{O\text{-}CH_3}{\diagup}}}}{} \quad (43)$$

$$\underset{BF_4^-}{}\qquad\qquad\qquad\qquad\qquad\qquad\qquad$$

$$\qquad \mathbf{128} \qquad\qquad\qquad\qquad\qquad \mathbf{129} \qquad\qquad\qquad\qquad \mathbf{130}$$

$$\xrightarrow[]{HF\text{-}BF_3} \qquad \rightleftharpoons \qquad \longrightarrow \qquad (44)$$

$$\qquad \mathbf{131} \qquad\qquad\qquad\qquad\qquad \mathbf{132} \qquad\qquad\qquad\qquad \mathbf{133}$$

This ring opening may likewise involve the gitonic superelectrophile **132**.

Superelectrophilic intermediates have further been proposed in the reactions of some esters. For example, a recent report describes the conversions of methyl benzoate to benzophenone products (70–93% yields) in reactions with superacid (eq 45).[52]

$$ (45) $$

82%

134   135

To explain the high level of electrophilic reactivity of this system, the protosolvated species **134** and **135** are proposed as probable intermediates.

Like the diprotonated esters, gitonic superelectrophiles are thought to arise from the reactions of some carboxylic acids in superacidic media. Protosolvation of the monoprotonated carboxylic acids has been suggested in the superacid-catalyzed formation of acyl cations from the corresponding carboxylic acids. For example, tetrafluoromethane is formed in the reaction of trifluoroacetic acid (CF_3CO_2H) with FSO_3H–SbF_5 (50 mol % SbF_5; eq 46).[53]

136   137   138   139

$$ (46) $$

With decreasing mol % of SbF_5 and thus decreasing acidity, the yield of CF_4 diminishes, and at 10 mol % SbF_5 no CF_4 is detectable. These results are interpreted as the protonation of CF_3CO_2H to form the carboxonium ion **136**, and subsequent protosolvation involving the *gitonic* dication **137**. Formation of **137** then leads to cleavage to the protiotrifluoroacetyl cation **138**, which ultimately leads to CF_4. Theoretical calculations found the dication **137** to be a stable minimum structure at the MP2/6-31G* level.[53] However calculations at the same level of theory could not find a stable minimum for the protiotrifluoroacetyl dication **138** because of its spontaneous dissociation into CF_3^+ and protonated carbon monoxide (COH^+).

This suggests that the formation of the $CF_3{}^+$ cation and fluoride abstraction from $SbF_6{}^-$ leads to the product CF_4.

In a similar respect, protosolvation has been suggested in the superacid-induced formation of the acetyl cation from acetic acid and the formation of the formyl cation from formic acid.[54] In theoretical studies related to such processes, both diprotonated acetic acid and diprotonated formic acid were found at a minimum on the potential energy surfaces (MP2/6-31G* level). In the case of acetic acid, the global minimum structure involves protonation at both oxygen atoms (**140**) and the structure is characterized by a long $C-OH_2$ bond and relatively short $C-OH$ bond.^{13}C and ^{17}O NMR chemical shifts were also calculated using IGLO, GIAO-SCF, and GIAO-MP2 methods. The gitonic superelectrophile (**140**) can be considered as a donor-acceptor complex of H_2O and the protonated acetyl cation (CH_3COH^{2+}). A similar structure (both C_s point group) is found at the global minimum for diprotonated formic acid (**141**).

Although monoprotonated acetic and formic acid can be directly observed as persistent species in superacids at low temperature,[55] the diprotonated ions **140** and **141** have not been detectable in the condensed phase as they cleave to their acyl cations

6.2.3 Oxonium 1,3-Dications

Like gitonic 1,2-dicationic species, analogous superelectrophiles having two oxonium cationic centers in a 1,3-dicationic structure are so far virtually unknown. There have been no reports of persistent bis-oxonium 1,3-dicationic species such as **142–143**. This may be due to the facile cleavage pathways that are available to such systems. In the case of 1,3-dioxane, reaction in superacid does not lead to the gitonic superelectrophile **143**, but rather ring opening products such as **145** and **146** are observed.[56] These are thought to arise from monocationic intermediates like **144** (eq 47).

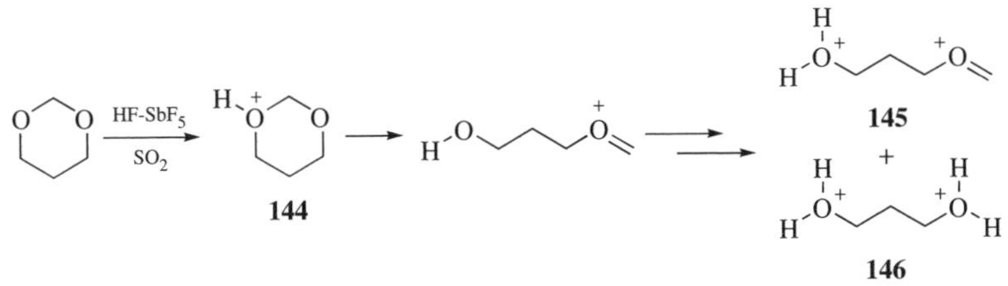

$$
\text{(47)}
$$

Several gitonic superelectrophiles have been reported having closely oriented oxonium and carboxonium ion centers, some of which may be considered 1,3-dications. A series of hydroxy-substituted carboxylic acids were studied in FSO_3H-SbF_5 in solution and the oxonium-carbonium dications could be directly observed at low temperature.[57] In the case of lactic acid, dication **147** is a persistent ion at $-80°C$, but at temperatures above $-60°C$, formation of the diprotonated lactide (**148**) is observed (eq 48).

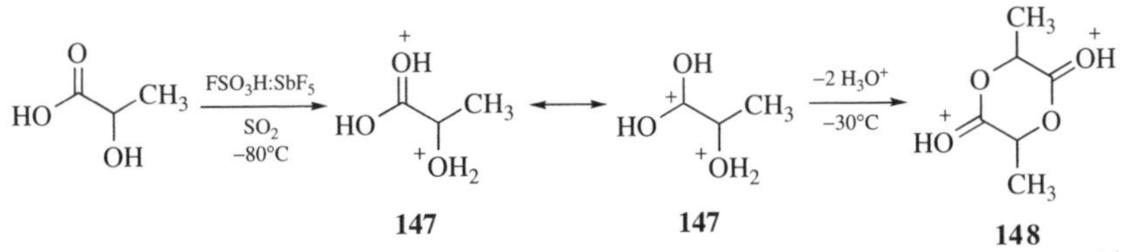

$$
\text{(48)}
$$

Although no mechanism is proposed for the dimerization of lactic acid in the FSO_3H-SbF_5 solution, the process may very well involve the superelectrophile **147**. Other oxonium-carbenium 1,3-dications have been suggested in superacid promoted pinacolone rearrangments of diprotonated aliphatic glycols and alkoxy alcohols (eqs 49–50).[58]

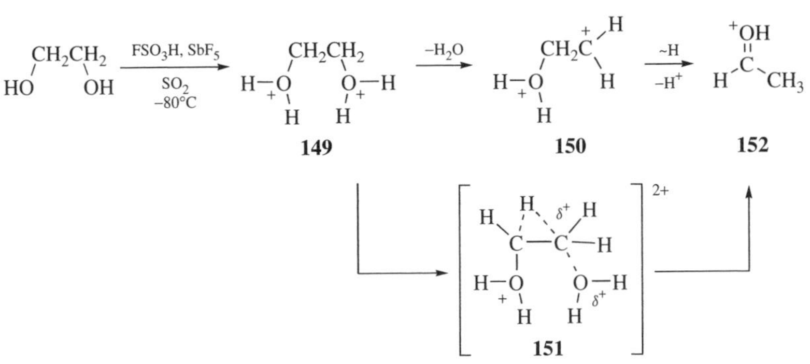

$$
\text{(49)}
$$

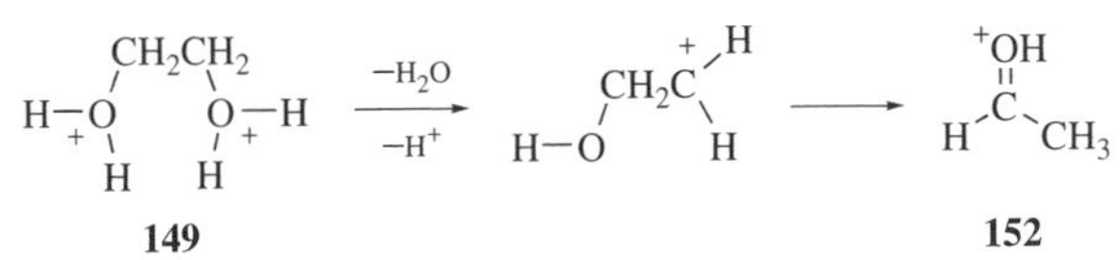

$$
\text{(50)}
$$

Ethylene glycol is shown to be diprotonated in FSO_3H-SbF_5 solution at $-80°C$, giving the distonic superelectrophile **149**. When the solution of **149** is warmed to $25°C$, NMR indicates that protonated acetaldehyde (**152**) is generated. This conversion can be understood as the loss of water with an accompanying shift of a hydride and subsequent proton loss. Although the gitonic superelectrophile **150** may only be a short-lived species, or a transition state structure (i.e., **151**), this conversion is novel with respect to the formation of a gitonic superelectrophile (**150** or **151**) from a distonic dicationic species (**149**). An alternative mechanism may also be proposed involving equilibration with the monoprotonated species in the dehydration step (eq 50). Similar conversions have been reported for the superacid-promoted rearrangements of alkoxy alcohols.[59] For example, 2-methoxy-ethanol is also diprotonated in FSO_3H-SbF_5 solution, forming the observed bis-oxonium dication (**153**) at $-80°C$ (eq 51).

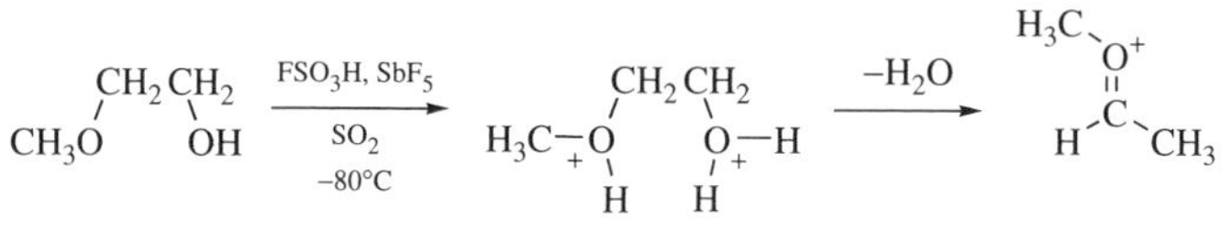

$$\text{(51)}$$

When the solution is warmed to $70°C$, dication **153** rearranges to the carboxonium ion **154** ($t_{1/2} = 30$ min). This conversion likely follows a mechanistic pathway similar to the superacid-promoted conversion of ethylene glycol to acetaldehyde (eq 49).

Further insight into the above reactions may be found in the previously discussed conversion of pivaldehyde (**155**) to methyl isopropyl ketone (**160**, Scheme 3) and in the closely related superacid-catalyzed preparation of methyl isopropyl ketone from isobutane and carbon monoxide. These conversions involve formation of the O,O-diprotonated species (**156**) which triggers a methyl shift to produce a oxononium-carbenium 1,3-dication (**157**).[60] Theoretical calculations indicate that 1,3-dication **157** is the global minimum on the potential energy surface, and that a direct hydride shift producing **158** is highly improbable energetically ($\Delta H = 23.7$ kcal/mol). Charge-charge repulsive effects lead to the destabilization of **158** relative to **157**. To explain the formation of methyl isopropyl ketone (**160**), it is suggested that the monocationic intermediate (**159**) is formed and the hydride shift occurs in a rapid, energetically favorable step ($\Delta H = -19.7$ kcal/mol). This suggests that the superacid-promoted rearrangements involving glycols and alkoxy alcohols described above may likewise involve formation of a oxonium-centered superelectrophile (i.e., **150**) followed by the deprotonation step.

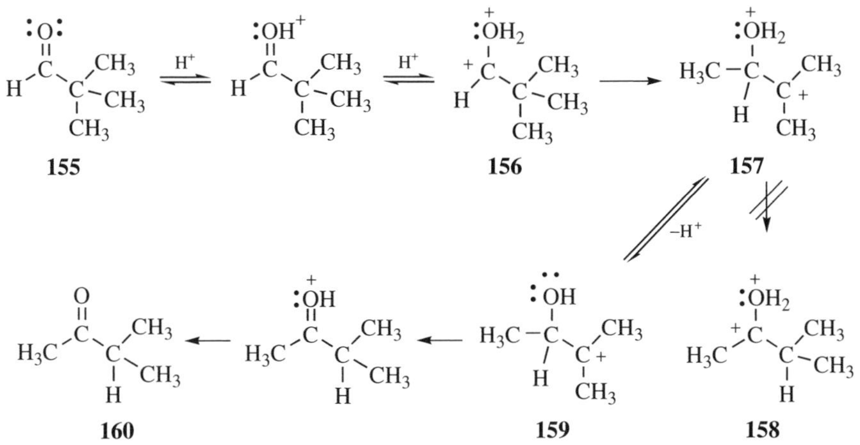

Scheme 3. Proposed mechanism for the superacid-catalyzed isomerization of pivaldehyde (**155**) to methyl isopropyl ketone (**160**).

Another novel class of oxonium dications involves the superacid-promoted ring opening reactions of oxazolines and related conversions.[61] Oxazolines are well known for their ability to react with strong nucleophiles (i.e. alcohols) when protonated at the ring nitrogen.[62] It was shown that the dicationic species (**161** or **162**) are capable of reacting with weak nucleophiles such as benzene (eq 52).

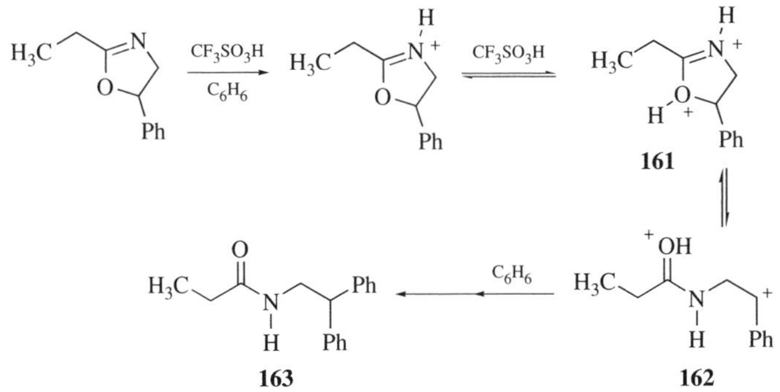

(52)

In this conversion, protonation at both the ring nitrogen and oxygen atoms leads initially to the 1,3-dication (**161**). Although benzene may attack the superelectrophile **161** directly to give the ring opened product, it seems more likely that ring opening precedes nucleophilic attack. Ring opening effectively separates the two positive charge centers. A similar conversion was reported in which isoxazolidine **164** was reacted with an excess of aluminum chloride in benzene to give product **167** in good yield (eq 53).[63]

$$(53)$$

It was proposed that product **167** arises from coordination of the isoxazolidine (**164**) to AlCl₃ to generate the dicationic, superelectrophilic intermediate (**165**), which undergoes ring-opening to give product **167** by a Friedel-Crafts type reaction.

6.2.4 Acyl-dications

There have been several types of gitonic superelectrophiles having acyl cationic groups as part of a 1,3-dicationic system. Monocationic acyl cations (**168**) can be prepared as persistent species in superacidic media and salts have even been studied by X-ray crystallography.[1b] Much of the interest in superelectrophilic species from acyl cations has focused on the protioacyl dications (**169**, *vide supra*).[67]

In principle, gitonic superelectrophiles may also be possible in structures that have two acyl cationic groups or in structures having an acyl cation adjacent to another cationic center. Like the superelectrophilic carboxonium dications, there is also some ambiguity here in distinguishing between gitonic and distonic superelectrophiles, as there are two important representations of acyl ions (**168a,b**). For the purposes of the present discussion, it is assumed that both the acyl oxygen and carbon atoms have partial positive charge.

Diacyl dications have been examined in both experimental and theoretical studies. Attempts to directly observe the oxalyl dication (**174**) were not successful as ionization of oxalyl chloride (**170**) with SbF₅ leads to the chlorocarbonyl cation (**172**, Scheme 4).[65] The initially formed chloroxalyl cation **171** immediately decarbonylates under the reaction conditions. Interestingly, complexation of the chloroxalyl cation **171** with excess Lewis acid is expected to generate superelectrophile **173**. Theoretical calculations (MP2/6-31G* level) indicate that the oxalyl dication (**174**) is at a minimum on the potential energy surface.[66] Calculated bond lengths for

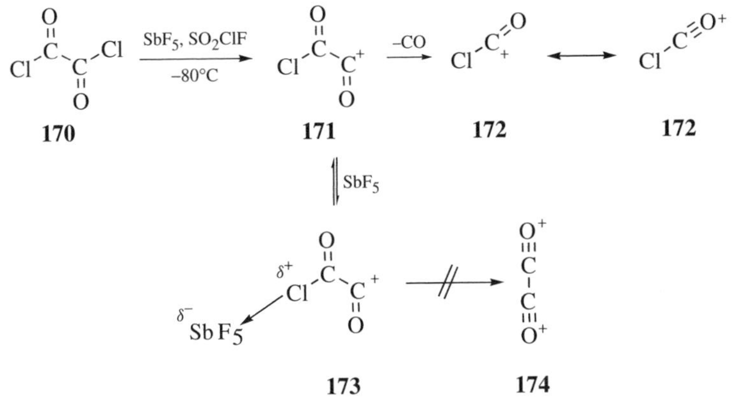

Scheme 4. Reaction of oxalyl chloride with SbF$_5$ and the attempted generation of the oxalyl dication **174**.

the species (**174**) are estimated to be 1.442 Å for the carbon-carbon bond and 1.145 Å for the carbon-oxygen bonds.

As expected, separation of the acyl groups leads to increasing stability of the diacyl dications. Malonyl fluoride (**175**) reacts with excess SbF$_5$ to give the donor-acceptor complex (**176**), which is in equilibrium with the dicationic species (**177**, eq 54).[67]

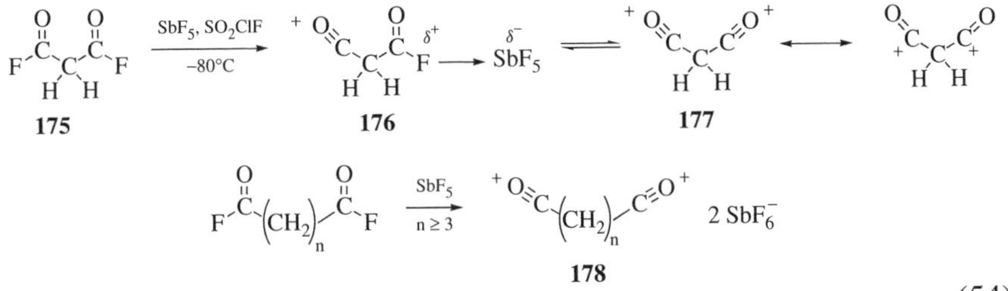

(54)

Larger systems (**178**) produce persistent diacyl ions that may be observed directly by NMR and IR spectroscopy and in some cases are considered distonic superelectrophiles (Chapter 7). Increasing distances between the acyl ion centers lead to structures with electrophilic reactivities similar to monocationic acyl ions.

Several types of onium dications have been studied in which a single acyl cationic center has been part of 1,3-dicationic superelectrophiles. For example, pyruvic acid has been studied in FSO$_3$H–SbF$_5$ solution at low temperatures.[34] Initially the diprotonated species is observed in equilibrium with some of the monocation (eq 55).

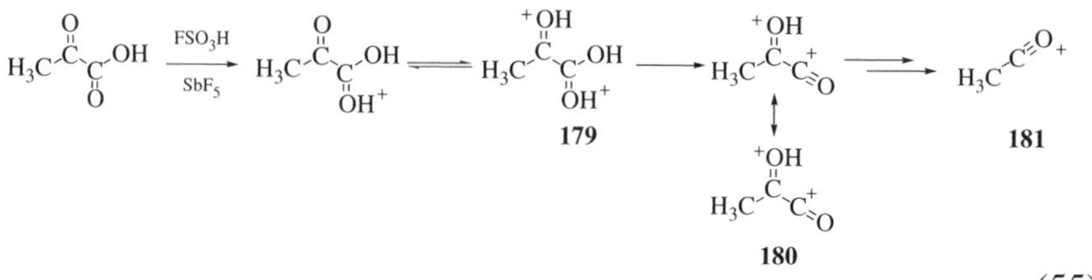

(55)

The diprotonated species (**179**) is then observed slowly cleaving to the acetyl cation (**181**) in a process thought to involve dication (**180**). A number of β-carbenium-acyl dications have also been studied by experiment and theory. Reaction of the 4-chloro-butanoyl cation (**182**) in superacidic HF–SbF$_5$ or HSO$_3$F–SbF$_5$ leads to formation of the 2-butenoyl cation (**185**, eq 56).[68]

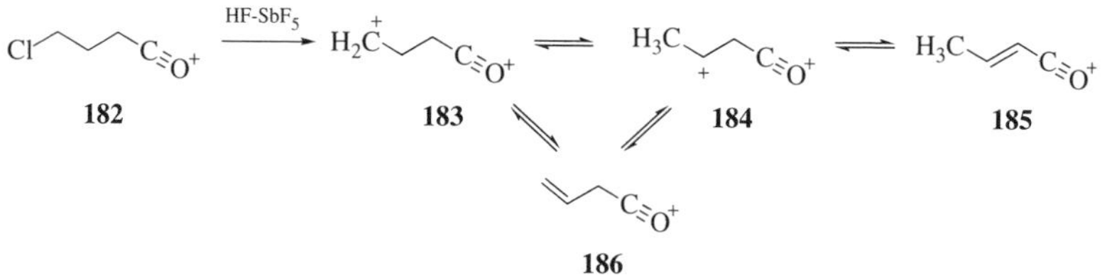

(56)

One of the proposed intermediates in this transformation is the superelectrophilic species (**184**), which undergoes deprotonation to give the 2-buten-oyl cation. Further evidence for the superelectrophile **184** is obtained from experiments in which the 2-butenoyl cation (**185**) is generated in DSO$_3$F–SbF$_5$. Significant deuterium incorporation is found at the α and γ positions, suggesting equilibria involving **184–186**. In a similar respect, formation of the 4-chloro-3-methylbutanoyl cation (**187**) in superacid leads to the two acyl dications (**188–189**, eq 57).[69]

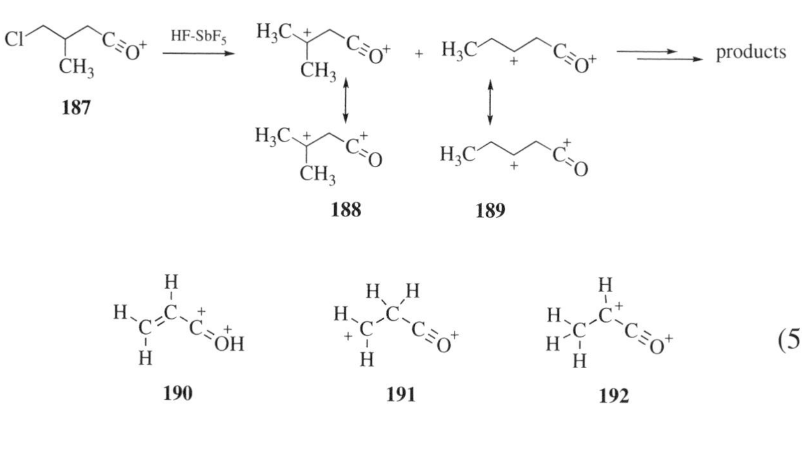

(57)

Carbenium-acyl dications have also been investigated by theoretical methods.[70] The structures of the propenoyl ($CH_2=CH-CO^+$) and the isopentenoyl (($(CH_3)_2C=CH-CO^+$) cations and their superelectrophilic, protonated dications were calculated at the MP2/6-311+G** and MP2/cc-pVTZ levels. In the case of the propenoyl cation, calculations find three structures for the protonated dications (**190–192**) at minima on the potential energy surface. The global minimum is found to be the C_α protonated structure (**191**), while the oxygen protonated (**190**) structure is 15.3 kcal/mol higher in energy. Gas-phase proton loss from **191** is calculated to be endothermic by 3.0 kcal/mol, and the transition state for deprotonation is estimated to be 77.8 kcal/mol higher in energy than structure **191**. Thus, dication **191** is predicted to be a kinetically stable species in the gas-phase. Structure **191** may be visualized as a product from the reaction of carbon monoxide with the ethylene dication ($CH_2CH_2^{2+}$). Calculations of NBO charges indicate that the CO group bears +0.90 of charge (including +1.08 on carbon) and the terminal CH_2 group bears +1.02 of charge. With the isopentenoyl (($(CH_3)_2C=CH-CO^+$) cation, three isomeric protonated dications were located at minima, the oxygen protonated species (**193**), the C_α protonated structure (**187**), and the methyl C–H protonated species (**193**). In the case of C_β protonation (analogous to **191**), all attempts to locate a stationary point for the product dication were unsuccessful and the ion spontaneously rearranges to dication **195**. The structure **195** is less stable than the global minimum (**188**) by only 10.4 kcal/mol. Again the most stable product from protonation of the alkenoyl cation corresponds to the one producing the carbenium-acyl dication (**188**). It is estimated to be 32.8 kcal/mol more stable than dication **193** and 31.4 kcal/mol more stable than dication **194**.

| 193 | 188 | 194 | 195 | 196 |

In dication **188**, the NBO charge at the carbenium ion center is +0.69 and at the acyl carbon is +1.09. The *tert*-butyl cation has been found to have NBO charge at its carbenium center of +0.67, suggesting a modest superelectrophilic activation of the carbenium ion center in **188**, compared with the *tert*-butyl cation. When charges on the methyl groups are also considered, structure **188** is similar to the protosolvated *tert*-butyl cation **196**.

6.2.5   Aza-carbo Dications

As seen in the gitonic and vicinal systems, ammonium and related cationic centers may be components of superelectrophiles and reactive dications having the 1,3-dicationic structure. Several types of superelectrophilic aza-carbo dications have been studied in which protonated nitro groups are involved. For example, it was found that nitroethylene reacts with benzene in the presence of 10 equivalents of CF_3SO_3H to give deoxybenzoin oxime in 96% yield (eq 58).[71] Since the reaction does not occur with only one equivalent of CF_3SO_3H, the formation of the *N,N*-dihydroxyiminium-methyl dication **197** was proposed. In spectroscopic studies, the stable dication (**199**) can be directly observed by ^1H and ^{13}C NMR spectroscopy from solutions of 1-nitro-2-methyl-1-propene (**198**) in CF_3SO_3H (eq 59).

It was further proposed that dications **197** and **199** are stabilized by Y-delocalization involving six π-electrons (four lone pair electrons from the oxygen and olefinic electrons). This stabilization is thought to be similar to that of the guanidinium ion **200**. *Ab initio* calculations were carried out to estimate the Y delocalization stabilization energies of **197** and **200**. By studying the energies of conformational isomers, the total π-stablization energy was estimated to be 89.50 kcal/mol for **197** and 113.17 kcal/mol for the guanidinium ion **200** at the 4-31G level of theory. Similarly, 2-nitropropene (**201**) reacts with C_6H_6 in CF_3SO_3H, and following a methanol then water quench, α-phenylacetone is formed in 85% yield (Scheme 5).[72] It is proposed that the superelectrophilic intermediate **202** is formed by diprotonation, and arylation gives the *aci*–nitro

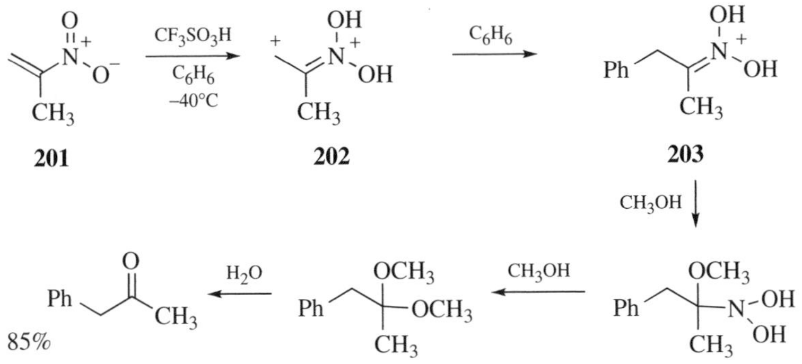

Scheme 5.

species **203**. Quenching of the reaction mixture yields the ketal, and with water, the ketone product is obtained. The arylated product is also formed in high yield if chlorobenzene is used, indicating that the electrophilic intermediate (**202**) is highly reactive. When nitro-substituted cyclic olefins are reacted with benzene and CF_3SO_3H at $-40°C$, similar products are formed. 1-Nitrocyclohexene (**204**) reacts to give 2-phenylcyclohexanone in 72% (eq 60).[72]

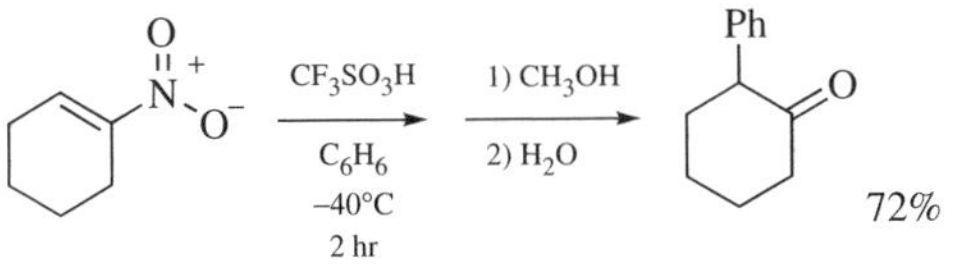

If the nitro olefins are reacted with benzene at higher temperatures, the product mixtures are dominated by the formation of $4H$-1,2-benzoxazines. For example, compound **204** reacts in CF_3SO_3H at 40°C to give the $4H$-1,2-benzoxazine (**206**) in 87% yield (Scheme 6). It was proposed that the cyclization involves the dicationic intermediate (**207**), with positive charge substantially delocalized into the aryl ring. Intramolecular reaction of the hydroxy group then provide the novel heterocyclic product **208**. Further evidence for the proposed dicationic intermediates comes from a study in which β-nitrostyrenes are found to generate stable, diprotonated species in CF_3SO_3H.[73] When (E)-β-nitrostyrene **209** is dissolved in CF_3SO_3H at low temperature, the spectroscopic data are consistent with the formation of O,O-diprotonated species (**210**, eq 61).

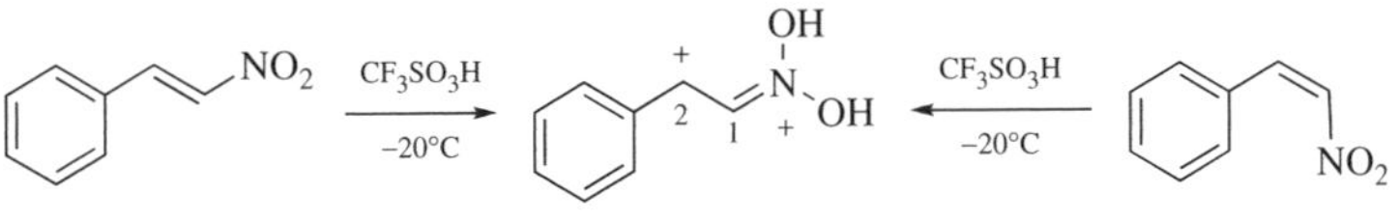

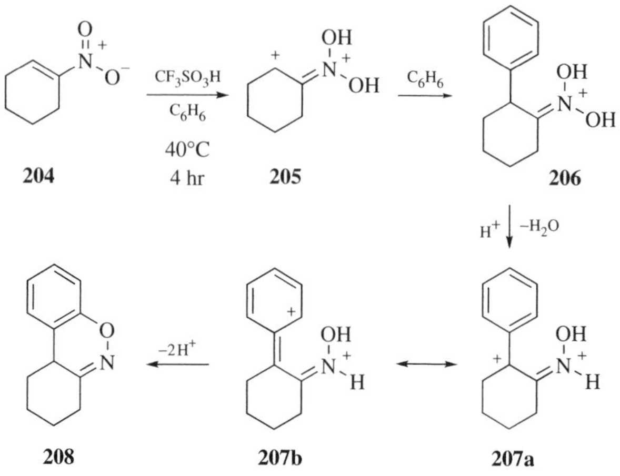

Scheme 6.

^{13}C NMR spectroscopy shows the C2 resonance at δ^{13}C 165.1 and C1 resonance at 128.3. As expected with the formation of dication **210**, (Z)-β-nitrostyrene **211** gives the identical spectra from CF$_3$SO$_3$H. Cryoscopic experiments also confirmed the formation of the dicationic species in the superacid.

An analogous series of dicationic species have been proposed in the reactions of nitro-substituted arenes in superacid. For example, 2-nitro-naphthalene (**212**) reacts in superacid to give the arylated product (**214**) in good yield (eq 62).[74]

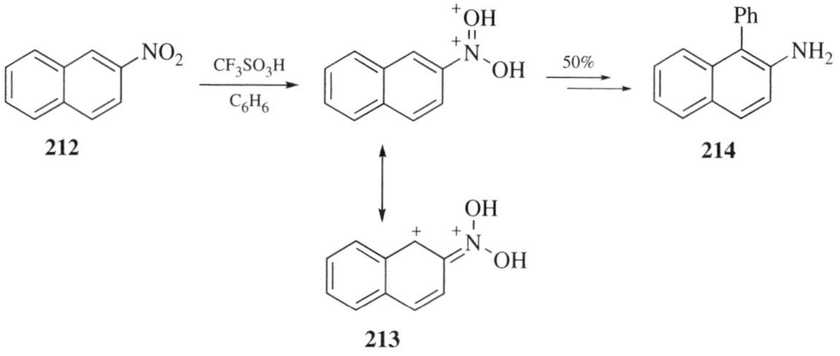

(62)

Although no detailed mechanism for the conversion was proposed, the initial step likely involves the formation of the superelectrophilic, dipro-tonated species (**213**), which reacts with benzene by electrophilic attack. There is some 1,3-dicationic character in **213**, however it is understood that the positive charge is delocalized throughout the naphthalene ring-system. Evidence for the dicationic species comes from cryoscopic measurements, as well as, ^{1}H,^{13}C, and ^{15}N NMR spectroscopy. All of these data confirm the formation of **213** in CF$_3$SO$_3$H. In a similar respect, 1-nitronapthalene (**215**) is found to produce the dicationic species (**216**) in CF$_3$SO$_3$H and the superelectrophile **216** can be directly observed by spectroscopic studies. When 1-nitronapthalene (**215**) is dissolved in CF$_3$SO$_3$D, no deuterium is incorporated onto the ring positions, indicating that diprotonation occurs at the nitro group. Intermediates like **213** and **216** have also been found to be involved in the superacid-promoted reactions of hydroxylanilines and aniline-*N*-oxides (eq 64).

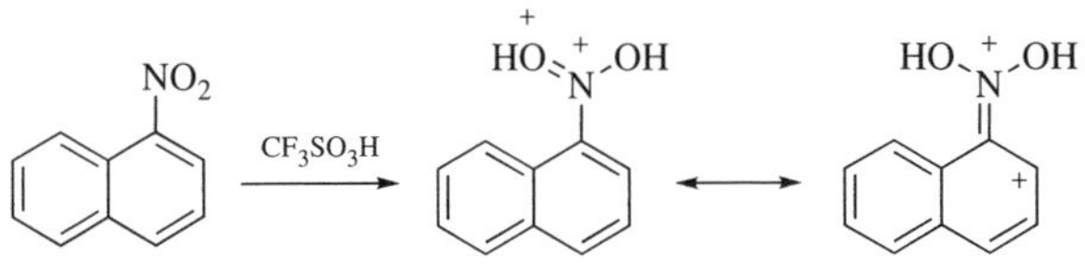

(63)

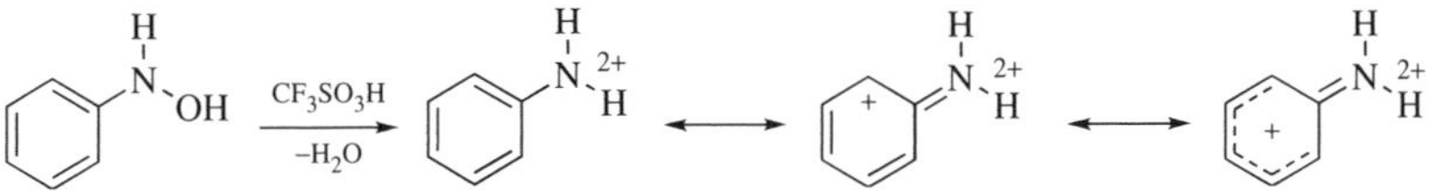

(64)

The chemistry of these species is discussed in Chapter 4.

It has been demonstrated that nitronic acids and α-carbonyl nitromethanes can form superelectrophilic intermediates in strong acids and the resulting species are capable of reacting with benzene (eqs 65–67).[23]

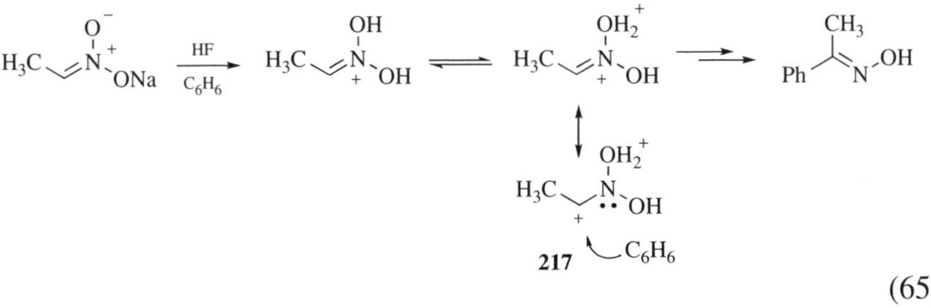

(65)

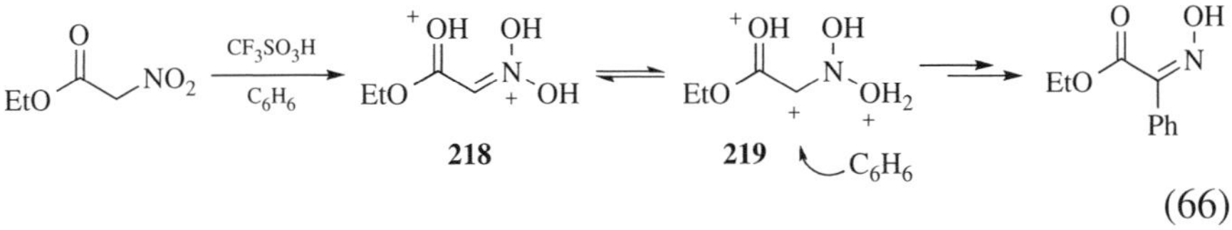

(66)

$$
\underset{Ph}{\overset{O}{\|}}\!\!-\!\!NO_2 \quad \xrightarrow[\text{C}_6\text{H}_6]{\text{CF}_3\text{SO}_3\text{H}} \quad \underset{\textbf{220}}{Ph\!\!-\!\!\overset{^+OH}{}\overset{OH}{\underset{|}{N}}\!\!-\!\!OH} \quad \rightleftharpoons \quad \underset{\textbf{221}}{Ph\!\!-\!\!\overset{^+OH}{}\overset{OH}{\underset{|}{N}}\!\!-\!\!OH_2} \quad \xrightarrow{} \quad \underset{Ph}{\overset{O}{\|}}\!\!-\!\!\overset{OH}{\underset{|}{N}}\!\!-\!\!Ph
$$

$$(67)$$

It has been proposed that the gitonic superelectrophiles (**217**, **219**, **221**) arise from double protonation at the nitro group, and in the case of α-carbonyl nitromethanes (eqs 66–67), protonation also occurs at the carbonyl group to form highly reactive tricationic superelectrophiles (**219** and **221**). Because weaker acids are capable of forming the diprotonated species but do not lead to arylated products, tricationic reactive superelectrophiles are thought to be involved in limited equilibria concentrations. The dicationic species **218** can be directly observed by NMR spectroscopy.

It has been shown that ammonium-carbenium 1,3-dications can be generated from superacid-promoted reactions of some amino-alcohols.[76] Reactions of compounds such as **222** in CF_3SO_3H lead to the formation of ammonium-carbenium dications, which have been shown to possess superelectrophilic reactivities. Dication **223** reacts in high yield with benzene (eq), while the analogous monocationic electrophile, 1,1-diphenylethyl cation (**59**), does not react with benzene. When compound **222** is reacted in FSO_3H–SbF_5 at low temperature, the dication **223** can be observed by NMR spectroscopy. The *para*-carbon atoms are deshielded in the ^{13}C NMR (by about 10 ppm from the alcohol **222**) indicating significant delocalization of the carbocation charge into the phenyl rings. Interestingly, a number of biologically important compounds possess the phenethylamine substructure, and it has been shown that some of these compounds can ionize to the dicationic electrophiles in superacid. For example, adrenaline (**224**) leads to dication (**225**) in FSO_3H–SbF_5. NMR studies suggest that the dication is best represented as the charge separated, distonic superelectrophile (**225a**), although it is expected that some 1,3-dicationic character is also present in the structure (i.e., **225b**).

$$
\underset{\textbf{222}}{\overset{CH_3}{\underset{OH}{\overset{|}{Ph}\!\!\diagup\!\!Ph}\!\!-\!\!NH_2}} \quad \xrightarrow[\text{C}_6\text{H}_6]{\text{CF}_3\text{SO}_3\text{H}} \quad \underset{\textbf{223}}{\overset{CH_3}{\underset{Ph}{\overset{|}{Ph}\!\!-\!\!}\!\!+\!\!NH_3}} \quad \xrightarrow{} \quad \underset{80\%}{\overset{CH_3}{\underset{Ph}{\overset{|}{Ph}\!\!-\!\!}NH_2}} \quad (68)
$$

$$
\boxed{\underset{\textbf{59}}{\overset{Ph}{\underset{Ph}{\diagdown}}\!\!\overset{+}{C}\!\!-\!\!CH_3}}
$$

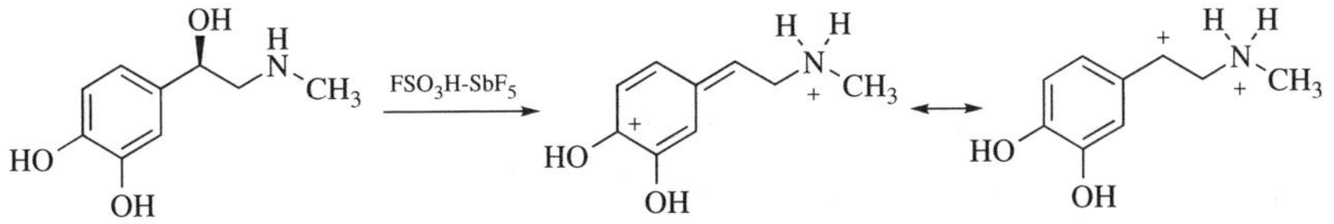

224 225a 225b

(69)

Ammonium-carbenium dications and related species are also generated readily from olefinic precursors.[77] For example, the tetrahydropyridine (**226**) leads to the 1,3-dication (**227**) and vinyl-substituted *N*-heteroaromatics can give dications (i.e., **228**) in superacid, both of which show high electrophilic reactivities (eqs 70–71).

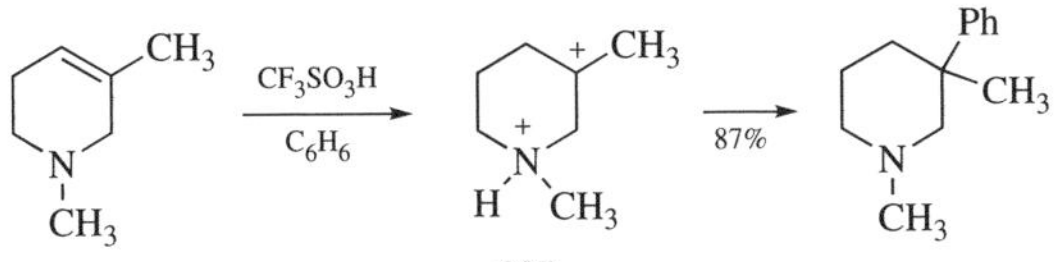

226 227

(70)

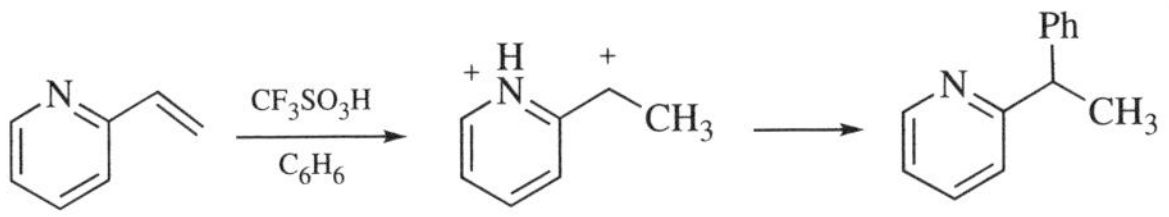

228

(71)

It was previously noted that superelectrophilic carboxonium ions may be generated from suitable precursors, including amino-ketones, *N*-heteroaromatic ketones and aldehydes, amino-acetals, and other substrates.[45] In their superacid-promoted condensation reactions, these compounds often produce ammonium-carbenium superelectrophiles as intermediates in the reactions. As an example, the amino-acetal (**229**) reacts with arenes in the presence of superacid to give the arylated product (**231**, eq 72).[43]

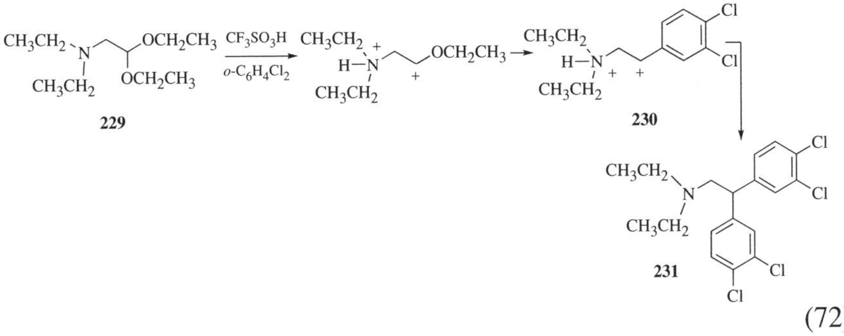

229 230 231

(72)

Initial ionization gives an ammonium-carboxonium dication, which then produces the ammonium-carbenium dication (**230**). A variety of dicationic electrophiles like **230** have been proposed.

Several studies have examined the possibility of generating carbenium-nitrilium dications. Ionization of benzophenone cyanohydrin (**232**) in $FSO_3H-SbF_5-SO_2ClF$ at $-78°$ C leads to the formation of the monocationic species (**233**), which was characterized by NMR spectroscopy (eq 73).[78] Despite the superacidic conditions, no direct evidence for the gitonic superelectrophile (**234**) was obtained. When compound **232** is reacted with benzene in CF_3SO_3H, the phenylated product (**236**) is obtained in good yield (eq 74).

$$(73)$$

$$(74)$$

Based on related observations, it was concluded that the superelectrophile **234** is not involved in the phenylation reaction. However, weak interaction (solvation) of the nitrile lone pair (i.e., **235**) with the superacid may generate increasing dipositive character and the observed superelectrophilic reactivity.

REFERENCES

(1) (a) G. A. Olah *Angew. Chem. Int. Ed. Engl.* **1993**, *32*, 767. (b) G. A. Olah; K. K. Laali; Q. Wang; G. K. S. Prakash *Onium Ions*, Wiley, New York, **1998**, Chapter 1. (c) G. A. Olah; G. K. S. Prakash; K. Lammertsma *Res. Chem. Intermed.* **1989**, *12*, 141.

(2) (a) G. A. Olah in *Stable Carbocation Chemistry*, G. K. S. Prakash and P. v. R. Schleyer, Eds., Wiley, New York, **1997**, Chapter 1. (b) G. A. Olah *J. Org. Chem.* **2001**, *66*, 5843.

(3) G. A. Olah; J. L. Grant; R. J. Spear; J. M. Bollinger; A. Serianz; G. Sipos *J. Am. Chem. Soc.* **1976**, *98*, 2501.

(4) G. A. Olah; G. K. S. Prakash; J. G. Shi; V. V. Krishnamurthy; G. D. Mateescu; G. Liang; G. Sipos; V. Buss; J. M. Gund; P. v. R. Schleyer *J. Am. Chem. Soc.* **1985**, *107*, 2764.

(5) G. Rasul; G. A. Olah; G. K. S. Prakash *Proc. Nat. Acad. Sci. USA* **2004**, *101*, 10868.

(6) G. A. Olah; V. Prakash Reddy; G. Rasul; G. K. S. Prakash *J. Am. Chem. Soc*. **1999**, *121*, 9994.

(7) N. J. Head; G. A. Olah; G. K. S. Prakash *J. Am. Chem. Soc*. **1995**, *117*, 11205.

(8) G. A. Olah; Q. Wang; A. Orlinkov; P. Ramaiah *J. Org. Chem*. **1993**, *58*, 5017.

(9) N. S. Dokunikhin; S. L. Solodar; L. M. Vinogradov *J. Org. Chem. U.S.S.R. (Engl. Tran.)* **1979, ** *15*, 2137.

(10) G. A. Olah; G. K. S. Prakash; G. Liang; P. W. Westerman; Klaus Kunde; J. Chandrasekhar; P. v. R. Schleyer *J. Am. Chem. Soc*. **1980**, *102*, 4485.

(11) M. Bremer; P. v. R. Schleyer; K. Schoetz; M. Kausch; M. Schindler *Angew. Chem. Int. Ed. Engl*. **1987**, *26*, 761.

(12) H. Hogeveen; P. W. Kwant *Acc. Chem Res*. **1975**, *8*, 413.

(13) G. A. Olah; J. S. Staral; G. Liang; L. A. Paquette; W. P. Melega; M. J. Carmody *J. Am. Chem. Soc*. **1977**, *99*, 3349.

(14) G. Maas; P. Stang *J. Org. Chem*. **1983**, *48*, 3038.

(15) J. A. Barltrop; J. C. Barrett; R. W. Carder; A. C. Day; J. R. Harding; W. E. Long; C. J. Samuel *J. Am. Chem. Soc*. **1979**, *101*, 7510.

(16) (a) R. M. Pagni *Tetrahedron* **1984**, *40*, 4161. (b) G. K. S. Prakash; T. N. Rawdah; G. A. Olah *Angew. Chem. Int. Ed. Engl*. **1983**, *22*, 390.

(17) (a) G. A. Olah; N. Hartz; G. Rasul; G. K. S. Prakash *J. Am. Chem. Soc*. **1993**, *115*, 6985. (b) G. A. Olah; N. Hartz; G. Rasul; G. K. S. Prakash; M. Burkhart; K. Lammertsma *J. Am. Chem. Soc*. **1994**, *116*, 3187.

(18) G. A. Olah; G. K. S. Prakash; G. Rasul *J. Org. Chem*. **2001**, *61*, 2907.

(19) (a) D. Farcasiu; A. Ghenciu *J. Org. Chem*. **1991**, *56*, 6050. (b) K. Y. Koltunov; I. B. Repinskaya *Zhurnal Organich. Khimii* **1994**, *30*, 90. (c) D. M. Brouwer; J. A. Van Doorn; A. A. Kiffen *Rec. Trav Chim Pays-Bas* **1975**, *94* (8), 198. (d) D. M. Brouwer; J. A. Van Doorn *Rec. Trav Chim Pays-Bas* **1970**, *89* (6), 553. (e) M. Juhasz; S. Hoffmann; E. Stoyanov; K.-C. Kim; C. A. Reed *Angew. Chem. Int. Ed*. **2004**, *43*, 5352.

(20) D. M. Brouwer *Rec. Trav Chim Pays-Bas* **1968**, *87*, 1295.

(21) S. Walspurger; A. V. Vasilyev; J. Sommer; P. Pale *Tetrahedron* **2005**, *61*, 3559.

(22) K. Y. Koltunov; S. Walspurger; J. Sommer *Chem. Commun*. **2004**, 1754.

(23) T. Ohwada; N. Yamagata; K. Shudo *J. Am. Chem. Soc*. **1991**, *113*, 1364.

(24) K. Y. Koltunov; L. A. Ostashevskaya; I. B. Repinskaya *Russ. J. Org. Chem. (Engl. Transl.)* **1998**, *34*, 1796.

(25) I. B. Repinskaya; K. Y. Koltunov; M. M. Shakirov; V. A. Koptyug *Russ. J. Org. Chem. (Engl. Transl.)* **1992**, *28*, 1013.

(26) (a) R. Rendy; Y. Zhang; A. McElrea; A. Gomez; D. A. Klumpp *J. Org. Chem*. **2004**, *69*, 2340. (b) G. K. S. Prakash; P. Yan; B. Torok; G. A. Olah *Catal. Lett*. **2003**, *87*, 109.

(27) D. A. Klumpp; R. Rendy; Y. Zhang; A. Gomez; A. McElrea *Org. Lett*. **2004**, *6*, 1789.

(28) K. Y. Koltunov; G. K. S. Prakash; G. A. Olah *Heterocycles* **2004**, *62*, 757, and references cited therein.

(29) K. Y. Koltunov; S. Walspurger; J. Sommer *J. Mol. Catal., A: Chem*. **2006**, *245*, 231.

(30) (a) I. Iwai; T. Hiraoka *Chem. Pharm. Bull*. **1963**, *11*, 638. (b) Reference 22.

(31) (a) Y. Zhang; S. A. Aguirre; D. A. Klumpp *Tetrahedron Lett*. **2002**, *43*, 6837. (b) D. A. Klumpp; R. Rendy; Y. Zhang; A. Gomez; A. McElrea; H. Dang *J. Org. Chem*. **2004**, *69*, 8108. (c) See also: A. V. Vasilyev; S. Walspurger; P. Pale; J. Sommer *Tetrahedron Lett*. **2004**, *45*, 3379.

(32) (a) G. A. Olah; M. Calin *J. Am. Chem. Soc*. **1968**, *90*, 4672.

(33) D. Bruck; A. Dagan; M. Rabinovitz *Tetrahedron Lett*. **1978**, *19*, 1791.

(34) G. A. Olah; A. T. Ku; J. Sommer *J. Org. Chem*. **1970**, *35*, 2159.

(35) D. M. Brouwer *Rec. Trav Chim Pays-Bas* **1968**, *87*, 225.

(36) J. W. Larsen; P. A. Bouis *J. Am. Chem. Soc*. **1975**, *97*, 6094.

(37) G. A. Olah; A. M. White *J. Am. Chem. Soc*. **1967**, *89*, 4752.

(38) K. Y. Koltunov; G. K. S. Prakash; G. Rasul; G. A. Olah *Eur. J. Org. Chem*. **2006**, 4861.

(39) G. K. S. Prakash; T. Mathew; D. Hoole; P. M. Esteves; Q. Wang; G. Rasul; G. A. Olah *J. Am. Chem. Soc*. **2004,** *126*, 15770.

(40) D. A. Klumpp; K. Y. Yeung; G. K. S. Prakash; G. A. Olah *Syn.Lett*. **1998**, 918.

(41) G. A. Olah; D. N. Brydon; R. D. Porter *J. Org. Chem*. **1970**, *35*, 317.

(42) M. R. Seong; H. N. Song; J. N. Kim *Tetrahedron Lett*. **1998**, *39*, 7101.

(43) D. A. Klumpp; G. V. Sanchez Jr.; Y. Zhang; S. L. Aguirre; S. de Leon *J. Org. Chem*. **2002**, *67*, 5028.

(44) D. A. Klumpp; M. Garza; A. Jones; S. Mendoza *J. Org. Chem*. **1999**, *64*, 6702.

(45) (a) D. A. Klumpp; A. Jones; S. Lau; S. DeLeon; M. Garza *Synthesis*, **2000**, 1117. (b) D. A. Klumpp; M. Garza; G. V. Sanchez; S. Lau; S. DeLeon *J. Org. Chem*. **2000**, *65*, 8997. (c) D. A. Klumpp; Y. Zhang; P. J. Kindelin; S. Lau *Tetrahedron* **2006**, *62*, 5915. (d) Y. Zhang; D. A. Klumpp *Tetrahedron Lett*. **2002**, *43*, 6841. (e) K. Y. Koltunov; G. K. S. Prakash; G. Rasul; G. A. Olah *J. Org. Chem*. **2002**, *67*, 4330. (f) K. Y. Koltunov; G. K. S. Prakash; G. Rasul; G. A. Olah *J. Org. Chem*. **2002**, *67*, 8943.

(46) B. E. Maryanoff; H.-C. Zhang; J. H. Cohen; I. J. Turchi; C. A. Maryanoff *Chem. Rev*. **2004**, *104*, 1431, and references cited therein.

(47) R. Schmidt; E. Schlipf *Chem. Ber.* **1970**, *103*, 3783.

(48) G. A. Olah; Q. Wang; G. Sandford; A. B. Oxyzoglou; G. K. S. Prakash *Synthesis* **1993**, 1077.

(49) D. A. Klumpp; Y. Zhang; P. J. Kindelin *Abstract of Papers*, 230th National Meeting of the American Chemical Society, Washington, DC, August 2005; American Chemical Society: Washington, DC, 2005.

(50) G. A. Olah; N. Hartz; G. Rasul; A. Burrichter; G. K. S. Prakash *J. Am. Chem. Soc.* **1995**, *117*, 6421.

(51) H. Hogeveen *Rec. Trav Chim Pays-Bas* **1970**, *89*, 1303.

(52) J. P. Hwang; G. K. S. Prakash; G. A. Olah *Tetrahedron* **2000**, *56*, 7199.

(53) G. K. S. Prakash; G. Rasul; A. Burrichter; K. K. Laali; G. A. Olah *J. Org. Chem.* **1996**, *61*, 9253.

(54) (a) G. A. Olah; A. Burrichter; G. Rasul; R. Gnann; K. O. Christe; G. K. S. Prakash *J. Am. Chem. Soc.* **1997**, *119*, 8035. (b) N. Hartz; G. Rasul; G. A. Olah *J. Am. Chem. Soc.* **1993**, *115*, 1277.

(55) G. A. Olah; A. M. White; D. H. O'Brien *Chem. Rev.* **1970**, *70*, 561, and references cited therein.

(56) R. T. Akhmatdinov; I. A. Kudasheva; E. A. Kantor; D. L. Rakhmankulov *Zhurnal Organich. Khimii* **1993**, *19*, 1965.

(57) G. A. Olah; A. T. Ku *J. Org. Chem.* **1970**, *35*, 3913.

(58) G. A. Olah; J. Sommer *J. Am. Chem. Soc.* **1968**, *90*, 927.

(59) G. A. Olah; J. Sommer *J. Am. Chem. Soc.* **1968**, *90*, 4323.

(60) G. A. Olah; T. Mathew; E. R. Marinez; P. M. Esteves; M. Etzkorn; G. Rasul; G. K. S. Prakash *J. Am. Chem. Soc.* **2001**, *123*, 11556.

(61) D. A. Klumpp; R. Rendy; A. McElrea *Tetrahedron Lett.* **2004**, *45*, 7959.

(62) T. G. Gant; A. I. Meyers *Tetrahedron* **1994**, *50*, 2297.

(63) (a) Y. Seo; K. R. Mun: K. Kim *Synthesis* **1991**, 951. (a) Y. Seo; K. Kim *Bull. Korean Chem. Soc.* **1995**, *16*, 356.

(64) G. A. Olah; A. Germain; H. C. Lin; D. Forsyth *J. Am. Chem. Soc.* **1975**, *97*, 2928.

(65) G. K. S. Prakash; J. W. Bausch; G. A. Olah *J. Am. Chem. Soc.* **1991**, *113*, 3203.

(66) (a) R. C. Haddon; D. Poppinger; L. Radom *J. Am. Chem. Soc.* **1975**, *97*, 1645. (b) Reference 65.

(67) G. A. Olah; M. B. Comisarow *J. Am. Chem. Soc.* **1966**, *88*, 3313.

(68) D. Farcasiu; G. Miller *J. Org. Chem.* **1989**, *54*, 5423.

(69) D. Farcasiu; G. Miller; S. Sharma *J. Phys. Org. Chem.* **1990**, *3*, 639.

(70) G. Rasul; G. K. S. Prakash; G. A. Olah *J. Phys. Chem. A* **2006**, *110*, 1041.

(71) T. Ohwada; A. Itai; T. Ohta; K. Shudo *J. Am. Chem. Soc.* **1987**, *109*, 7036.

(72) T. Ohwada; K. Okabe; T. Ohta; K. Shudo *Tetrahedron* **1990**, *46*, 7539.

(73) T. Ohwada; T. Ohta; K. Shudo *J. Am. Chem. Soc*. **1986**, *108*, 3029.

(74) T. Ohta; K. Shudo; T. Okamoto *Tetrahedron Lett*. **1984**, *25*, 325.

(75) T. Ohwada; N. Yamagata; K. Shudo *J. Am. Chem. Soc*. **1991**, *113*, 1364.

(76) D. A. Klumpp; S. L. Aguirre; G. V. Sanchez, Jr.; S. J. de Leon *Org. Lett*. **2001**; *3*, 2781.

(77) (a) D. A. Klumpp; P. S. Beauchamp; G. S. Sanchez, Jr.; S. Aguirre; S. de Leon *Tetrahedron Lett*. **2001**, *42*, 5821. (b) Y. Zhang; A. McElrea; G. V. Sanchez, Jr.; D. A. Klumpp; D. Do; A. Gomez; S. L. Aguirre; R. Rendy *J. Org. Chem*. **2003**, *68*, 5119.

(78) T. Ohwada; K. Shudo *J. Am. Chem. Soc*. **1988**, *110*, 1862.

7

DISTONIC SUPERELECTROPHILES

7.1 STRUCTURAL CONSIDERATIONS

In distonic superelectrophiles, the two (or more) electrophilic charge centers are separated by at least two carbon or other heavy atoms. As discussed earlier, increasing separation of charge leads to decreasing electrophilic activation. Depending on the electrophilic system and the distance between charge centers, distant onium dications may exhibit chemistry no different than isolated onium monocations. For example, it was shown that 4-acetylpyridine will condense with benzene through the distonic superelectrophile (**1**, eq 1), but the analogous more charge-separated species (**2**) is unreactive towards weakly nucleophilic benzene.[1]

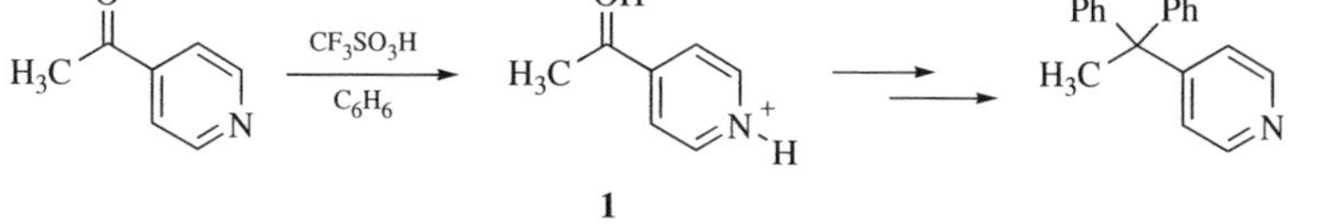

$$
\text{(1)}
$$

1

Reactive to C_6H_6

Superelectrophiles and Their Chemistry, by George A. Olah and Douglas A. Klumpp
Copyright © 2008 John Wiley & Sons, Inc.

2

Unreactive to C_6H_6

While structure **2** is an onium dication, it can not be considered a super-electrophile. Only if the electrophilic site(s) exhibit significantly increased reactivities due to interaction of the onium charge centers can the species be classified as distonic superelectrophiles.

Several examples of superelectrophiles have already been described in which delocalization of charge may lead to structures that could be considered gitonic or distonic superelectrophiles, depending on which resonance form is being considered predominant. While the bis-oxonium structure (**3a**) is formally a distonic superelectrophile, the bis-carbenium structure (**3c**) is considered a gitonic superelectrophile (eq 2). These types of systems have been previously discussed and therefore will not be included in this chapter. In a similar respect, there are examples of equilibration between gitonic (**4**) and distonic (**5**) superelectrophilic systems (eq 3).[2]

$$\text{(2)}$$

3a   **3b**   **3c**

$$\text{(3)}$$

4   **5**

Although there have been almost no systematic studies of these equilibria, it is expected that for many of these systems the distonic superelectrophile should be preferred in the equilibria, due to favorable separation of positive charge. Besides the number of atoms separating charge centers, another important consideration is the actual distance between charges. If a particular conformational or structural effect leads to charge centers being forced into closer proximity, then this may lead to superelectrophilic activation. Several examples of this effect are described in subsequent sections.

7.2 DISTONIC SYSTEMS

7.2.1 Carbodicationic Systems

There has been considerable experimental and theoretical work related to carbodications, many of which may be considered distonic superelectrophiles.[3] In the absence of aryl stabilization of the positively charged centers, long-lived acyclic carbodications can be formed only if the charge bearing carbons are separated by at least two carbon atoms while the carbenium centers are tertiary.[4] For example, the 2,2'-ethylenediisopropyl dication **6** has been prepared by the ionization of 2,5-dichloro-2,5-dimethylhexane in SbF_5/SO_2ClF (eq 4).

The ^{13}C NMR spectrum shows a characteristic absorption of $\delta^{13}C$ 331.3 for the carbenium centers.

The instability (and superelectrophilic nature) of aliphatic 1,4-dicationic systems can be seen in the failure to prepare some analogous cyclopropyl and 2,5-dimethyl-substituted 2,5-norbornadiyl dications (eqs 5 and 7).[4]

Despite the stabilizing effects of the cyclopropyl ring, diol **7** ionizes in superacid to give only the heptadienyl cation (**8**) from ring opening and proton loss.[4] It is only with further stabilization by cyclopropyl groups that the 1,4-dication was found to be persistent (eq 6).[5] Likewise, ionization of diol (**11**) does not provide the 2,5-dimethyl-2,5-norbornyldiyl dication (**12**, eq 7).[6] Only with more powerful electron donating groups, such as aryl or hydroxy groups, is the 2,5-norbornadiyl dication persistent (*vida infra*). The 1,4-dialkyl-1,4-cyclohexyl dications are also found to be unstable.[6] For example, diol (**13**) does not give the expected 1,4-dication **15**, but instead forms the allylic cation **14** (eq 8), and 1,4-di-*n*-butyl-1,4-cyclohexanediol (**16**) leads to the rearranged dication (**18**, eq 9).[6] It is not clear if the distonic superelectrophile (**17**) is formed initially, but nevertheless, rearrangement leads to the more stable charge separated species (**18**). Stabilization of the carbocationic centers with cyclopropyl groups, however, leads to a persistent 1,4-dication (eq 10).[7]

The above examples of the 2,5-norbornadiyl dication (**12**) and the 1,4-dialkyl-1,4-cyclohexyl dications (**15** and **17**) illustrate the importance of distance between charge centers implicating stability. Another interesting example of this effect is in the isomerization of the 1,5-manxyl dication (**19**, eq 11).

It was observed that dication **19** was stable at $-105°C$, but upon warming to $-60°C$ the 3,7-dimethylbicyclo[3.3.1]nona-3,7-diyl dication (**20**) is cleanly formed.[8,9] The ^{13}C NMR of dication **20** shows a resonance at δ^{13} C 323 for the cationic carbons. Calculations at the B3LYP/6-31G* level indicate that the isomerization increases the distance between charge centers from 2.80 Å in **19** to 3.58 Å in **20** and this leads to a concomitant decrease in potential energy of more than 26 kcal/mol. The increasing distance between the charge centers is thought to be the driving force for the isomerization and it is consistent with the superelectrophilic character of dication **19**. Interestingly, bicyclic dications (**19–20**) are shown to have charges in closer proximity than the 2,6-dimethyl-2,6-heptadiyl dication (**21**, charge-charge distance: 5.21 Å). Though each of these systems (**19-20**) are considered 1,5-dications, the structural constraints of the bicyclic framework force the positive charge centers closer to each other. This enhances their distonic superelectrophilic character. A sulfur-stabilized dication (**22**) has also been reported from the superacid-promoted reaction of tetramethylhexathiaadamantane (eq 12).[10]

$$(12)$$

Like the structurally similar dication **20**, dication **22** has been observed by ^{13}C NMR in superacidic solutions ($FSO_3H–SbF_5$ or CF_3SO_3H) and the carbocationic centers are found at $\delta^{13}C$ 230. This large shielding compared to that of dication **20** ($\delta^{13}C$ 323) is indicative of the sulfoxonium-type interaction. AM1 calculations estimate the distance between the charge centers to be 3.88 Å, which is slightly longer than the distance (3.58 Å) found in dication **20**.

In addition to the 1,5-manxyl dication (**19**), the 1,4-bicyclo[2.2.2] octadiyl dication (**25**) has been claimed to have been observed (eq 13).[11c]

$$(13)$$

Subsequent NMR experiments indicate that the monocationic species (**24**) is formed initially as a donor-acceptor complex and then a second chloride abstraction could provide the dication (**25**). At temperatures above $-60°C$

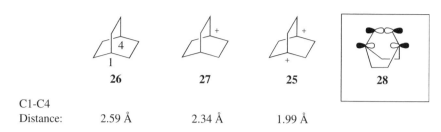

C1-C4 Distance:	2.59 Å	2.34 Å	1.99 Å

Figure 1.

dication **25** is found to be unstable. The carbenium ion centers are found at δ^{13} C 307. MINDO calculations[11c] estimate the C1- C4 distance to be 1.99 Å. Interestingly, calculations show a shortening of this distance upon progressing from the parent uncharged bicyclo[2.2.2]octane (**26**), to the monocation **27**, and to the dication **25** (Figure 1). Although it might be expected that strong coulombic repulsion would lead to an *increasing* distance between C1 and C4, this is not observed. The shortening of the C1-C4 distance in the dication **25** is attributed to hyperconjugative transfer of electron density to the carbocationic centers, which leads to symmetry allowed 1,4 bonding (i.e. **28**). Calculations indicate that more than half of the positive charge is delocalized over the 12 hydrogen atoms as a result of hyperconjugative effects.

The 2,6-adamantadiyl dication systems have likewise been studied and the results are consistent with the superelectrophilic character of some these 1,5-carbodications. When diol **29** was reacted in superacid the 2,6-adamatanediyl dication **31** was not formed (eq 14).[12] Despite the super acidic conditions, only the dioxonium ion (i.e. diprotonated diol, **30**) could be observed. Structure **31** and other adamantadiyl dications ($C_{10}H_{14}^{2+}$) have been also been studied by theoretical calculations.[13] As described in Chapter 6, dication **31** is found to be 3.3 kcal/mol less stable than the 1,3-dication having the two carbocationic centers located at the bridgehead carbons. Even the ionization of the *tertiary*-diol **32** gives only a single carbocationic center, while the other hydroxy group produces an oxonium center (eq 15).[12]

(14)

(15)

Water elimination in the superacidic solution is a highly exothermic step, but nevertheless the 2,6-adamantadiyl dication **34** is not formed. This observation suggests that structures like **34** can be distonic superelectrophiles. As in the case of other 1,4- and 1,5-carbodications, the 2,6-adamantadiyl dications are stabilized and persistent when the carbenium centers bear an aryl substitutent (*vide infra*).

Other adamantane-based dications have also been prepared. Although systems such as diadamanta-4,9-diyl dication (**35**) and 1,1′-bisadamanta-3,3′-diyl dication (**36**) have been prepared (both 1,6-dications), experimental data suggests that these systems are more related to the monopositive adamantyl cations rather than distonic superelectrophilic systems.[14] However, the adamanta-1,3-dimethyldiyl dication (**38**) has been prepared in superacid media (eq 16).[15]

(16)

The ^{13}C NMR data for **38** clearly indicate that, due to the close proximity of the cationic centers, the positive charges are highly delocalized into the substituents and the adamantyl cage.

Many of the unstable distonic superelectrophiles discussed here have been prepared as stabilized species by incorporating aryl substituents into them (Table 1).[4,6,12,15] In the cases of **39** and **44**, however, the analogous methyl-substituted dications (**6** and **38**) are also persistent. In comparing the ^{13}C NMR data, the aryl-stabilized dications (**39** and **44**) show large shieldings of the carbocationic centers when compared to **6** and **38**. For example, the 2,2′-ethylenediisopropyl dication **6** has a resonance of δ $^{13}C^+$ 331.3 while dication **39** has the resonance of $\delta^{13}C^+$ 222.4. This is indicative of the significant charge delocalization into the phenyl rings in

Table 1. Aryl-stabilized dicationic systems (39–44) and their
$\delta^{13}C^+$ data

Structure	^{13}C NMR, C^+	Structure	^{13}C NMR, C^+
39	222.4	**42**	245.1
40	220.9	**43**	252.3
41	247.9	**44**	245.6

dication **39**. Charge delocalization and stabilization results in diminishing
superelectrophilic character.

There are a number of other aryl-substituted carbodicationic systems
that can be properly described as distonic superelectrophiles. For example,
dication **45** has been generated from 2,2'-*p*-phenylenedi-2-propanol in
SbF$_5$ at $-78°$C.[4] When compared to the dimethyl(phenyl)carbenium ion
(cumyl cation) **46**, NMR data indicate that the positive charges are dis-
persed to a considerable extent into the neighboring methyl groups in the
dication **45**.

$\delta^{13}C$ NMR Signals:

$-CH_3$	47.1
C^+	240.8
C_{ipso}	149.1
C_{ring}	137.9

45   **46**

$\delta^{13}C$ NMR Signals:

$-CH_3$	33.9
C^+	255.7
C_{ipso}	140.7
C_{ortho}	142.4

Due to coulombic repulsion, there is also less charge delocalization into
the phenyl group. The diminished neighboring group participation (stabilization) is one of the characteristics of superelectrophilic activation.[16] In
the case of dication **45**, the aryl group is less capable of donating electron
density to either carbenium center.

Other distonic superelectrophiles arise from aryl-substituted carbodicationic systems in which the positive charge centers are forced into close

proximity due to structural effects. A series of aryl-substituted phenyl enediyl dications (**47–49**)

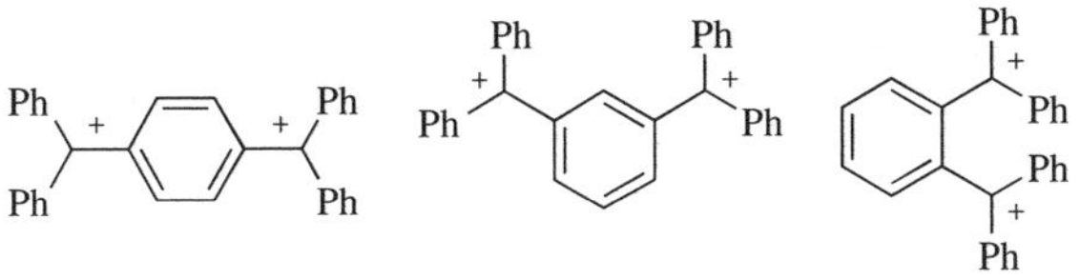

	47	**48**	**49**
$pK_{R+} \left(\begin{array}{c} \text{second} \\ \text{ionization} \end{array} \right)$	-10.5	-9.9	-16.6

were prepared in sulfuric acid solutions.[17] Within this series, it was found that the pK_{R+} value (for the ionization producing the dication) is significantly higher for the *ortho*-substituted system (**49**). This higher pK_{R+} value is attributed to the close proximity of the two charge centers and it suggests some distonic superelctrophilic character in **49**. The 1,8-bis (diarylmethyl)naphthalene dications have been extensively studied and experimental observations suggest partial distonic superelectrophilic character. The 1,8-bis(diphenylmethyl)naphthalene dication (**50**) has been prepared by several methods, including Ichikawa and coworker's deoxygenation method using strong silylating agents (eq 17).[18]

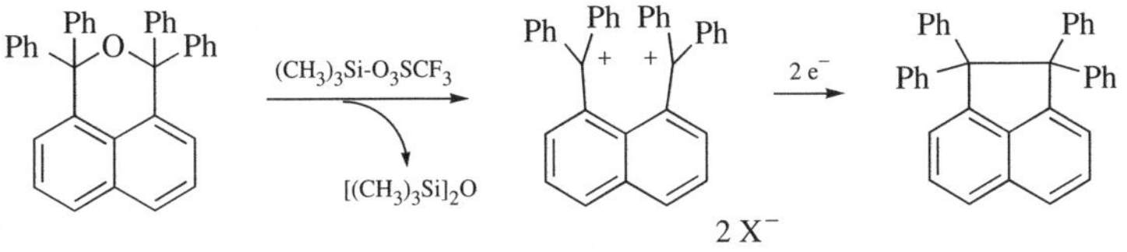

(17)

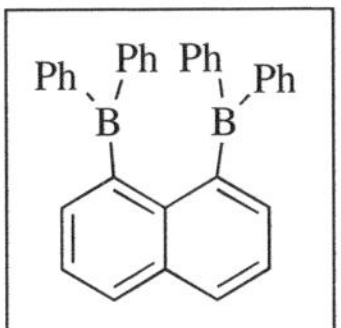

52

Although formally considered a 1,5-dication, **50** possesses a structure in which the carbenium centers are constrained at a distance of separation of 3.11 Å. NMR studies show the carbenium ion centers at $\delta^{13}C$ 207.7, consistent with the carbocationic structure **50**. In cyclic voltamographic analysis, the compound **51** shows an especially high oxidation potential (two-electron oxidation peak at 1.10 V), when compared to analogous dications and triarylmethyl monocations.[19] It has also been shown that the oxidation potential varies in a predictable manner based on the nature of the aryl rings. The high oxidation potentials are considered to be the result of close proximity of the charge centers. Despite the close proximity of the charges, dication **50** has been reported as stable in acetonitrile solutions. This suggests a decreased degree of electrophilic reactivity. An isoelectronic neutral boron-containing compound (**52**) has likewise been prepared and studied.

An interesting application of the dicationic species has been reported in which **50** has been used as a two-electron oxidant to couple *N,N*-dialkylanilines (eq 18).[19]

$$(18)$$

The oxidative coupling is thought to involve single electron transfer steps. The use of **50** in this coupling reaction is shown to superior to the use of other oxidizing agents, such as Ceric ammonium nitrate (CAN) and $PhI(O_2CCF_3)_2$. This type of conversion is not possible with triphenyl-methyl (trityl) cation salts, indicating that dications like **50** could be considered distonic superelectrophiles.

More highly stabilized 1,8-bis(diarylmethyl)naphthalene dications have been prepared, including the *p*-methoxyphenyl derivative **53**.[20] This dica-tion is generated from ionization of the diol in HBF_4 and $(CF_3CO)_2O$.[20a] Dication **53** has been characterized by experimental studies (single crystal X-ray analysis and NMR) and theoretical calculations. The carbenium ion centers are found to be separated by just 3.076 Å (X-ray and *ab initio* results) and show ^{13}C NMR resonances at $\delta^{13}C$ 191.8. Two electron reduction is also shown to give the acenaphthene derivative **54**.

53 **54**

$$(19)$$

Other 1,8-naphthalene or acenaphthene dication systems have been described and some have been shown to have useful electrochromic properties.[20b,c]

Closely related systems have been studied which are composed of 1,1'-biphenyl-2,2'-diyl dications and 1,1'-binaphthalene-2,2'-diyl dications. In the case of the 1,1'-biphenyl-2,2'-diyl dications, these have been prepared by the ionizationof related diols (i.e., **55**) in HBF_4 and by the oxidation of compound **57** (electrochemically or with reagents such as $(p\text{-}BrC_6H_4)_3N^{+\bullet}SbCl_6^-$).[21] Conversely, dication **56** is converted back to **57** by Mg or by electrochemical reduction.

$$\text{(20)}$$

Dication **56** has been isolated and studied by crystallography, revealing a separation of the carbenium ion centers by 3.66 Å.

Among the binaphthyl-systems, Suzuki and co-workers have reported two methods to prepare these dications (eq 21–22).[22]

$$\text{(21)}$$

The diol or diether (**58**) can be ionized in HBF_4. Optical resolution has been achieved to produce the chiral distonic superelectrophilic dication (**59**). X-ray analysis shows an interplanar separation of the charged rings by about 3.53 Å. Due to its helical structure and the exciton coupling of the dye components, there is a very high amplitude of the circular dichroism (CD) signals. Along with its redox chemistry, the chiroptic response makes these compounds promising candidates for chiral redox memory systems, or electrochiroptic materials. Another system (**62**) has been generated from the oxidation of the binaphthylic diolefin (**61**) by iodine (eq 22).[23]

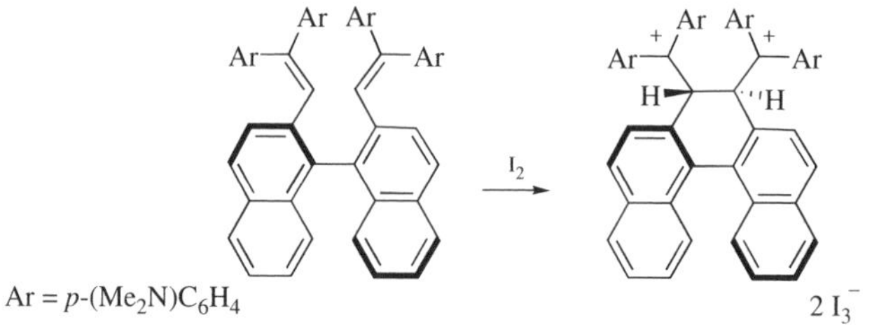

(22)

Ar = p-(Me$_2$N)C$_6$H$_4$

61   **62**

Oxidation leads to formation of the new σ-bond and aryl-stabilized carbocationic centers in **62**. Interestingly, there is no evidence of proton loss from the dihydro[5]helicene dication **62** and the dication is stable in the presence of the reasonably nucleophilic counter ion, I$_3^-$. This again suggests that these stabilized systems are considered only as weakly distonic superelectrophiles.

There is also the possibility of distonic, superelectrophilic bis-carbonium ions. Despite the fact that such species may be important in the superacid-catalyzed cracking reactions of aliphatic hydrocarbons, there have been very few studies of such systems. The structures and energies of small distonic alkonium dications have been studied using *ab initio* calculations.[24] For diprotonated *n*-butane (C$_4$H$_{12}^{2+}$) two structures were located as stable minima on the potential energy surface. Structure **63** is formed by a protonation of the two terminal C-H bonds, resulting in a pair of two electron-three center bonds. The other structure (**64**) arises from protonation of the terminal C-H bond and the most distant C-C bond.

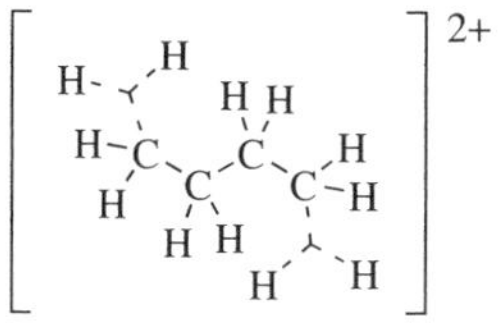

63 (C_2) 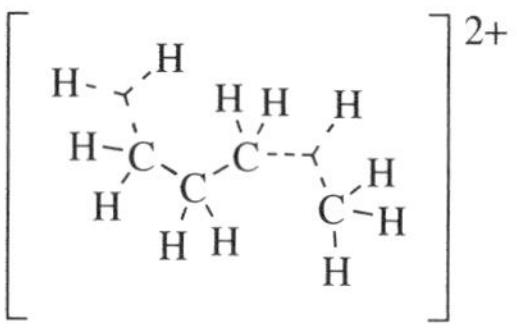   **64** (C_1)

The gas phase structures are found to be within 2 kcal/mol in energy, with **63** being more stable. It is notable that both **63** and **64** are the diprotonated structures possible from *n*-butane having the maximum charge-charge separation. Interestingly, gas-phase proton loss is estimated to be endothermic by 50.3 kcal/mol (MP4(SDTQ)/6-311G**//MP2/6-31G** level), while the carbon-carbon bond cleavage reaction is found to be exothermic by more than 50 kcal/mol (eq 23).

63

$$\tag{23}$$

The product ions are the most stable monocationic carbonium ions, those protonated at the secondary carbon.

Another class of distonic superelectrophiles are the carbenium-carbonium dications. As discussed, Olah and co-workers found experimental and theoretical evidence for the protosolvated *tert*-butyl dication ($[(CH_3)_2 CCH_4]^{2+}$) and 2-propyl dication ($[CH_3CHCH_4]^{2+}$), both gitonic superelectrophiles (*vide supra*).[25] However, analogous distonic superelectrophilic systems like **65** and **66** have not yet been studied.

65   **66**

7.2.2 Carbo-onium Dication

Among other distonic superelectrophiles described in the literature, there are carbo-onium dications. These include carbo-carboxonium dications, carbo-ammonium dications, and related ions. Despite the separation of charge in these superelectrophiles, some have been shown to have very high electrophilic reactivities. Like the carbodications described previously, the discussion here is limited to those systems that have been shown to have electrophilic reactivities greater than the related monocationic onium ions, as well as structural criteria supporting their designation as a distonic superelectrophilic species.

There have been a wide variety of carbo-carboxonium dications described in the literature. Some of the related distonic superelectrophiles can be used for remote functionalization of appropriate substrates. For

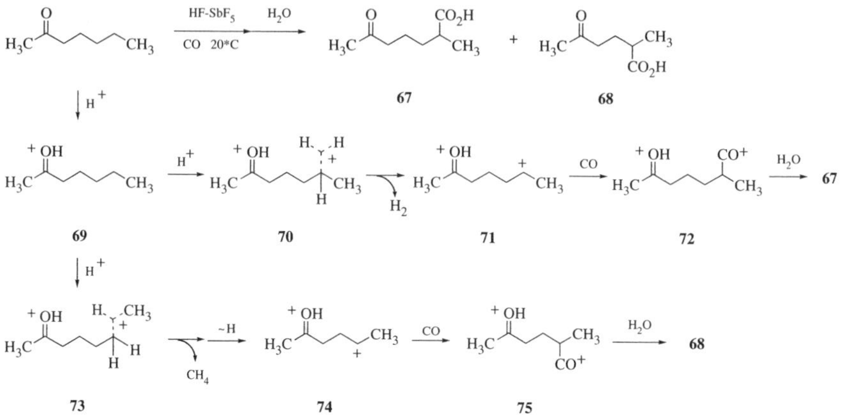

Scheme 1.

example, it was shown that aliphatic ketones are converted to the keto acids (**67** and **68**) by the reactions in superacid in the presence of carbon monoxide (Scheme 1).[26] The two products from 2-heptanone are thought to arise from protolytic cleavage of C-H and C-C bonds. Protonation of the carbonyl group leads to the carboxonium ion (**69**) and further protosolvation leads to the carboxonium-carbonium dications (**70** and **73**) and subsequently to the carboxonium-carbenium dications (**71** and **74**). Capture of the carboxonium-carbenium dications by carbon monoxide and water then gives products **67** (20% relative yield) and **68** (74% relative yield). The key step in this conversion is formation of the carboxonium-carbonium dications (**70** and **73**), and it is notable that these species are formed with the maximum possible charge-charge separation. It is not known to what extent each cationic charge influences the other charge center's electrophilic character. This chemistry has been shown to be selective and useful in functionalizing remote 3° alkyl carbons (eq 24).[26]

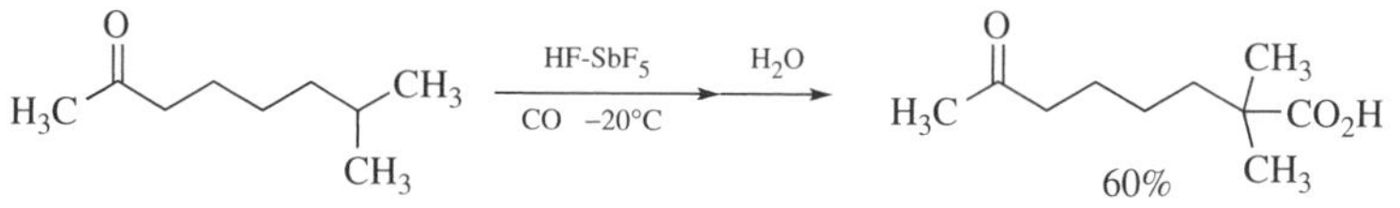

In chemistry involving conversion of camphor, two distonic superelectrophiles are proposed.[27] When ketone **76** is reacted with HF–SbF$_5$, the enone **80** is produced (eq 25).

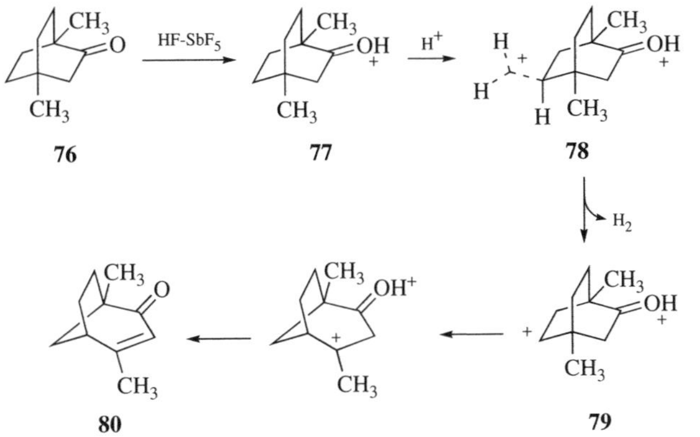

(25)

The conversion is thought to involve formation of the carboxonium ion (**77**) by protonation of the carbonyl oxygen, and subsequent protonation then occurs at the C-H bond. The resulting carboxonium-carbonium dication (**78**) possesses the maximum possible charge-charge separation for this bicyclic framework. Subsequently, an intermediate carboxonium-carbenium dication (**79**) is produced, which isomerizes to the *tertiary*-carbenium ion, and deprotonation provides the product enone (**80**). Similar distonic superelectrophiles are proposed in other rearrangements of terpenes in superacid.[28]

Another type of carboxonium-carbenium dication is obtained from remote functionalization of alkylamides (eq 26).[29]

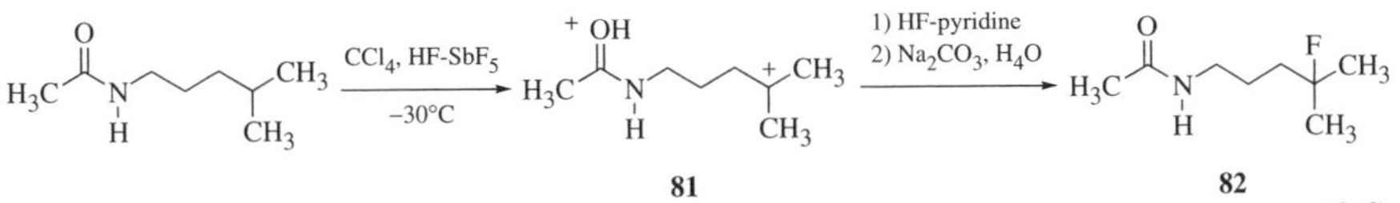

(26)

In the presence of superelectrophilic trihalomethyl cation, the carbocationic center is formed by hydride abstraction generating the distonic superelectrophile (**81**). Capture of the superelectrophile with fluoride leads to the fluorinated product (**82**).

A number of steroidal systems have also been shown to generate carboxonium-carbenium dications in superacid. These intermediates are known to lead to novel products and their chemistry has been recently reviewed.[28d]

Olah and associates demonstrated the oxyfunctionalization of aldehydes and ketones by superacid promoted reactions with ozone.[30] These conversions are thought to involve carbo-carboxonium dications and it has been shown to be an effective method for the preparation of bifunctional products. In the case of aldehydes and ketones, oxyfunctionalization produces dicarbonyl products (eqs 27–28). It has been found that the reaction requires a minimum separation of two carbons between the carboxonium and the developing carbonium ion centers in order for product formation to occur. Thus, pentanal gives dicarbonyl product (**84**), while no such reaction occurs with butanal (eq 29).

$$O = \underbrace{}_{} \quad \xrightarrow[{O_3 \quad -78°C}]{FSO_3H-SbF_5} \quad O = \underbrace{}_{} = O \qquad 60\% \tag{27}$$

83   **84** 80%

$$\tag{28}$$

85   not formed

$$\tag{29}$$

The mechanism is thought to involve the reaction of protonated ozone with the C-H σ-bond and formation of the carboxonium-carbonium dication (**83**). Evidently, formation of the analogous gitonic superelectrophile (**85**) from butanal is disfavored due to the close proximity of the two positive charges. Like most of the chemistry discussed preceedingly there is a strong preference for functionalization at the site most distant from the carboxonium ion group (assuming it is a 2° or 3° reaction center).

Carbo-carboxonium dications have also been generated by the direct ionization of appropriate functional groups by the action of Brønsted superacids. For example, unsaturated acids are shown to give the reactive distonic superelectrophiles, which are shown to be moderately reactive.[31] Protonation of the carboxyl and olefinic groups give the distonic superelectrophiles (**86** and **88**; eqs 30–31).

86   **87**   85% CH_3

$$\tag{30}$$

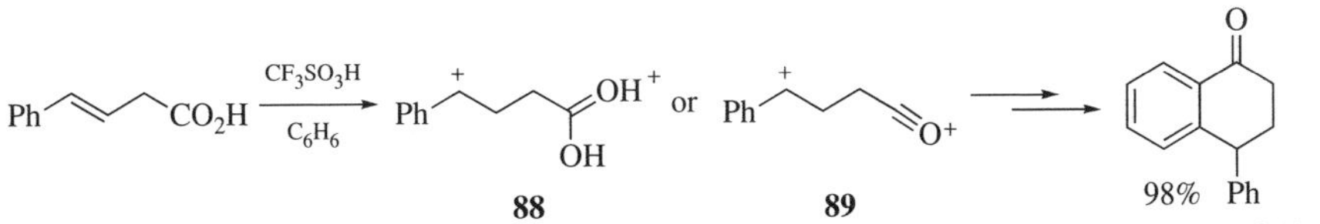

(31)

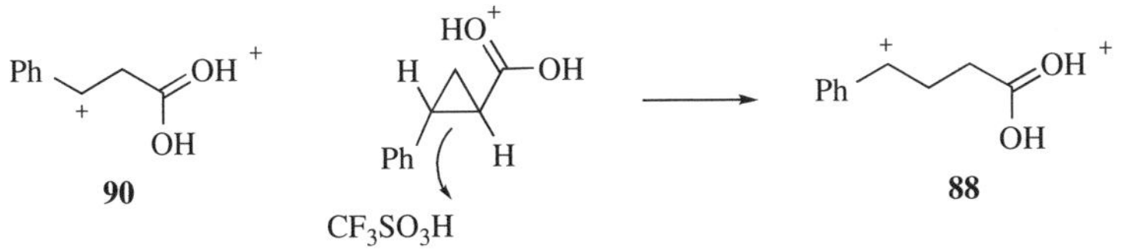

(32)

It has been shown in another study that protonated carboxylic acids tend to form acyl ions at temperatures above $10°C$, so it is possible that the acyl-carbenium dications (**87** and **89**) are the electrophiles that lead to the final cyclization products.[32] Based on the ability to react with deactivated arenes, it has been shown that there is a marked decrease in superelectrophilic character upon going from the gitonic superelectrophile **90** to the distonic superelectrophile **88**. In addition to protonation of the olefin, protonation of a cyclopropane derivative has also been shown to produce the distonic superelectrophile **88**.[32b] There is evidence that these types of distonic superelectrophiles are significantly more reactive than analogous monocationic electrophiles. For example, the distonic superelectrophile (**91**) is generated from a condensation reaction with α-ketoglutaric acid, and despite the stabilizing effect of the two phenyl groups, dication **91** reacts withbenzene and gives the phenyl-substituted tetralone (eq 33).[33]

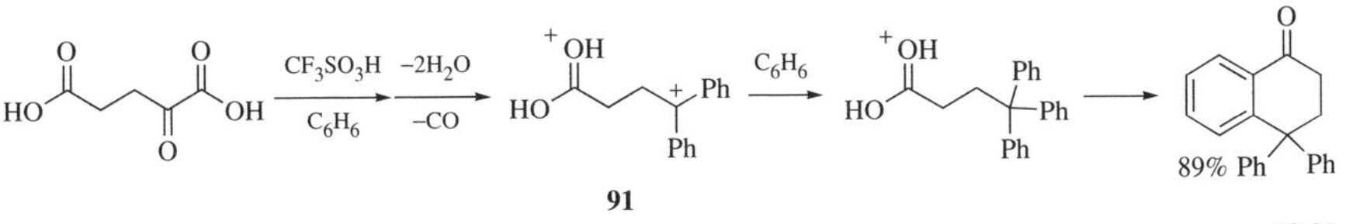

(33)

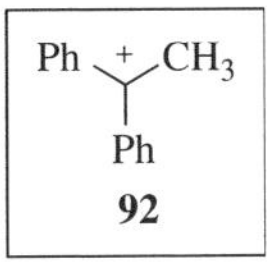

In contrast, the 1,1-diphenyl ethyl cation (**92**) is unreactive to benzene. This indicates that the carboxonium group (or the corresponding acyl ion) participates in the superelectrophilic activation of the adjacent carbocationic center.

Several reports have suggested that carbo-carboxonium superelectrophiles may also be produced from phenols, naphthols, and related species, by diprotonation in superacidic media.[34] For example, 1-naphthol is thought to form the distonic superelectrophile (**93**) with a variety of acids (excess AlCl₃, HF–SbF₅, and HUSY zeolite).[34a] In the presence of benzene, the substituted tetralone, and with cyclohexane, 1-tetralone is produced (eq 34).

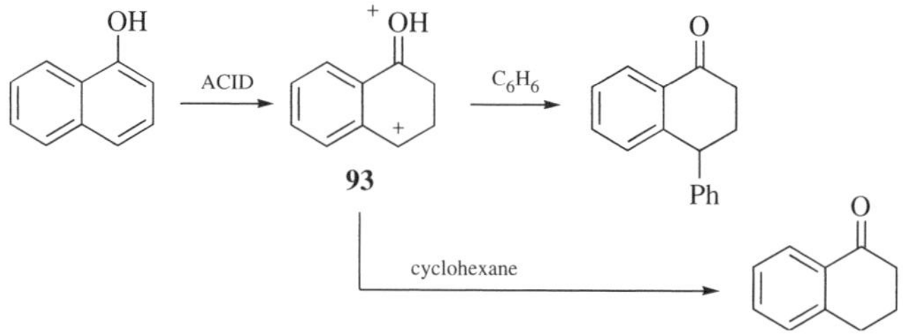

(34)

Since benzene and cyclohexane are both fairly weak nucleophiles, this chemistry indicates that such dicationic species (i.e., **93**) are indeed superelectrophilic.

As mentioned, 2-oxazolines may form a ring-opened distonic superelectrophile in reactions in superacid. These carboxonium-carbenium dications are capable of reacting with benzene and moderately deactivated substrates.[2] For example, the optically active oxazoline (**94**) reacts in CF₃SO₃H to generate the chiral dication (**95**) and this superelectrophilic species is capable of reacting with o-dichlorobenzene in fair to modest yield and diastereoselectivity (eq 35).

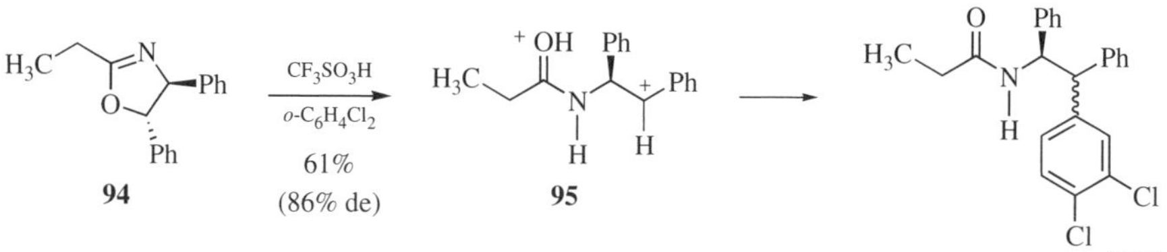

(35)

Carbo-acyl dicationic species have been proposed as intermediates in several reports, but these types of distonic superelectrophiles have not yet been sufficiently studied. Work by the Olah group has shown that protonated carboxylic acids cleave to the acyl ions in superacidic media at temperatures above $-10°C$.[32] In principle, ionization of a second group (such as hydroxyl or olefinic) can generate a carbocationic site adjacent to the acyl cation, producing the carbo-acyl dicationic species. Two types of distonic superelectrophiles are possible for the carbo-acyl dications: the acyl-carbenium dications (i.e., **97**) and acyl-carbonium dications (i.e. **99**). As described in Chapter 6, it has been proposed that the distonic superelectrophile (**97**) is formed as a short-lived and reactive intermediate from the 4-chlorobutanoyl cation (**96**, eq 36).[35] An acyl-carbonium dication (**99**) could be produced by protosolvation of an acyl cation (**98**, eq 37). It has been shown in a closely related study that proto(deutero)solvation of the propionyl cation (**100**) occurs in superacid, to provide deuterium incorporation at the methyl group.[36] Experimental and theoretical evidence demonstrated that formation of the carbonium ion center occurs preferentially at the most distant carbon, giving the gitonic superelectrophile (**101**, eq 38).

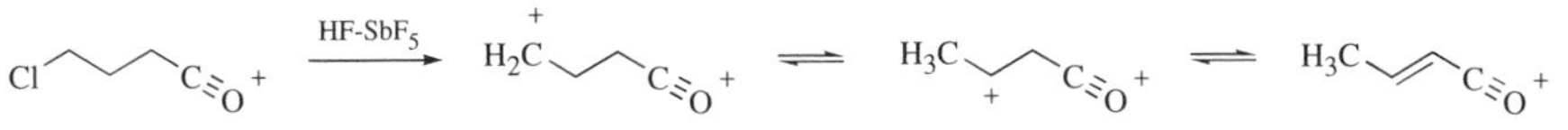

(36)

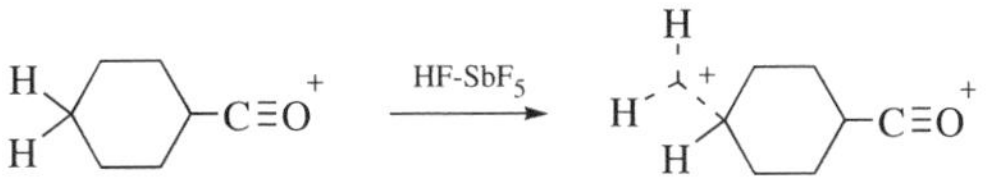

(37)

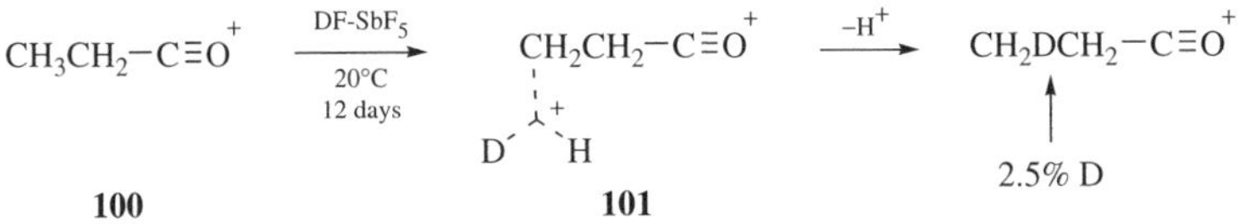

(38)

This provides maximum charge-charge separation in the dication. Thus, it is expected that analogous distonic superelectrophiles will likewise tend to undergo protosolvation at the carbon(s) most distant from the acyl cation center.

A fair number of carbo-ammonium dicationic species and related systems have been reported. These distonic superelectrophiles have been directly observed and shown to be useful in synthetic methodologies. For example, the acid-catalyzed Grewe-cyclization is a well-known reaction used in the preparation of morphine analogues.[37] The conversion involves formation of the distonic superelectrophile (**102**) from an appropriate tetrahydropyridine derivative (eq 39).

$$\text{(39)}$$

Related ammonium-carbenium distonic superelectrophiles (1,4-dications, **103–104**) have been shown to possess strong electrophilic reactivies, undergoing arylation with benzene and dichlorobenzene (eqs 40–41).[38]

$$\text{(40)}$$

$$\text{(41)}$$

The formation of distonic superelectrophile **104** provides a high-yield route to the antispasmodic drug, fenpiprane **105**. The 1,4-dication (**108**), 1,5-dication (**106**) and 1,6-dication (**107**) have likewise been shown to react with benzene in high yields.[39] The superelectrophilic vinyl-dications (**109–110**) have also been studied.[40,41]

A number of related distonic superelectrophiles have been generated from N-heteroaromatic compounds (Table 2). Vinyl-dications (**111–112**) have been produced from the ethynyl pyridines,[40] while N-alkenyl N-heterocycles provide dications (**113–115**).[42a] Vinyl-substituted N-heterocycles provide access to distonic superelectrophiles such as dication **116**.[39] Dications **117** and **118** are generated from their precursors, and both intermediates lead to efficient cyclization reactions with the adjacent

Table 2. Distonic superelectrophiles (111–118) formed with CF₃SO₃H.

Precursor	Distonic Superelectrophile	Precursor	Distonic Superelectrophile
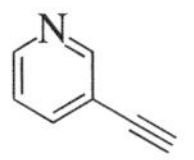	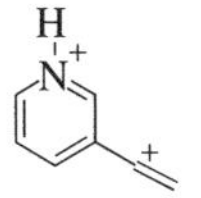 **111**	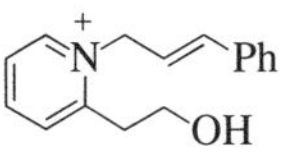	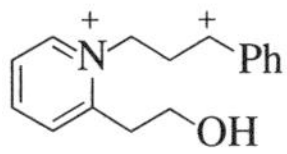 **115**
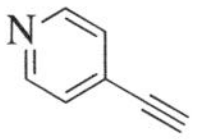	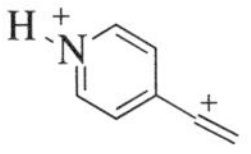 **112**	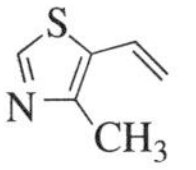	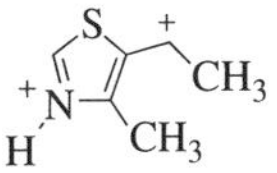 **116**
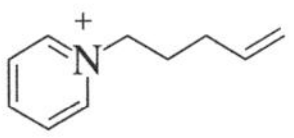	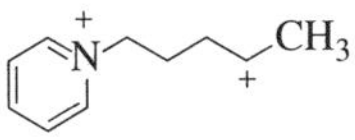 **113**	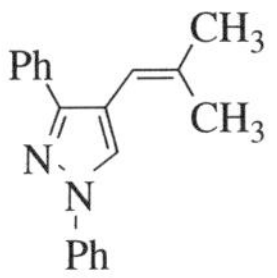	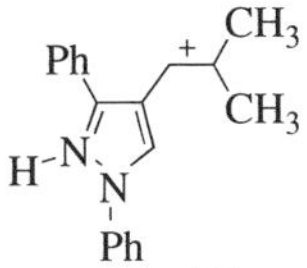 **117**
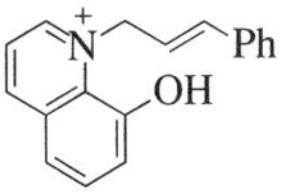	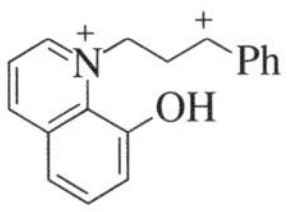 **114**	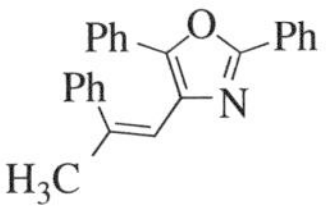	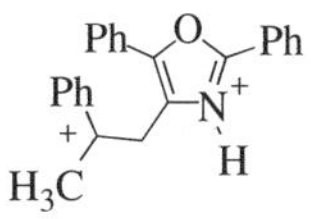 **118**

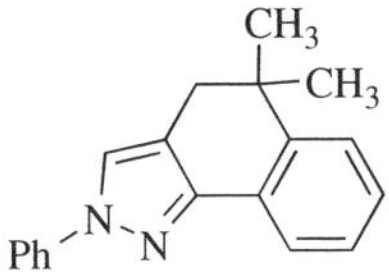

119

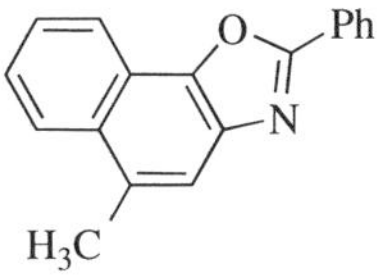

120

phenyl groups.[42b,43] Products **119** and **120** are obtained from the conversions. Distonic superelectrophiles **111–116** have all been shown to react with benzene in Friedel-Crafts type reactions, indicating their reactivities as electrophiles. As expected, the diprotonated species **111–118** are formed with the largest possible charge separation, with preference for highly substituted carbenium centers.

Some theoretical work has been done to estimate the relative stabilities of distonic superelectrophiles with respect to the distance between the charge centers. For example, compound **121** is ionized in CF₃SO₃H to give initially the 1,3-dication (**122**), which is found to undergo the kinetically controlled isomerization to the charge separated 1,4-dication (**123**, Figure 3).[43] Experiments using isotopic labeling indicate that the

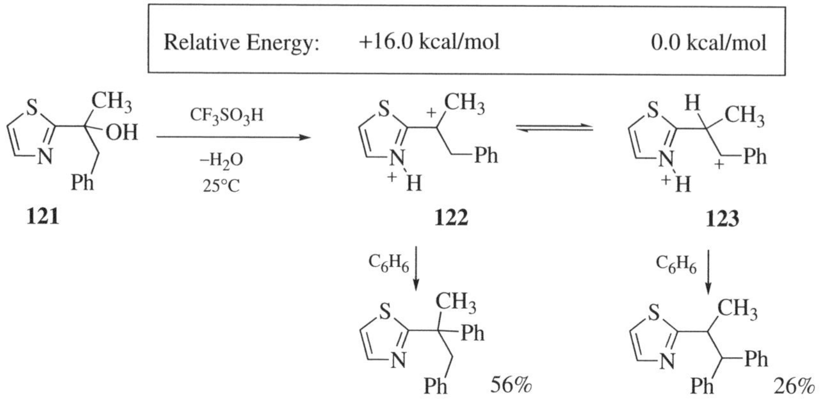

Figure 2.

charge migration occurs via successive deprotonation-reprotonation steps. Calculations at the B3LYP 6-311G**//B3LYP 6-311G** level of theory estimate that the 1,4-dication (**123**) is about 16 kcal/mol more stable than the 1,3-dication (**122**) for the gas-phase geometry-optimized structures. In another study, the olefinic quinoline (**124**) was found to preferentially give the charge separated 1,5-dication (**126**) rather than the 1,4-dication (**125**, Table 3).[44] As described subsequently, dication **125** leads by cyclization to a very efficient conversion to benz[c]acridine. The two dications (**125–126**) were studied using calculational methods to determine their relative stabilities (Table 3). Although their relative energies varied with the

Table 3. Calculated energies for dications 125 and 126.

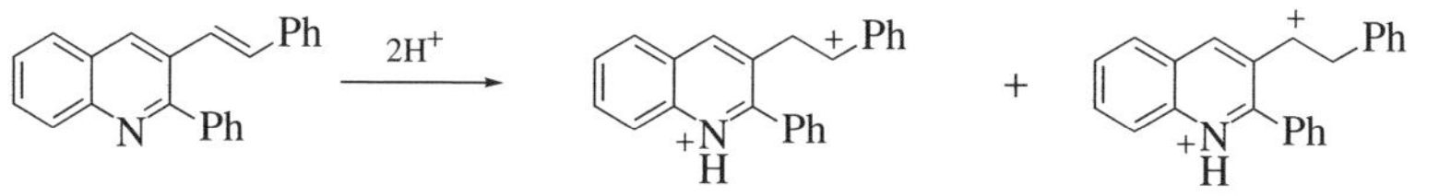

Level of Theory	Relative Energy, kcal/mol^{-1}	
	125	**126**
HF/3-21G	0.0	17.7
HF/6-31G (d)	0.0	17.1
HF/6-311G (d)	0.0	17.9
PBE1/6-311G (d)	0.0	10.0
MP2/6-311g (d)	0.0	10.3
MPW1/6-311G (d)//PCMsp	0.0	7.4

level of theory used in the calculations, all computational methods showed the charge-separated species (**125**) to be considerably more stable. Besides the charge-separation, dication **125** also benefits from favorable benzylic interaction. A calculation was also done using the solvation-sphere model MPW1/6-311G(d)//PCMsp, using sulfuric acid as the solvent. While the energy difference between the two dications becomes even less, this solution phase model still suggests more than a 7 kcal/mol difference in stabilities between the two dications.

Examination of the dication pairs **122/123** and **125/126** may also suggest a further useful definition of the distonic superelectrophiles: the distonic superelectrophiles may be distinguished from onium dications that are without superelectrophilic activation (i.e., two isolated cationic centers) by comparison of the energies, reactivities, or structural/electronic criteria, of two closely related species. For example, it is expected that ammonium-carbenium dications **127** and **128** should vary considerably in terms of energy, due to the proximity of the charges.

127

1,4-dication

1,5-dication

129

1,9-dication

1,10-dication

128

130

Thus, dication **127** can be called a distonic superelectrophile. However, there is very little difference in energy between the ammonium-carbenium dications **129** and **130**, due to the large distance between charges. Dication **129** is therefore not a distonic superelectrophile, but rather an onium dication with isolated electrophilic sites. This concept has been demonstrated experimentally for a number of onium dications, as in studies of carbodications having a large distance between the charge centers. In order to distinguish between distonic superelectrophiles and onium dications, an arbitrary energy criteria may be set. For example, if increasing charge-charge separation by a single carbon atom leads to a stabilization greater than 1.0 kcal/mol, then the species may be considered as a distonic superelectrophile. Likewise, the energy levels of LUMO orbitals could be compared and used as a basis for distinguishing between these two types of closely related dicationic species.

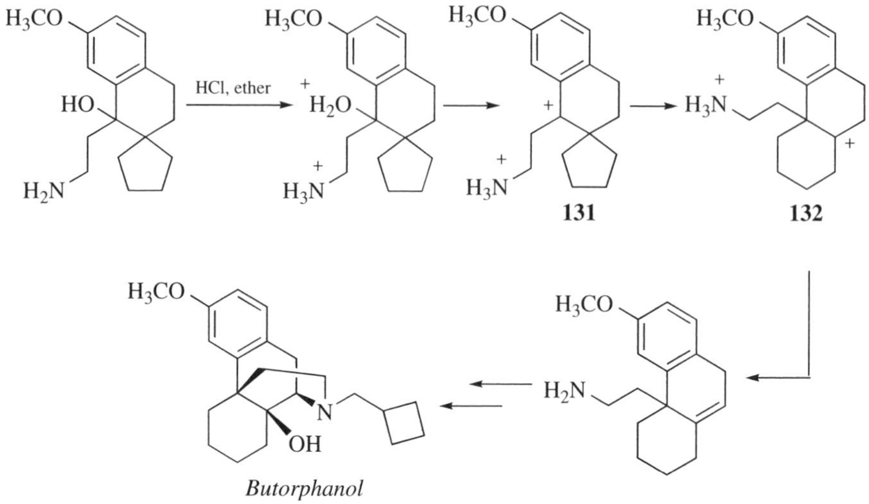

Butorphanol

Scheme 2.

Charge-charge separation has been shown to be a useful driving force in the reactions of carbo-ammonium dications. For example, in the commercial synthesis of the analgesic drug *butorphanol*, a key step involves the ring expansion step of dication **131** to dication **132** (Scheme 2).[45] The carbocationic center in **131** is located at a benzylic charge position, being adjacent to an electron rich aryl ring. It is somewhat unexpected that migration should occur to produce dication **132** because the resulting carbocationic center loses its benzylic stabilization. Clearly, an important driving force for this conversion is the charge-charge separation, where the 1,4-dication (**131**) produces a 1,5-dication (**132**). In other studies, it was demonstrated that distonic superelectrophiles **133** and **135** undergo charge migration (eqs 42–43).[43]

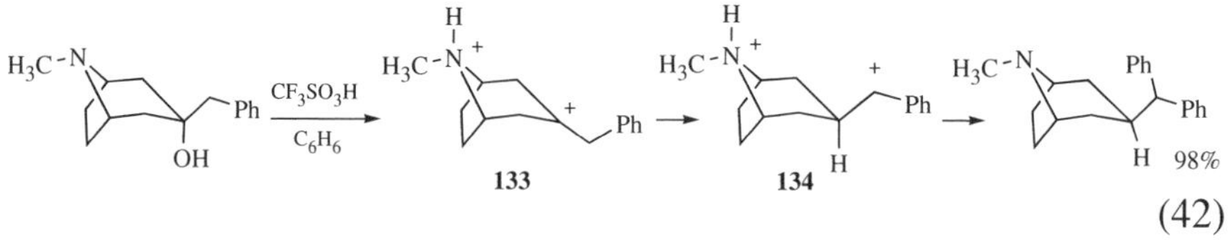

$$(42)$$

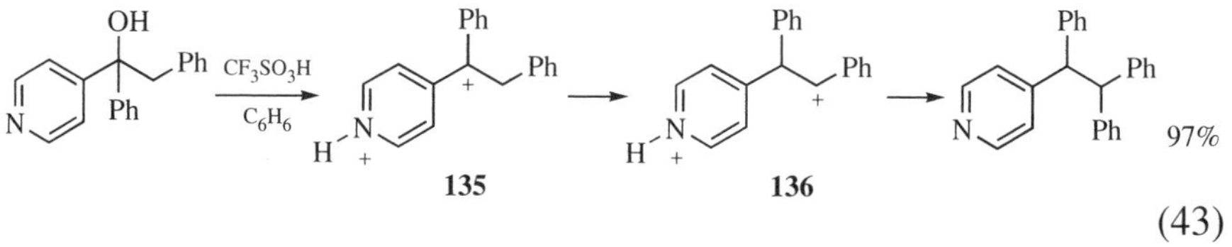

$$(43)$$

The charge-separated species (**134** and **136**) are then quantitatively trapped with benzene. Interestingly, the conversion of **135** to **136** involves the migration of charge from one benzylic position to another benzylic position. This indicates that charge-charge repulsion drives the conversion. The charge migration chemistry has been exploited in the synthesis of a variety aza-polycyclic aromatic compounds (eqs 44–45).[43]

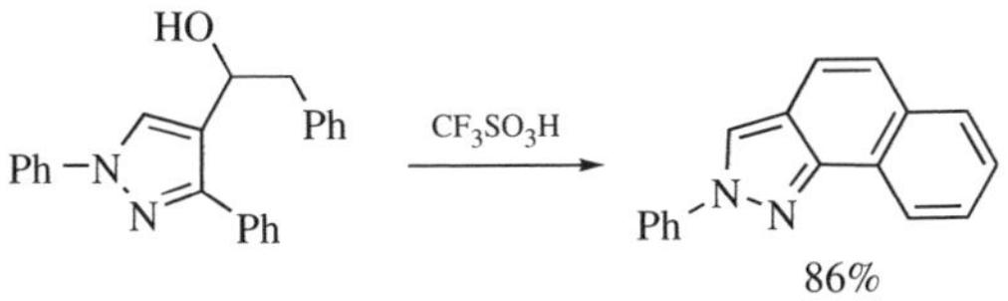

$$\text{(44)}$$

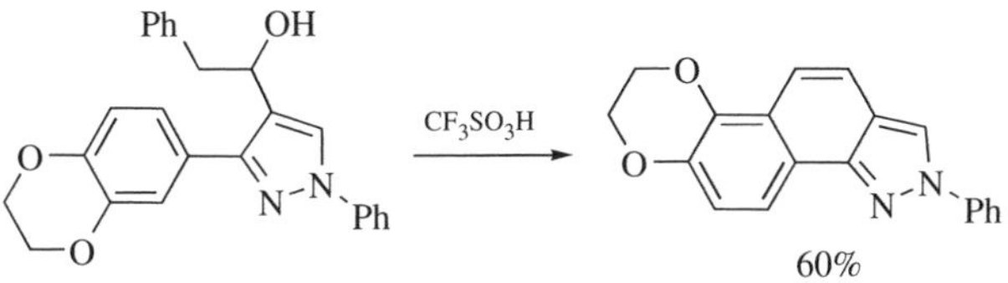

$$\text{(45)}$$

As shown in Scheme 3, these conversions involve two key reaction steps. From the pyridine derivative (**137**), ionization of the starting material involves protonation of the N-heterocycle and the hydroxyl group, with loss of water producing the distonic superelectrophile (**138**). Charge migration then provides the dicationic species (**139**) which is able to undergo ring closure. *Ipso*-protonation of the phenyl group leads to benzene elimination and formation of the condensed aromatic system.

Besides carbo-ammonium dicationic systems, there have been studies related to carbo-phosphonium dication systems. Some of the reported chemistry suggests that superelectrophilic activation is involved. Olefinic phosphonium salts are protonated in superacid to generate dications like **140** and these species have been shown to react with benzene in good

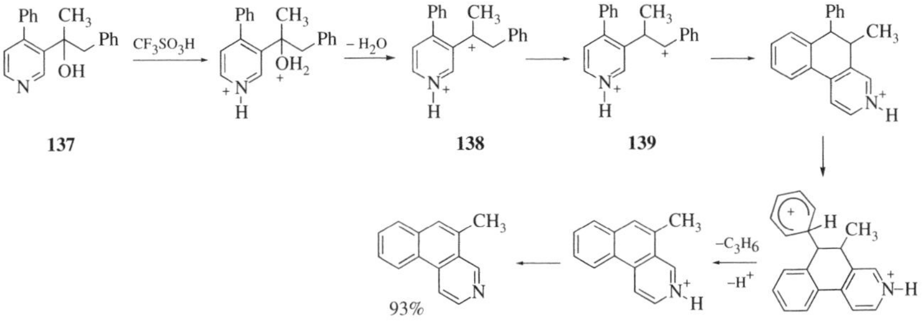

Scheme 3. Proposed mechanism for aza-polycyclic aromatic compound formation.

yields (eq 46).[46] In the reaction of the 2-pentenyl system (141), the superacid catalyzed addition reaction gives a product arising from a charge migrationstep (eq 47).

$$\tag{46}$$

$$\tag{47}$$

Initial protonation gives the 1,4-dication (142), but charge migration provides the more stable 1,5-dication (143). Reaction of dication 143 with benzene then gives the product.[46]

There have been few reports of carbo-halonium dicationic species in the literature. In an attempt to prepare an adamantadiyl dication, Olah and co-workers reported NMR evidence for the carbenium ion (144),

$$\tag{48}$$

which possesses significant donor-acceptor interaction between the bromine atom and the Lewis acid (eq 48).[12] This species is expected to also exhibit modest superelectrophilic activation from the partial positive charge on the bromine.

7.2.3 Carboxonium-Centered Distonic Dication

A number of studies have reported the formation of carboxonium-centered distonic dications. These systems include diprotonated diketones, diprotonated dicarboxylic acids, their derivatives, and others. A wide variety of mixed dications have also been reported such as ammonium-carboxonium dications, oxonium-carboxonium dications, acyl-carboxonium dications, and other species. Many of these carboxonium-centered dications exhibit properties indicative of distonic superelectrophiles.

7.2.4 Bis-carboxonium Dication

As described in the previous chapter, a number of diketones have been diprotonated in superacid to generate distonic superelectrophilic species. Several types of 1,4-diketones yield the respective distonic superelectrophiles (eq 49).[47]

(49)

These distonic superelectrophiles (**145–147**) have been characterized by low-temperature ^1H NMR (and ^{13}C NMR in the case of **147**) from FSO_3H–SbF_5 solution. Dication **146** was also studied by calorimetric studies to determine the heat of diprotonation of 2,5-hexanedione.[48] It was found that the heat of diprotonation for the γ-diketones (like 2,5-hexanedione) is about 5 kcal/mol less than expected, when compared to twice the heat of protonation of acetone or other monoketones. The destabilization of dication **146** by 5 kcal/mol can be the result of electrostatic effects, and it can be considered evidence for the superelectrophilic character of such dications. When 2,6-admantanedione is reacted in FSO_3H–SbF_5 solution, the dication **148** is formed as a persistent species, observable by ^1H and ^{13}C NMR.[12] The carboxonium carbons of **148** are observed at $\delta^{13}C$ 247.7, while the monocationic species (**149**) has a carboxonium carbon at $\delta^{13}C$ 267.1. These ^{13}C NMR data were interpreted as evidence for the increasing importance of the carboxonium-type resonance structure (**148a**) due to electrostatic repulsive effects. Some examples of aromatic diketones (i.e., diacetylbenzenes) have also been reported to produce bis-carboxonium dications in their protonation reactions in superacids.[47]

A homologous series of aliphatic ketoacids were studied in FSO_3H–SbF_5 solution by low-temperature NMR. As discussed previously, the gitonic superelectrophiles (**150a,b**) have been generated, but the distonic superelectrophiles (**150c-e**) were also observed.[49] Among the distonic superelectrophiles, the two systems with the greatest distance between charge centers (**150 d,e**) undergo cleavage to the acyl-carboxonium dications (**151 d,e**) at 0°C by loss of water (eq 50). Diprotonated levulinic acid (**150c**), however, is found to be stable to dehydrative cleavage up to temperatures as high as 60°C (eq 51).

The resistance to cleavage is an indication of the superelectrophilic character of dication **150c**. Several aromatic compounds have likewise been shown to produce dicationic species upon the protonation of carboxyl and carbonyl functional groups. Other bis-carboxonium dications have been described involving protonation of carbonyl, amide, and other groups.[50] These distonic superelectrophiles (**152–153**) have been shown to be useful in condensation reactions (eqs 52–53).

A number of bis-carboxonium distonic superelectrophiles have been generated from dicarboxylic acids and diesters. Early studies of the protonation of the aliphatic diacids were done by cryoscopic techniques using sulfuric acid media.[51] More recent work was also involved using low-temperature NMR methods. The aliphatic dicarboxylic acids are converted to their diprotonated products (**154a-f**) in FSO_3H-SbF_5-SO_2 solution and the dications **154a-f** have been characterized by ^1H NMR.[52] Like the dications from the ketoacids discussed previously, it was shown that the bis-carboxonium dications may undergo dehydration to form acyl cation center(s). This cleavage reaction is also shown to be sensitive to the distance between the carboxonium charge centers (eq 54).

154 a: n = 0   d: n = 3
    b: n = 1   e: n = 4
    c: n = 2   f: n = 5

Diprotonated pimelic, adipic, and glutaric acids (154 d-f) cleave to the acyl-carboxonium dications (155 d-f) at a rate that is similar to that of monoprotonated aliphatic carboxylic acids. However, succinic acid (154c) only cleaves to an extent of 50% in the $FSO_3H-SbF_5-SO_2$ solution. Loss of the second water molecule from 155 d-f occurs readily, but even with increased acidity the acyl-carboxonium dication 155c only cleaves to an extent of about 50% to the biacyl ion (156c). These data are consistent with a significant degree of superelectrophilic activation in the case of 154c and 155c, although the results also show a diminishing superelectrophilic activation in the charge separated species (154 d-f) and (155 d-f). There are also close parallels between these results and the formation of bis-acyl dications from the aliphatic diacid chlorides (vide infra).

The dicationic species arising from phthalic acid and its esters have also been studied. Diprotonated phthalic acid (157) is observed by low temperature ^1H and ^{13}C NMR, with its carboxonium carbons found at $\delta^{13}C$ 181.0.[53a] Warming the solution of 157 leads to the appearance of a new set of signals that were assigned to the cleavage product, the acyl-carboxonium dication 158 (eq 55).

NMR evidence also suggests the degenerate rearrangement via the anhydride derivative (159). A similar process was described for the chemistry of succinic and glutaric anhydrides in superacid. The methyl ester of

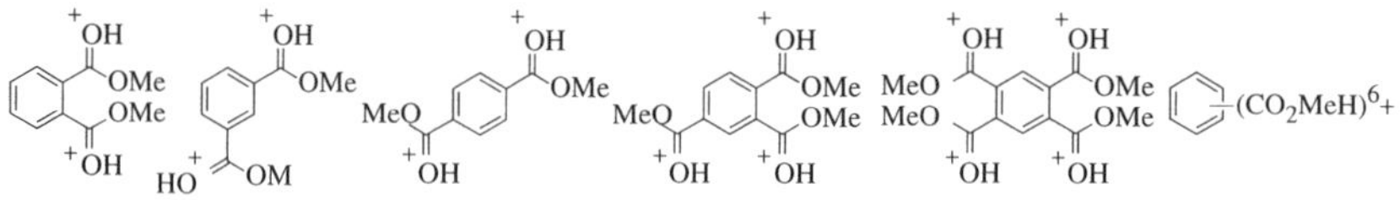

160   161   162   163   164   165

Scheme 4.

phthalic acid was also shown to form the diprotonated species (**160**) in
$FSO_3H–SbF_5–SO_2ClF$ at low temperature (Scheme 4). Other multiply
protonated species (**161–165**) are formed by the reactions of their respective esters in superacid.[53b]

Diprotonated aliphatic esters have been studied by low-temperature
NMR experiments and by calorimetric techniques.[48,54] A series of diesters
(**166a-e**) were studied in superacid by low-temperature NMR, and in each
case the diprotonated species (**167a-e**) could be observed as a persistent
species at $-60°C$ (eq 56).[54]

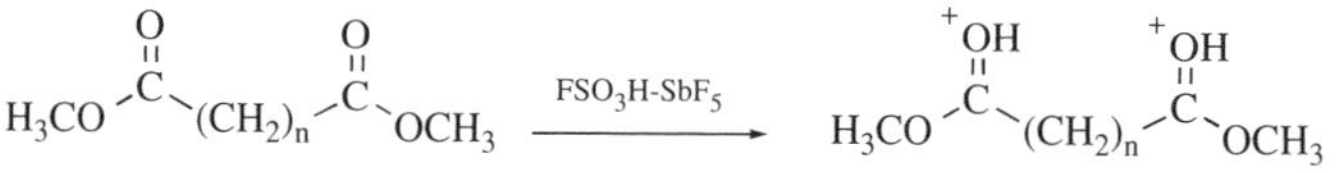

$$(56)$$

166 a: n = 0 d: n = 3
         b: n = 1 e: n = 4
         c: n = 2

167 a: n = 0 d: n = 3
         b: n = 1 e: n = 4
         c: n = 2

In general, the diprotonated diesters (**167a-e**) are more easily obtained than
analogous diprotonated diketones. It has been suggested that this reflects
the effective charge delocalization, or dispersal, in the ester-based carboxonium ions.[48] This is evident by the acid-strength required to form the
diprotonated species, with diketones requiring much stronger superacids.
Based on the results from calorimetric studies of diester (di)protonation, it
was observed that the effects of charge-charge repulsion is negligible with
separation of the two positive charges by three or more methylene units.[48]
Thus, **167c** can be properly described as a distonic superelectrophile, while
the higher homologs **167 d-e** have rapidly diminishing superelectrophilic
activation.

Another interesting type of bis-carboxonium dications is the bis(1,3-
dioxolanium) dication series (**168–170**).

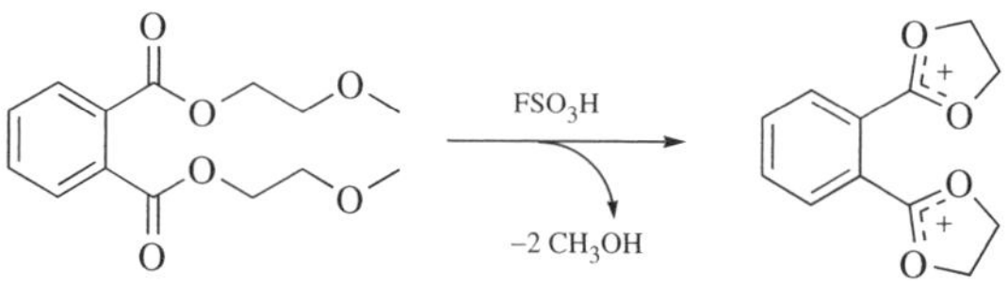

168

$$(57)$$

169   170

These ions were prepared from the appropriate 2-methoxyethyl esters (eq 57).[55] Reaction in superacid leads to the formation of the 1,3-di oxolanium cation centers and the respective dicationic species. Dications **168–170** were directly observed by NMR spectroscopy. Their structures, energies, and GIAO/DFT $\delta^{13}C$ values were determined by *ab initio* computational methods. There is generally good agreement between the observed ^{13}C NMR spectra and the calculated spectra. Dication **168** gives a ^{13}C NMR signal at $\delta^{13}C$ 182.2 for the carboxonium carbons in FSO$_3$H solution, while the calculated value (B3LYP/6-311G*//B3LYP/6-31G* level) is at $\delta^{13}C$ 188.5. When the energies of the dications are compared, the *ortho*-isomer (**168**) is found 15.0 kcal/mol higher in energy then the *meta* and *para* isomers (**169–170**). This increasing energy is thought to be due to the steric effects of the *ortho* substituents, as well as the electrostatic repulsive effects of the two closely oriented cationic centers. Clearly, species **168** should be considered a distonic superelectrophile, due in part to energy considerations. Using the same chemistry, the tricationic species (**171**) is produced in superacid (eq 58).

$$(58)$$

171

When comparing the ^{13}C NMR data from the dications **168–170** and the trication **171**, there is evidence for increasing superelectrophilic activation with the trication. For example, the methylene carbons are found at $\delta^{13}C$ 77.5 for dication **169** and $\delta^{13}C$ 86.9 for trication **171**. This increased deshielding of the methylene carbons can be understood in terms of a greater delocalization of the positive charges (in the trication) onto the ring oxygen atoms, leading to deshielding of the methylene carbons. The enhanced resonance interaction with the oxygen atoms is thought to be the result of the electrostatic repulsive effects of the three positive charges. The optimized geometries also reveal a structural effect from the third positive charge. The bond length between the carboxonium carbon and the *ipso*-carbon increases from 1.453 Å in dication **169** to 1.471 Å in the trication **171**. This increase in bond length is due to coulombic repulsive effects.

7.2.5 Onium-carboxonium dication

Among the reported distonic superelectrophiles, a significant number of ammonium-carboxonium dications and related species have been studied. It has been shown that these electrophiles show enhanced reactivities compared with monocationic carboxonium ions. For example, 4-piperidone (**172**) is diprotonated in superacidic CF_3SO_3H to give the distonic superelectrophile (**173**), which condenses with benzene in high yield (eq 59).[56] In contrast, cyclohexanone forms the monocationic carboxonium (**174**) ion, but ion **174** is not sufficiently electrophilic to react with benzene (eq 60).

$$(59)$$

$$(60)$$

The observed electrophilic reactivity is indicative of superelectrophilic activation in the dication **173**. Other ammonium-carboxonium dications have also been reported in the literature, some of which have been shown to react with benzene or other weak nucleophiles (Table 4).[1,42b,57–60] Besides ammonium-carboxonium dications (**175–179**), a variety of *N*-heteroaromatic systems (**180–185**) have been reported. Several of the dicationic species have been directly observed by low-temperature NMR, including **176**, **178–180**, **183**, and **185**. Both acidic (**175**, **180–185**) and non-acidic carboxonium (**176–177**) dicationic systems have been shown to possess superelectrophilic reactivity. The quinonemethide-type dication (**178**) arises from the important biomolecule adrenaline upon reaction in superacid (entry 4). The failure of dication **178** to react with aromatic compounds (like benzene) suggests only a modest amount of superelectrophilic activation. An interesting study was done with aminobutyric acid

Table 4. Distonic superelectrophiles based on ammonium-carboxonium dications and related heterocyclic systems.

Entry	Precursor	Dication	Entry	Precursor	Dication
(1)		**175**	(7)		**181**
(2)		**176**	(8)		**182**
(3)		**177**	(9)		**183**
(4)		**178**	(10)		**184**
(5)		**179b**	(11)		**185**
(6)		**180**			

derivatives.[60] The series of diprotonated species (**179a-c**) were observed as persistent species in FSO$_3$H-SbF$_5$ solution at low temperature. The diprotonated α- and β-aminobutyric acids (**179a,b**) were found to be stable at temperatures as high as 45°C, but the diprotonated γ-aminobutyric acid was found to undergo cleavage to the amino-acylium dication (**186**, eq 61).

179a   179b   179c   186   (61)

As described in the previous chapter, **179a,b** are thought to be stable towards cleavage due to the close proximity of the charge centers. Diprotonated 4-pyridinecarboxaldehyde (**180**) has been shown to react with nitrobenzene and saturated hydrocarbons, demonstrating its superelectrophilic character (eqs 62–63).[58a]

$$(62)$$

$$(63)$$

Other diprotonated acyl-pyridines have likewise been studied.[61] In studies of 5-, 6-, 7-, and 8-hydroxyquinolines and 5-hydroxyisoquinoline, dicationic intermediates like **185** (Table 4) were found to be involved in superacid catalyzed reactions with benzene and cyclohexane.[59] For example, 8-hydroxyquinoline (**187**) reacts in $CF_3SO_3H–SbF_5$ to generate dications (**188** and **189**) and undergoes ionic hydrogenation in the presence of cyclohexane (eq 64). Compound **187** also reacts with benzene in suspensions of aluminum halides (eq 65).

$$(64)$$

$$(65)$$

190

Interestingly, the comparable monocation (**190**) is not reactive towards benzene or cyclohexane. This is an indication of the superelectrophilic character of dication **188**. The isomeric hydroxyquinolines and 5-hydroxy-isoquinoline react with 5–7 molar excess of aluminum chloride and cyclohexane at 90°C to give ionic hydrogenation products, and the corresponding distonic superelectrophiles (**191–193**) are proposed as intermediates.

185   191   192   188   193

When the reactions with benzene were compared, it was found that the distonic superelectrophiles **185**, **188**, and **193** are the most reactive electrophiles.[59] Based on the results from computational studies (MNDO and DFT), the relative reactivities of the isomeric hydroxyquinolines and 5-hydroxyisoquinoline can be correlated with the energies of the lowest unoccupied molecular orbital (ε_{LUMO}, the square of the coefficients (c^2) at the reactive carbon atoms, and the NBO charges (q) on CH groups. For example, 8-hydroxyquinoline (**187**) is found to be more reactive than 6-hydroxyquinoline in the superacid catalyzed reactions with benzene and cyclohexane. When the ε_{LUMO} for dication **188** is compared with the ε_{LUMO} for dication **191**, dication **188** has a LUMO of much lower energy. It was proposed that lower ε_{LUMO} values correspond to increased electrophilic reactivities. It is also noted that typical dicationic structures are found to have much lower ε_{LUMO} values ($\sim -12\,eV$) compared to analogous monocation ε_{LUMO} values ($\sim -7\,eV$). This study also showed evidence for the importance of atomic charge, q, at a reaction center in determining the reactivity from a kinetic point of view. Given that the energy levels for the highest occupied molecular orbitals (HOMOs) for benzene ($-9.782\,eV$) and cyclohexane ($-9.109\,eV$) are considerably higher (less negative) than the ε_{LUMO} values for the dications, the superelectrophilic character of these species can be understood. To further characterize the intermediates involved in the conversions, dicationic species **185** and **191–193** were also studied by ^1H and ^{13}C NMR from 5-hydroxyquinoline, 6-hydroxyquinoline, 7-hydroxyquinoline, and 5-hydroxyisoquinoline, respectively, in CF_3SO_3H-SbF_5 solutions. Despite the high acidity of the media, however, dications **191–193** were found to be in equilibrium with the monoprotonated species.

In a related report, 5-amino-1-naphthol (**194**) was shown to react with weak nucleophiles (cyclohexane and benzene) and the distonic superelectrophile **196** is proposed as the key intermediate (Figure 3).[62] NMR

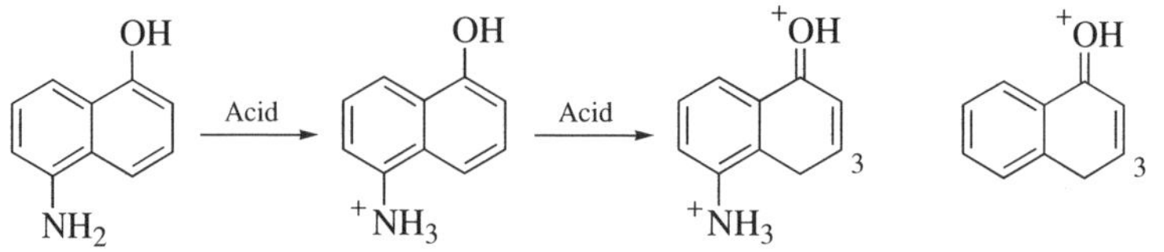

| | | ε_{LUMO} | -10.979 eV | -7.277 eV |
| | | q_3 | $+0.3$ | $+0.25$ |

Figure 3. 5-Amino-1-naphthol, its protonated ions **195-196**, and comparison to monocation **190**.

studies indicate that the monocationic species **195** is formed in CF_3CO_2H (H_0 -2.7), and the dicationic species **196** from CF_3SO_3H (H_0 -14.1) or CF_3SO_3H-SbF_5 (H_0 -20), although with CF_3SO_3H both the monocation (**195**) and dicationic (**196**) are present in appreciable concentrations. A dicationic species similar to **196** is also thought to arise from solutions with excess anhydrous $AlCl_3$ or $AlBr_3$ (eq 66). The superelectrophilic reactivity of dication **196** is consistent withcomputation results, the energy level of the lowest unoccupied molecular orbital (ε_{LUMO}) and the atomic charge of the reaction center (q_i). In order to show the effect of a second charge center, dication **196** is compared to the monocationic 4-hydroxy-1-naphthalenonium ion **190**. Dication **196** is found to have a much lower ε_{LUMO} and a greater positive charge at carbon 3, when compared to the monocation **190**. In reactions with excess (4 equivalents) $AlCl_3$ or $AlBr_3$, 5-amino-1-naphthol undergoes ionic hydrogenation with cyclohexane while the reaction with benzene gives 5-amino-3- phenyl-1-tetralone (**197**) (eq 66).

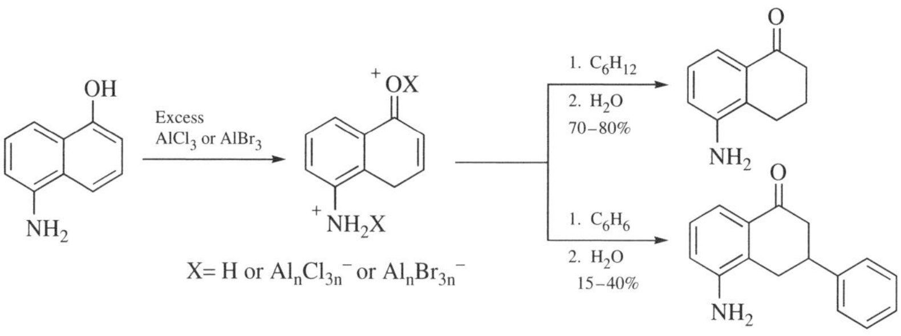

(66)

Since monocation **190** does not react with either benzene or cyclohexane, the results show that the protonated amino group significantly enhances the electrophilicity of the carboxonium ion. These results are also consistent with the predictions based on ε_{LUMO} and q_i.

The activating effects of ammonium groups on carboxonium electrophiles has also been exploited in the Friedel-Crafts acylations with amides.[50] For example, in comparing the superacid-catalyzed reactions of acetanilide, the monoprotonated species (**198**) is found to be unreactive towards benzene (eq 67), while the diprotonated, superelectrophilic species (**199**) reacts with benzene to give the acyl transfer product in reasonably good yield (eq 68).

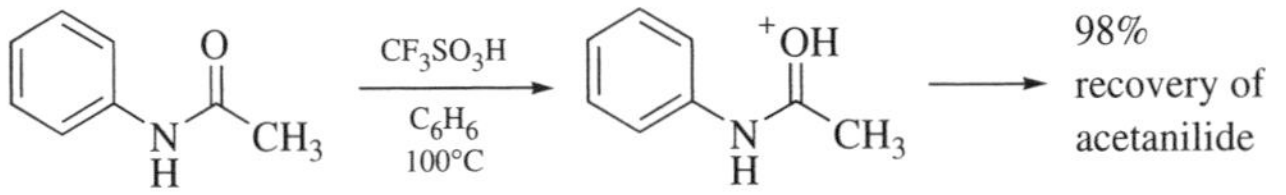

(67)

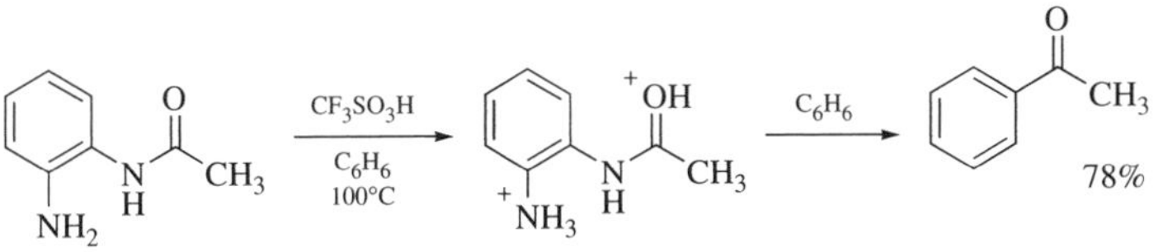

(68)

Other related distonic superelectrophiles (**200–201**) were also shown to provide acyl-transfer products with benzene, and in the case of the aminopyridine derivative (**202**) intramolecular reaction produces the indanone (eq 69).

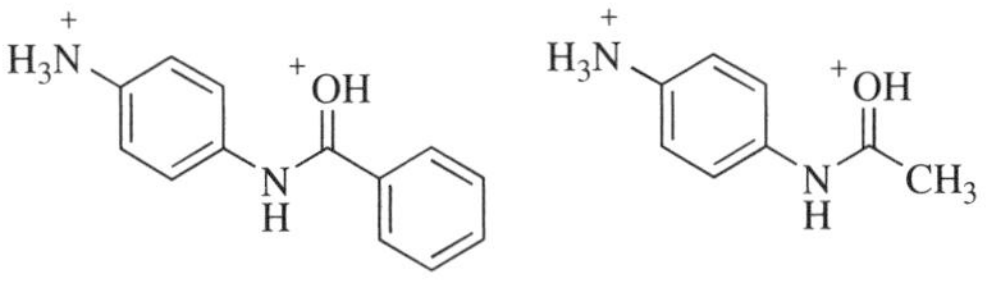

(69)

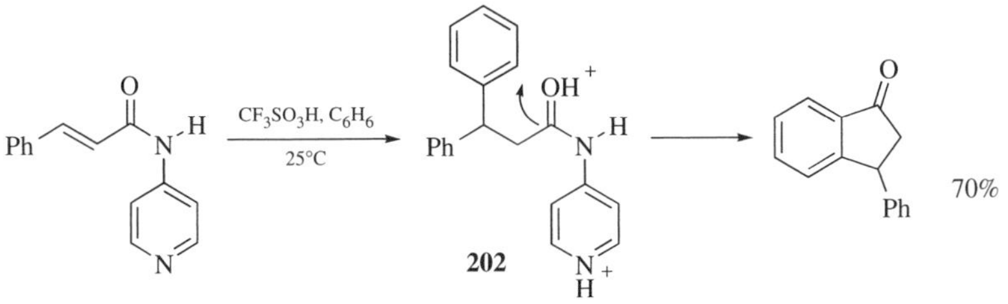

Although Friedel-Crafts acylation is well known with carboxylic acids, anhydrides, and acid halides, there are virtually no reports of Friedel-Crafts acylations being done with amides.[63] These results demonstrate the application of distonic superelectrophiles to accomplish such a difficult synthetic goal as the Friedel-Crafts acylation of aromatic compounds with amides.

The diprotonated benzoquinone monooximes have also been studied. Using low temperature NMR, dications such as **203** can be directly observed (eq 70)[64]. Little work has been done to study the electrophilic chemistry of these ionic species, although Shudo and Okamato generated dication **204** in superacid and found it capable of reacting with phenol (eq 71).[65]

(70)

(71)

Carbo-oxonium dications have also been described in the literature and some of these may be considered distonic superelectrophiles. An interesting example has been described by Sommer and co-workers in the case of diprotonated p-anisaldehyde (**206**, Figure 4).[66] The protonation equilibria of p-anisaldehyde in superacids were studied using dynamic NMR techniques. It is known that O-protonated benzaldehydes and related compounds exhibit significant rotational barriers (compared to the neutral benzaldehyde) due to π-delocalization (i.e., **205b**). For monoprotonated p-anisaldehyde (**205**), the rotational barrier was estimated to be 18.6 kcal/mol. In strong superacid systems however, the oxonium center is formed producing the diprotonated p-anisaldehyde (**206**). This is found to have a decreased rotational barrier of about 12.6 kcal/mol. The decreasing rotational barrier is the result of diminished neighboring aryl-group stabilization of the carboxonium group. Electron delocalization from the aryl group towards the carboxonium group is disfavored due to the interaction of the two positive charges (i.e. **206b**). Other oxonium-carboxonium dications have been reported, such as the diprotonated products (**207–208**) from hydroxycarboxylic acids,[67] but little is known about the electrophilic character and superelectrophilic activation in such dications.

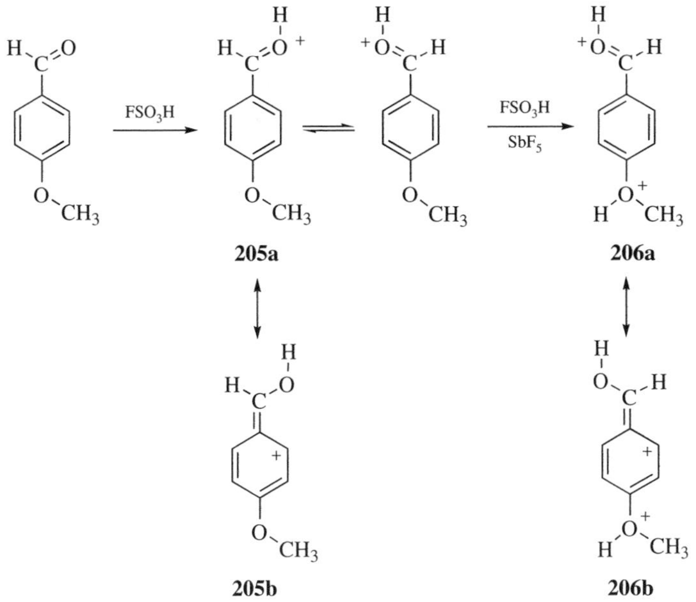

Figure 4.

7.2.6 Acyl-centered Distonic Dications

Monocationic acyl ions are readily prepared as persistent species in solutions of low nucleophile strength.[68] These acyl ions have been thoroughly characterized by IR and NMR spectroscopy, and several acyl ion salts have been characterized by X-ray crystallography. The monocationic acyl ions are often prepared in situ from carboxylic acids, esters, or anhydrides, by the action of a strong Brønsted acid, or the ions can be prepared from ionization of an appropriate acid halide with a strong Lewis acid. Both methods have been used to prepare acyl-centered dications, some of which can be considered distonic superelectrophiles. As described previously, dicarboxylic acids cleave to the bis-acyl ions in superacid ($FSO_3H–SbF_5$) provided that the acyl cations are separated by at least three methylene units (eq 54).[55] The first bis-acyl dications were reported by Olah and Comisarow, being prepared by the reactions of dicarboxylic acid fluorides with superacidic SbF_5 (eq 72).[69]

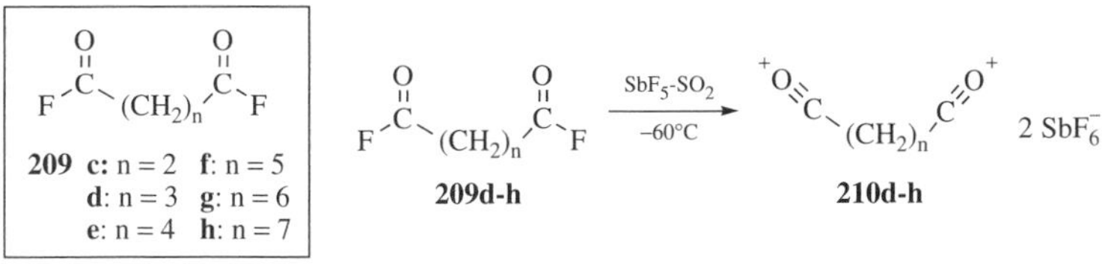

$$209 \quad \begin{array}{ll} \text{c: } n = 2 & \text{f: } n = 5 \\ \text{d: } n = 3 & \text{g: } n = 6 \\ \text{e: } n = 4 & \text{h: } n = 7 \end{array}$$

(72)

Within the series of dicarboxylic acid fluorides, ionization provided the bis-acyl dications in all cases except with succinyl fluoride. It was suggested that the electrostatic repulsive effects inhibit formation of the bis-acyl dication from succinyl fluoride (**209c**). Dications **210d-h** were characterized by IR and ^{1}H NMR in this study and they were also shown to be reactive towards benzene to give the expected diketones. Using calorimetric techniques to study the ionizations, similar results were obtained by Larson and Bouis.[49] It was found in these studies that systems with greater than four methylene groups (i.e., **210f-g**) exhibit similar heats of reaction. Thus, bis-acyl dications (**210f-h**) may be considered onium dications with separated acyl groups. However based on heats of reaction, it is estimated that dication **210e** is ca. 4 kcal/mol less stable than **210f**, and consequently bis-acyl dications separated by four or less methylene groups (**210c-e**) can be considered superelectrophilic.

Bis-acyl dications have also been generated from terephthalic acid derivatives. Ionization of terephthaloyl fluorides (**211a-c**) with SbF$_5$ has been used to prepare bis-acyl cations (**212a-c**, eq 73),[69]

211a: X = H
b: X = Cl
c: X = Br

212a: X = H
b: X = Cl
c: X = Br

213b: X = Cl
c: X = Br

$$(73)$$

which have been characterized by IR, UV, and ^{1}H NMR spectroscopy.[70] The halogenated derivatives **212b,c** can also be generated from tetrahaloterephthalic acids with SO$_3$. These bis-acyl dications have been studied by mass spectroscopy, being generated from the terephthaloyl chlorides. They fragment to the highly abundant $C_6X_4^{2+}$ ions (**213b,c**).

7.2.7 Varied Distonic Superelectrophilic Systems

Besides the discussed distonic superelectrophiles, there are a variety of other dicationic systems that may be considered as such. A significant number of sulfur-centered systems have been shown to form such persistent dicationic systems. For example, the dicationic carbosulfonium species (**215**) was obtained by the oxidation of compound **214** (eq 74).[71]

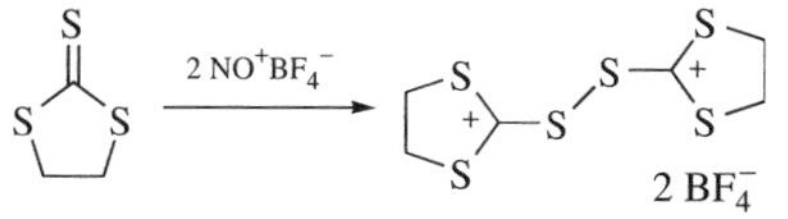

(74)

214   215

Many types of bis-sulfonium dications have been described in the literature, although little is known about the extent of superelectrophilic activation in these species. As an example, the dithioniabicyclo[2,2,2]octane dication (**217**) is produced by reaction of the bicyclic dithioether dication (**216**) with styrene derivatives (eq 75).[72] Tertiary sulfonium dications (useful synthetic intermediates) have been prepared (eq 76).[73]

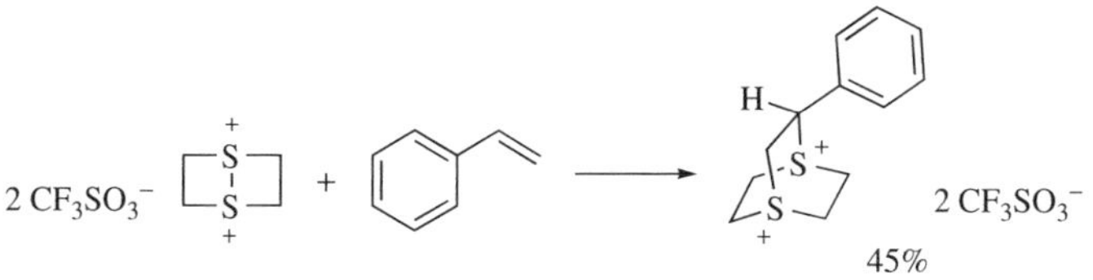

(75)

216   217

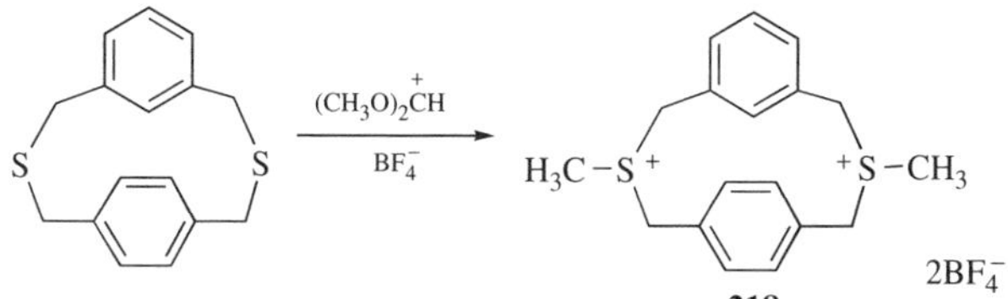

(76)

218

The chemistry of dicationic sulfur-centered systems has been recently reviewed,[74] and is therefore not further discussed here.

Due to the inherent stability of most ammonium ion centers, structures with two well-separated ammonium centers are considered onium dications without superelectrophilic activation.[68] In rare cases, however, distonic superelectrophiles are formed containing two ammonium cationic centers. One of the most important examples is the electrophilic fluorinating agent called Selectfluor® (**220**). It is a very useful reagent to prepare varied fluorinated products and its chemistry has been recently reviewed.[75] In a study comparing several types of electrophilic fluorinating agents, Selectfluor® **220** was found to have a reduction potential significantly more positive than the analogous monocationic salt **221** (Table 5).[76] Other than N-fluorobis(trifluoromethylsulfonyl)imide (**219**), **220** has the most positive reduction potential. Moreover, it has been shown that the electrophilic reactivity parallels the reduction potential

Table 5. Electrophilic fluorinating agents and their electrochemical reduction potentials (E_p).

	E_p, Reduction		E_p, Reduction
219	+0.18	**222**	−0.47
220	−0.04	**223**	−0.78
221	−0.37	**224**	−2.10

data. **219** reacts with benzene to give fluorobenzene. However **220** does not react with benzene, but fluorinates anisole in good yield. However, in CF_3SO_3H medium **220** fluorinates benzene, chlorobenzene, etc. Likewise, N-fluorobenzenesulfonimide (**223**) is incapable of fluorinating anisole, but does fluorinate the more nucleophilic, 1,3-dimethoxybenzene.[76] Based on these results, it is clear that the adjacent ammonium group enhances the electrophilic reactivity of the N-fluoro group in Selectfluor®, and thus **220** can be properly considered as a distonic superelectrophile.

Phosphonium and arsonium dications have been studied and their chemistry has also been recently reviewed.[74]

There have been several distonic superelectrophiles described in the literature that are oxonium-centered dications. For example, a series of diols were solvated in FSO_3H-SbF_5-SO_2 and the dicationic species were persistent at $-80°C$.[77] Upon warming to $25°C$, the bis-oxonium ions undergo rearrangement to the more stable carboxonium monocations (eq 77). Barring a concerted mechanism, the transformations are thought to involve the carbo-oxonium dications (i.e., **225**) and concomitant hydride shifts. Interestingly, 2,5-hexanediol ionizes in superacid, and with warming the cyclic oxonium ion (**229**) is formed (eq 78).

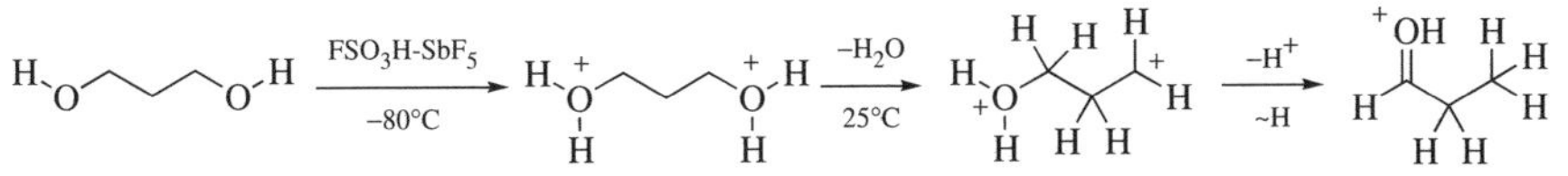
225 226
(77)

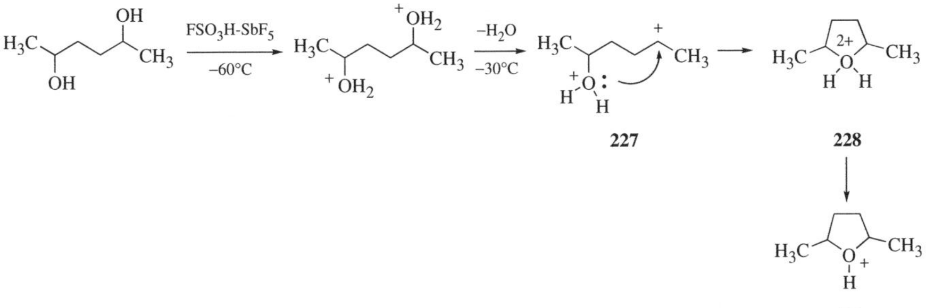
227 228 229
(78)

In superacid, it is possible to even generate gitonic superelectrophiles such as diprotonated water (H_4O^{2+}).[78] This suggests the possibility that the gitonic superelectrophile (**228**) may be involved in the cyclization step, although it is unclear if it is a transition state in the conversion or a discrete high lying intermediate. Some of the bis-oxonium ions lead to cleavage reactions (eq 79).

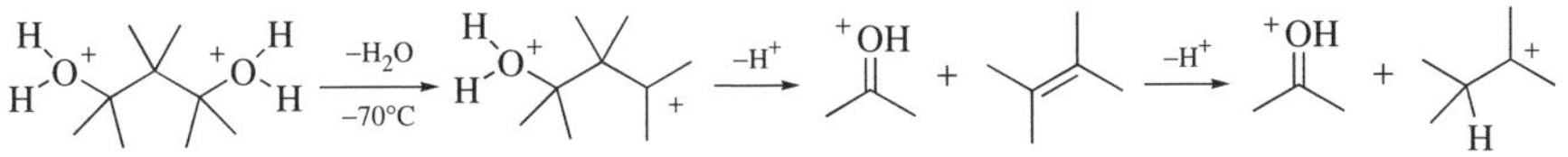
230
(79)

2,3,3,4-Tetramethyl-2,4-pentanediol forms the oxonium dication **230** in superacid, and dehydration is followed by carbon- carbon bond cleavage. Although such cleavage reactions can occur with monocationic onium ions, in this case the cleavage reaction is likely an indication of the super-electrophilic nature of the dicationic intermediate(s).

Similar distonic superelectrophiles have been described as products from the superacidic reactions of hydroxy ethers. Olah and Sommer prepared and observed (by low temperature NMR) both 1,4- and 1,5-dioxonium dications (**231–234**).[79]

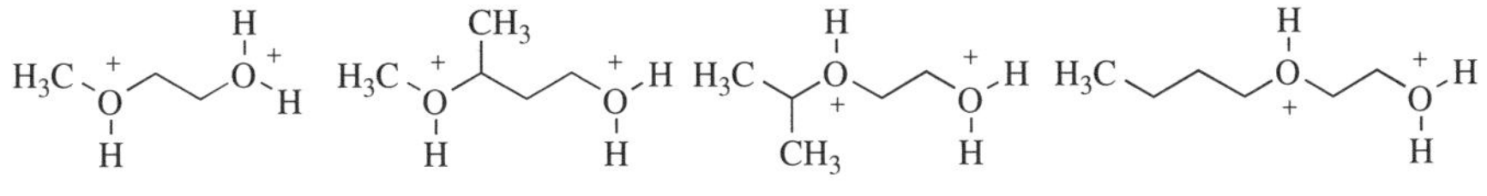
231 232 233 234

These species were prepared from the appropriate ethers in FSO_3H-SbF_5-SO_2 solution at $-60°C$. Like the protonated diols, these oxonium dications (i.e., **235**) were observed to undergo cleavage reactions leading to monocationic fragments or dehydration-rearrangement to give the oxonium ion **236** (eq 80).

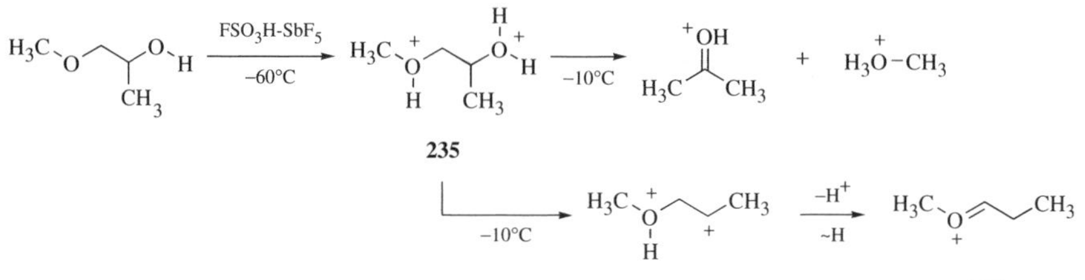

(80)

Another interesting oxonium-based dication arises from *p*-methoxy-benzene-diazonium ion in FSO_3H-SbF_5 solution (eq 81).[80] Dication **237** was observed by low-temperature ^{13}C NMR spectroscopy. Interestingly, the *o*-methoxybenzenediazonium ion does not form the oxonium dication **238** (eq 82),

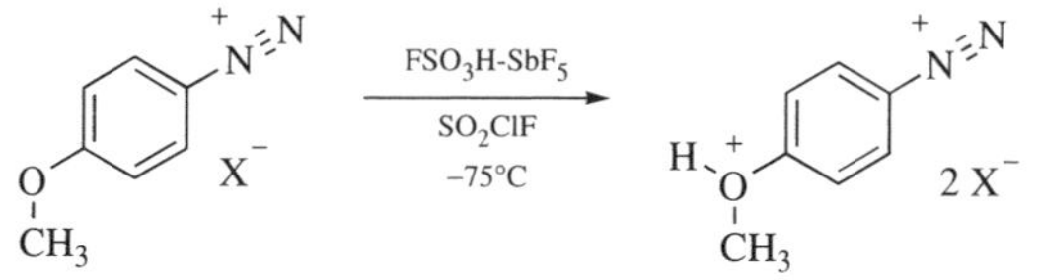

(81)

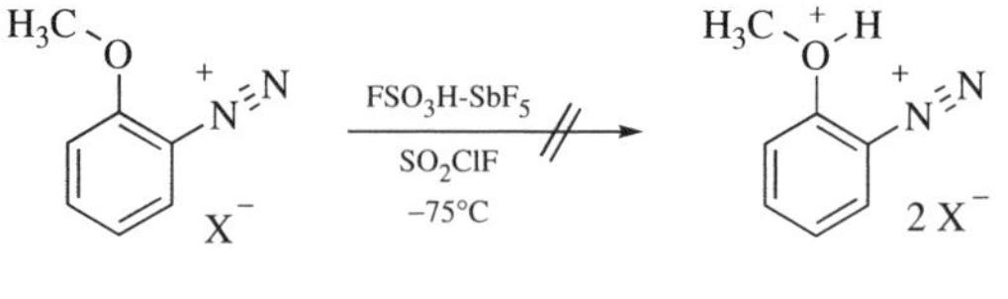

(82)

presumably due to the proximity of the developing charge on the oxonium and diazonium groups. This may be an indication that oxonium-based dications like **237** should be considered distonic superelectrophiles.

A number of distonic halonium-based dications are also known and their properties are often superelectrophilic.[81] The chemistry of halonium ions was thoroughly reviewed, thus only a few aspects are described here. Alkylation of dihaloalkanes with methyl and ethyl fluoroantimonate (CH_3F-SbF_5-SO_2 and CH_3CH_2F-SbF_5-SO_2) gives the monoalkylated halonium ions and/or the dialkylated dihalonium ions, depending on the reaction conditions. Iodine shows an unusual ability to stabilize positive charge, as seen in the formation of dialkyl alkylene diiodonium ions (**239**, eq 83). Remarkably, the dihalonium ion **239a** has been prepared with only a single methylene group separating the iodonium centers, formally a gitonic superelectrophile. However, in the case of dibromoalkanes, the dihalonium ion **240** is formed with three methylene groups separating the halonium centers (eq 84).

$$I\text{-}(CH_2)_n\text{-}I \xrightarrow[-78°C]{\substack{\text{excess} \\ CH_3F\text{-}SbF_5}} \overset{+}{CH_3}\text{-}I\text{-}(CH_2)_n\text{-}\overset{+}{I}\text{-}CH_3 \tag{83}$$

239 **a**: n = 1 **d**: n = 4
       b: n = 2 **e**: n = 5
       c: n = 3 **f**: n = 6

$$Br\diagdown\diagup Br \xrightarrow[-78°C]{\substack{\text{excess} \\ CH_3F\text{-}SbF_5}} H_3C\diagdown\underset{+}{Br}\diagup\diagdown\underset{+}{Br}\diagup CH_3 \tag{84}$$

240

In contrast, the 1,2-dibromoalkanes only give the monohalonium ions and cleavage products. Dialkyl alkylenedichloronium ions were not observed. In the case of dialkyl alkylenedibromonium ions, it is clear that charge-charge repulsive effects lead to a significant destabilization of the systems having only two methylene groups between the bromonium groups when compared to the charge separated species **240**. This may indicate the superelectrophilic character of **240** and other related species. The dialkyl phenylenedihalonium ions exhibit similar trends.[81] For example, all isomeric diiodobenzenes give the dihalonium ions **241–243** and even the triiodonium ions (**244**) can be prepared. *p*-Dibromobenzene can be dialkylated to give the dibromonium ion **245**, however, the reactions of *m*- or *o*-dibromobenzene with excess alkylating agent only provides the monoalkylation products (**246–247**).

241   242   243   244

245   246   247   248

Coloumbic repulsive effects are thought to destabilize the *o*- or *m*-dibro-
monium ions. Mixed dihalonium ions (**248**) have also been prepared.
Alkylarylchloronium ions however could not be prepared under these
conditions due to facile aromatic ring alkylation.

In a related study, bromoanisoles were reacted with excess CH_3F-SbF_5
and the dicationic product **250** was only formed with *p*-bromoanisole.[82]
NMR studies indicated that the dication is in equilibrium with the oxonium
monocation **249** (eq 85).

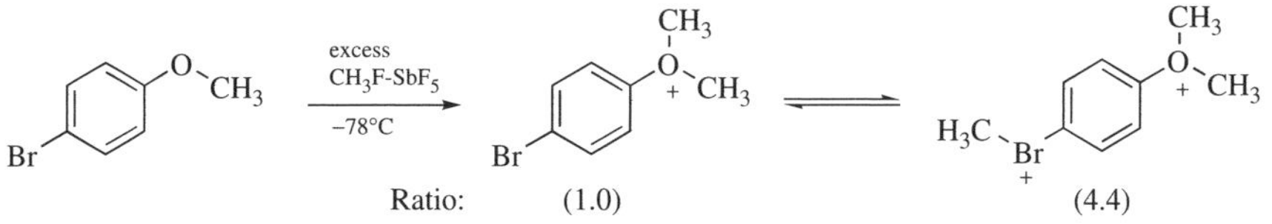

(85)

The isomeric *o*- or *m*-bromoanisole give only the oxonium monocations.
Presumably, dialkylation is inhibited by unfavorable charge-charge repulsive effects in the *m*- or *o*-isomers.

Interesting dihalonium ions have also been prepared from 1,4-dihalo-
cubanes (**251**, eq 86).[83] Both diiodo and dibromocubanes were dimethy-
lated in excess CH_3F-SbF_5 to give the dihalonium ions (**252a,b**) which
were stable at $-70°C$. 1,4-Dichlorocubane (**251c**) did not give the dihalonium ion **252c**. When the halocubanes (**253a-c**) were reacted with CH_3F-
SbF_5, the corresponding halonium cubanes (**254a-c**) could not be obtained
as stable species (eq 87).

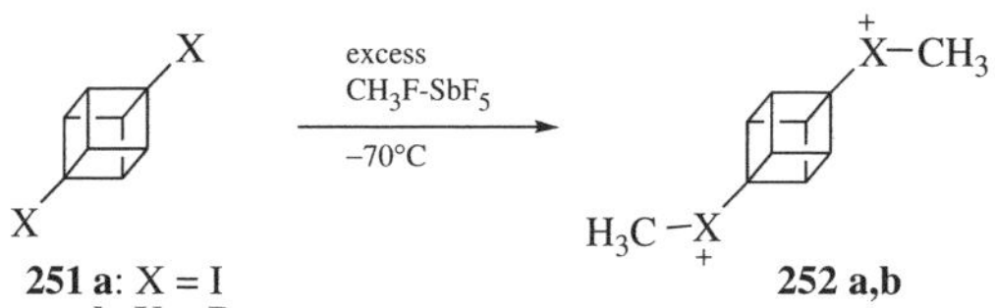

(86)

251 a: X = I
   b: X = Br
   c: X = Cl

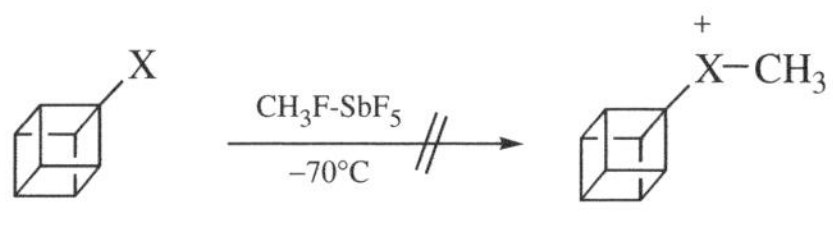

(87)

253a: X = I
   b: X = Br
   c: X = Cl

This unusual observation is explained with consideration of the superelectrophilic nature of the dihalonium ions (**252a,b**). It is thought that the
cubyl framework is stabilized by the pair of electron-withdrawing halonium groups, while in the case of the monohalonium cubanes (**254a-c**),

the cubyl framework does not have this stabilization. In this same study, cubyl diacyl and dicarboxonium dications were also prepared, as well as a novel tetracarbocationic species.

REFERENCES

(1) D. A. Klumpp; M. Garza; G. V. Sanchez; S. Lau; S. DeLeon *J. Org. Chem.* **2000**, *65*, 8997.

(2) D. A. Klumpp; R. Rendy; A. McElrea *Tetrahedron Lett.* **2004**, *45*, 7959.

(3) (a) R. M. Pagni *Tetrahedron* **1984**, *40*, 4161. (b) G. K. S. Prakash; T. N. Rawdah; G. A. Olah *Angew. Chem. Int. Ed. Engl.* **1983**, *22*, 390.

(4) G. A. Olah; J. L. Grant; R. J. Spear; J. M. Bollinger; A. Serianz; G. Sipos *J. Am. Chem. Soc.* **1976**, *98*, 2501.

(5) G. A. Olah; V. P. Reddy; G. Lee; J. Casanova; G. K. S. Prakash *J. Org. Chem.* **1993**, *58*, 1639.

(6) G. A. Olah; G. K. S. Prakash; T. N. Rawdah *J. Am. Chem. Soc.* **1980**, *102*, 6127.

(7) G. K. S. Prakash; A. P. Fung; T. N. Rawdah; G. A. Olah *J. Am. Chem. Soc.* **1985**, *107*, 2920.

(8) C. Taeschler; T. S. Sorensen *Tetrahedron Lett.* **2001**, *42*, 5339.

(9) A. d. Meijere; O. Schallner; P. Golitz; W. Weber; P. v. R. Schleyer; G. K. S. Prakash; G. A. Olah *J. Org. Chem.* **1985**, *50*, 5255.

(10) G. P. Miller; I. Jeon; G. Wilson; A. J. Athans *Org. Lett.* **1999**, *1*, 1771.

(11) (a) G. A. Olah; G. Liang; P. v. R. Schleyer; E. M. Engler; M. J. S. Dewar; R. C. Bingham *J. Am. Chem. Soc.* **1973**, *93*, 6829. (b) R. C. Bingham; D. H. Lo; M. J. S. Dewar *J. Am. Chem. Soc.* **1975**, *97*, 1285. (c) A. de. Meijere; O. Schallner; P. Golitz; W. Weber; P. v. R. Schleyer; G. K. S. Prakash; G. A. Olah *J. Org. Chem.* **1985**, *50*, 5255.

(12) G. K. S. Prakash; V. V. Krishnamurthy; M. Arvanaghi; G. A. Olah *J. Org. Chem.* **1985**, *50*, 3985.

(13) G. Rasul; G. A. Olah; G. K. S. Prakash *Proc. Nat. Acad. Sci. USA* **2004**, *101*, 10868.

(14) G. A. Olah; G. K. S. Prakash; J. G. Shih; V. V. Krishnamurthy; G. D. Meteescu; G. Liang; G. Sipos; V. Buss; J. M. Gund; P. v. R. Schleyer *J. Am. Chem. Soc.* **1985**, *107*, 2764.

(15) M. D. Heagy; Q. Wang; G. A. Olah; G. K. S. Prakash *J. Org. Chem.* **1995**, *60*, 7351.

(16) (a) G. A. Olah *Angew. Chem. Int. Ed. Engl.* **1993**, *32*, 767. (b) G. A. Olah; D. A. Klumpp *Acc. Chem. Res.* **2004**, *37*, 211.

(17) H. Hart; T. Salzberg; R. R. Rofos *J. Am. Chem. Soc.* **1963**, *85*, 1800.

(18) (a) T. Saitoh; S. Yoshida; J. Ichikawa *Org. Lett*. **2004**, *6*, 4563. (b) H. Wang; F. P. Gabbai *Org. Lett*. **2005**, *7*, 283.

(19) (a) T. Saitoh; S. Yoshida; J. Ichikawa *J. Org. Chem*. **2006**, *71*, 6414. (b) T. Saitoh; J. Ichikawa *J. Am. Chem. Soc*. **2005**, *127*, 9696.

(20) (a) H. Wang; F. P. Gabbai *Angew. Chem. Int. Ed*. **2004**, *43*, 184. (b) H. Kawai; T. Takeda; K. Fujiwara; T. Suzuki *Tetrahedron Lett*. **2004**, *45*, 8289. (c) H. Kawai; T. Nagasu; T. Takeda; K. Fujiwara; T. Tsuji; M. Ohkita; J. Nishida; T. Suzuki *Tetrahedron Lett*. **2004**, *45*, 4553.

(21) (a) T. Suzuki; J. Nishida; T. Tsuji *Angew. Chem. Int. Ed. Engl*. **1997**, *36*, 1329. (b) T. Suzuki; J. Nishida; T. Tsuji *Chem. Commun*. **1998**, 2193. (c) K. A. Carey; W. Clegg; M. R. J. Elsegood; B. T. Golding; M. N. Stuart Hill; H. Maskill *J. Chem. Soc., Perkin Trans 1*, **2002**, 2673.

(22) (a) J. Nishida; T. Suzuki; M. Ohkita; T. Tsuji *Angew. Chem. Int. Ed*. **2001**, *40*, 3251. (b) T. Suzuki; R. Yamamoto; H. Higuchi; E. Hirota; M. Ohkita; T. Tsuji *J. Chem. Soc. Perkin Trans 1*, **2002**, 1937.

(23) H. Higuchi; E. Ohta; H. Kawai; K. Fujiwara; T. Tsuji; T. Suzuki *J. Org. Chem*. **2003**, *68*, 6605.

(24) G. A. Olah; G. K. S. Prakash; G. Rasul *J. Org. Chem*. **2001**, *61*, 2907.

(25) (a) G. A. Olah; N. Hartz; G. Rasul; G. K. S. Prakash *J. Am. Chem. Soc*. **1993**, *115*, 6985. (b) G. A. Olah; N. Hartz; G. Rasul; G. K. S. Prakash; M. Burkhart; K. Lammertsma *J. Am. Chem. Soc*. **1994**, *116*, 3187.

(26) N. Yoneda; T. Fukuhara; Y. Takahashi; A. Suzuki *Chem. Lett*. **1983**, 17.

(27) J.-C. Jacquesy; R. Jacquesy; J. F. Patoiseau *Tetrahedron* **1976**, *32*, 1699.

(28) (a) J. P. Begue; M. Charpentier-Morize; C. Pardo; J. Sansoulet *Tetrahedron* **1978**, *34*, 293. (b) J. M. Coxon; G. J. Hydes; P. J. Steel *Tetrahedron* **1985**, *41*, 5213. (c) R. Jacquesy; J. F. Patoiseau *Bull. Soc. Chim. Fr*. **1978**, *2*, 225. (d) J.-C. Jacquesy in *Stable Carbocation Chemistry*, G. K. S. Prakash and P. v. R. Schleyer, Eds., Wiley, New York, **1998**, pp 549-574.

(29) A. Martin; M.-P. Jouannetaud; J.-C. Jacquesy *Tetrahedron Lett*. **1996**, *37*, 2967.

(30) G. A. Olah; N. Yoneda; R. Ohnishi *J. Am. Chem. Soc*. **1976**, *98*, 7341.

(31) (a) R. Rendy; Y. Zhang; A. McElrea; A. Gomez; D. A. Klumpp *J. Org. Chem*. **2004**, *69*, 2340. (b) G. K. S. Prakash; P. Yan; B. Torok; G. A. Olah *Catal. Lett*. **2003**, *87*, 109.

(32) G. K. S. Prakash; G. Rasul; A. Burrichter; K. K. Laali; G. A. Olah *J. Org. Chem*. **1996**, *61*, 9253.

(33) D. A. Klumpp; M. Garza; S. Lau; B. Shick; K. Kantardjieff *J. Org. Chem*. **1999**, *64*, 7635.

(34) (a) K. Y u. Koltunov; S. Walspurger; J. Sommer *Chem. Commun*. **2004**, 1754. (b) I. B. Repinskaya; K. Y u. Koltunov; M. M. Shakirov; L. N. Shchegoleva; V. A. Koptyug *Russ. J. Org. Chem*. **1993**, *29*, 803. (c) K. Y u. Koltunov; M. M. Shakirov; I. B. Repinskaya *Russ. J. Org. Chem*. **1998**, *34*, 595.

(35) D. Farcasiu; G. Miller *J. Org. Chem.* **1989**, *54*, 5423.

(36) G. A. Olah; A. Burrichter; G. Rasul; G. K. S. Prakash; M. Hachoumy; J. Sommer *J. Am. Chem. Soc.* **1996**, *118*, 10423.

(37) (a) D. C. Palmer; M. J. Strauss *Chem. Rev.* **1977**, *77*, 1. (b) R. Grewe *Angew. Chem.* **1947**, *59*, 194.

(38) D. A. Klumpp; P. S. Beauchamp; G. S. Sanchez, Jr.; S. Aguirre; S. de Leon *Tetrahedron Lett.* **2001**, *42*, 5821.

(39) Y. Zhang; A. McElrea; G. V. Sanchez Jr.; D. A. Klumpp; D. Do; A. Gomez; S. L. Aguirre; R. Rendy *J. Org. Chem.* **2003**, *68*, 5119.

(40) D. A. Klumpp; R. Rendy; Y. Zhang; A. Gomez; A. McElrea; H. Dang *J. Org. Chem.* **2004**, *69*, 8108.

(41) (a) J. R. Brooks; D. N. Harcourt; R. D. Waigh *J. Chem. Soc., Perkin Trans. 1* **1973**, 2588. (b) H. Takayama; T. Suzuki; T. Nomoto *Chem. Lett.* **1978**, 865. (c) K. R. Gee; P. Barmettler; M. R. Rhodes; R. N. McBurney; N. L. Reddy; L.-Y. Hu; R. E. Cotter; P. N. Hamilton; E. Weber; J. F. W. Keana *J. Med. Chem.* **1993**, *36*, 1938. (d) D. N. Brooks; D. N. Harcourt *J. Chem. Soc.* (C) **1969**, 626.

(42) (a) Y. Zhang; D. A. Klumpp *Tetrahedron Lett.* **2002**, *43*, 6841. (b) D. A. Klumpp; P. J. Kindelin; A. Li *Tetrahedron Lett.* **2005**, *46*, 2931.

(43) A. Li; P. J. Kindelin; D. A. Klumpp *Org. Lett.* **2006**, *8*, 1233.

(44) D. A. Klumpp; A. Li; D. J. DeSchepper *Abstract of Papers*, 232nd National Meeting of the American Chemical Society, San Francisco, CA, September 2006; American Chemical Society: Washington, DC, 2006; Abstract ORGN 561.

(45) A. Kleemann; J. Engel; B. Kutscher; D. Reichert *Pharmaceutical Substances*, 4th Ed., Thieme, Stuttgart, **2001**, p 308.

(46) Y. Zhang; S. A. Aguirre; D. A. Klumpp *Tetrahedron Lett.* **2002**, *43*, 6837.

(47) G. A. Olah; M. Calin *J. Am. Chem. Soc.* **1968**, *90*, 4672.

(48) J. W. Larsen; P. A. Bouis *J. Am. Chem. Soc.* **1975**, *97*, 6094.

(49) G. A. Olah; A. T. Ku; J. Sommer *J. Org. Chem.* **1970**, *35*, 2159.

(50) D. A. Klumpp; R. Rendy; Y. Zhang; A. Gomez; A. McElrea *Org. Lett.* **2004**, *6*, 1789.

(51) A. Wiles *J. Chem. Soc.* **1953**, 996.

(52) G. A. Olah; A. M. White *J. Am. Chem. Soc.* **1967**, *89*, 4752.

(53) (a) D. Bruck; M. Rabinovitz *J. Am. Chem. Soc.* **1977**, *99*, 240. (b) D. Bruck; M. Rabinovitz *J. Am. Chem. Soc.* **1976**, *98*, 1599.

(54) G. A. Olah; P. W. Westerman *J. Org. Chem.* **1973**, *38*, 1986.

(55) V. Prakas Reddy; G. Rasul; G. K. S. Prakash; G. A. Olah *J. Org. Chem.* **2003**, *68*, 3507.

(56) D. A. Klumpp; M. Garza; A. Jones; S. Mendoza *J. Org. Chem.* **1999**, *64*, 6702.

(57) (a) D. A. Klumpp; G. V. Sanchez, Jr.; Y. Zhang; S. L. Aguirre; S. de Leon *J. Org. Chem.* **2002**, *67*, 5028. (b) D. A. Klumpp; S. L. Aguirre; G. V. Sanchez, Jr.; S. J. de Leon *Org. Lett.* **2001**, *3*, 2781.

(58) (a) D. A. Klumpp; Y. Zhang; P. J. Kindelin; S. Lau *Tetrahedron* **2006**, *62*, 5915. (b) D. A. Klumpp; A. Jones; S. Lau; S. DeLeon; M. Garza *Synthesis*, **2000**, 1117.

(59) K. Y. Koltunov; G. K. S. Prakash; G. Rasul; G. A. Olah *J. Org. Chem.* **2002**, *67*, 4330.

(60) G. A. Olah; D. L. Brydon; R. D. Porter *J. Org. Chem.* **1970**, *35*, 317

(61) G. A. Olah; M. Calin *J. Am. Chem. Soc.* **1967**, *89*, 4736.

(62) K. Y. Koltunov; G. K. S. Prakash; G. Rasul; G. A. Olah *Tetrahedron* **2002**, *58*, 5423.

(63) H. Heaney in *Comprehensive Organic Synthesis*, B. M. Trost; I. Fleming, Eds.; Pergamon Press: Oxford, **1991**; Vol 2, pp 733-750. (a) G. A. Olah *Friedel-Crafts and Related Reactions*; Wiley-Interscience: New York, **1964**;Vol. 3, pp 16–36.

(64) G. A. Olah; D. J. Donovan *J. Org. Chem.* **1978**, *43*, 1743.

(65) K. Shudo; Y. Orihara; T. Ohta; T. Okamato *J. Am. Chem. Soc.* **1981**, *103*, 943.

(66) J. Sommer; P. Rimmelin; T. Drakenberg *J. Am. Chem. Soc.* **1976**, *98*, 943.

(67) G. A. Olah; A. T. Ku *J. Org. Chem.* **1970**, *35*, 3913.

(68) G. A. Olah; K. K. Laali; Q. Wang; G. K. S. Prakash *Onium Ions*, Wiley, New York, 1998.

(69) G. A. Olah; M. B. Comisarow *J. Am. Chem. Soc.* **1966**, *88*, 3313.

(70) (a) Reference 69. (b) J. O. Knobloch; F. Ramirez *J. Org. Chem.* **1970**, *40*, 1101.

(71) R. L. Blankespoor; M. P. Doyle; D. M. Hedstrand; W. H. Tamblyn; D. A. V. Dyke *J. Am. Chem. Soc.* **1981**, *103*, 3313.

(72) V. G. Nenajdenko; N. E. Shevchenko; E. S. Balenkova *J. Org. Chem.* **1998**, *63*, 2168.

(73) V. Boekelheide; P. H. Anderson; T. A. Hylton *J. Am. Chem. Soc.* **1974**, *96*, 1558.

(74) V. G. Nenajdenko; N. E. Shevchenko; E. S. Balenkova, I. V. Alabugin *Chem. Rev.* **2003**, *103*, 229.

(75) (a) G. S. Lal; G. P. Pez; R. G. Syvret *Chem. Rev.* **1996**, *96*, 1737. (b) P. T. Nyffeler; S. G. Duron; M. D. Burkart; S. P. Vincent; C.-H. Wong *Angew. Chem., Int. Ed.* **2005**, *44*, 192. (c) T. Shamma; H. Buchholz; G. K. S. Prakash; G. A. Olah, *Isr. J. Chem.* **1999**, *39*, 1029.

(76) A. G. Gilincinski; G. P. Pez; R. G. Syvret; G. S. Lal *J. Fluorine Chem.* **1992**, *59*, 157.

(77) G. A. Olah; J. Sommer *J. Am. Chem. Soc.* **1968**, *90*, 927.

(78) G. A. Olah; G. K. S. Prakash; M Barzaghi; K. Lammertsma; P. v. R. Schleyer; J. A. Pople *J. Am. Chem. Soc*. **1986**, *108*, 1032.

(79) G. A. Olah; J. Sommer *J. Am. Chem. Soc*. **1968**, *90*, 4323.

(80) K. Laali; G. A. Olah *J. Org. Chem*. **1985**, *50*, 3006.

(81) G. A. Olah; Y. K. Mo; E. G. Melby; H. C. Lin *J. Org. Chem*. **1973**, *38*, 367.

(82) G. A. Olah; E. G. Melby *J. Am. Chem. Soc*. **1973**, *95*, 4971.

(83) N. J. Head; G. Rasul; A. Mitra; A. Bashir-Heshemi; G. K. S. Praksh; G. A. Olah *J. Am. Chem. Soc*. **1995**, *117*, 12107.

8

SIGNIFICANCE AND OUTLOOK

The concept of superelectrophilic activation was first developed based on the increasing trend of electrophilic reactivities observed in superacidic media.[1] Varied electrophiles showed significantly enhanced reactivities in superacids and were capable of reacting with extremely weak nucleophiles. Many of the reported superelectrophiles have been generated in strong and superacidic media.[2] The need for a highly acidic media can be understood in terms of their low nucleophilicity and ability to protonate (or coordinate) available electron pair donor site(s) of the electrophiles decreasing neighboring group participation with the electrophilic center. This interaction with the acidic media leads to doubly electron-deficient, activated species or superelectrophiles. Fully formed cationic species, however, are the limiting case, and superelectrophilic activation more often involves only partial proton transfer or weak coordination with Lewis acids, i.e., electrophilic solvation.

Although superelectrophilic activation has been most often observed in superacid media, examples have also been reported involving reactions occurring in less acidic environments, including enzyme systems, zeolites, etc., as well as in the gas phase. It has also been shown that onium charge centers (ammonium, phosphonium, etc.) can be involved in superelectrophilic activation. These considerations suggest that superelectrophilic activation is a rather general phenomenon. Berkessel and Thauer suggested

Superelectrophiles and Their Chemistry, by George A. Olah and Douglas A. Klumpp
Copyright © 2008 John Wiley & Sons, Inc.

for example that locally highly electron deficient environments could be present at enzyme active sites.[3] Moreover, properly positioned acidic functional groups and Lewis acid sites may be able to participate in multidentate interaction with substrates, producing doubly electron-deficient superelectrophilic species. Besides superelectrophilic activation in biological systems, multidentate interaction may also be important in solid acid catalyzed reactions involving, for example, Nafion-H and H-ZSM-5 and their analogs.[4] Some solid acids are known to possess acidic sites clustered together in close proximity. In order to explain how these acids can sometimes display remarkable activities, for example, allowing the catalytic transformations of extremely low nucleophilicity alkanes (even methane), it has been suggested by Olah that bi- and multidentate interactions can lead to superelectrophiles.[4] In other types of zeolites, Corma noted that cationic charges in the zeolite framework "produce very high electrostatic fields (ϕ_{FN}) and field gradients (ϕ_{FQ}) in the channels and cavities."[5] This may also contribute to observed superelectrophilic activation.

There is now considerable experimental and theoretical evidence to support the concept of superelectrophilic activation discussed throughout our book. Several types of superelectrophilic dicationic species have been observed in the condensed phase using spectroscopic techniques, such as low-temperature NMR. Other superelectrophiles have been observed in the gas phase using mass spectrometric methods. Kinetic studies have likewise provided evidence for various superelectrophiles in superacid-catalyzed reactions. Superelectrophilic activation is also confirmed by varied observations, as trends of reactivity vs. acidity, electrophilic chemistry involving extremely weak nucleophiles, product studies, isotopic exchange, and structure-activity relationships. Calorimetric and kinetic studies have also provided thermodynamic data by which superelectrophiles can be studied. Theoretical calculations have been particularly useful. Many superelectrophiles were shown to be kinetically stable gas-phase species, often having considerable barriers to deprotonation or fragmentation reactions. By comparing usual electrophiles with the corresponding superelectrophiles, calculations have been able to correlate reactivity trends with molecular orbital energy levels and charge densities of the involved superelectrophiles. Moreover, thermodynamic activation parameters obtained from *ab initio* calculations have been compared with kinetically determined experimental values. Good agreement between the calculated and experimental data has been considered evidence for the involvement of superelectrophiles. Theoretical calculations have also provided useful structural and spectral data allowing comparison with experimentally determined data.

Superelectrophilic reactions have been shown to be particularly useful in various synthetic conversions particularly of unactivated σ and π-bonds, such as in alkanes and electron deficient arenes. Superelectrophiles have also been used in the synthesis of natural products and biologically active compounds. Superelectrophilic chemistry has also found varied applications in macromolecular chemistry and material science.[13-17]

Experimental and theoretical studies of superelectrophiles have revealed many interesting structure-activity relationships. Superelectrophilic activation was shown to be a consequence of decreasing neighboring group participation into electron deficient reactive cationic centers of the electrophilic reagent. Decreasing neighboring group participation often leads to enhanced reactivities, as a consequence of the inductive, electrostatic, and/or resonance effects of electron withdrawing substituents in the doubly electron-deficient species. For *de facto* dicationic superelectrophiles (and higher polycations), coulombic repulsion has been shown to significantly influence the chemistry of these species. Charge-charge separation is an important driving force for such superelectrophilic reactions. Onium (ammonium, pyridinium, phosphonium, etc.) substituents particularly enhance the reactivities of adjacent electrophilic sites. Similar activating effects have been demonstrated in organometallic and catalytic systems.

Significant structure-activity relationships in superelectrophiles relate to the distance between charge centers. To an approximation, superelectrophilic activation decreases with increasing separation of charge centers. This leads to the distinction between gitonic and distonic superelectrophiles, the latter being characterized by three or more atoms separating the charge centers. It was shown, however, that superelectrophilic activation may also occur by suitable geometrical arrangement forcing the charge centers into close proximity. It has been suggested that superelectrophilic activation in enzymatic systems may be accomplished by closely oriented cationic charges.[6]

Since the concept of superelectrophilic activation was proposed 30 years ago, there have been many varied superelectrophiles reported both in experimental and theoretical studies. Superelectrophiles can be involved in both gas and condensed phase reactions, ranging from interstellar space down to the active sites of certain enzymes. Moreover, synthetic conversions involving superelectrophiles are increasingly used in the synthesis of valuable products. Superelectrophilic activation has also been useful in the development of a number of new catalytic processes. It is our belief that superelectrophilic chemistry will continue to play an increasing role in both synthetic and mechanistic chemistry.

REFERENCES

(1) G. A. Olah; A. Germain; H. C. Lin; D. Forsyth *J. Am. Chem. Soc*. **1975**, *97*, 2928; G. A. Olah, G. K. S. Prakash, M. Barzaghi, K. Lammertsma, P. R. Schleyer, *J. Am. Chem. Soc*., **1986**, *108*, 1062; G. A. Olah, G. K. S. Prakash, M. Marcelli, K. Lammertsma, *J. Phys. Chem*. **1988**, *92*, 878; G. A. Olah, G. K. S. Prakash, K. Lammertsma, *Res. Chem. Intermed*. **1989**, *12*, 14.

(2) G. A. Olah; D. A. Klumpp *Acc. Chem. Res*. **2004**, *37*, 211.

(3) A. Berkessel; R. K. Thauer *Angew. Chem. Int. Ed. Engl*. **1995**, *34*, 2247.

(4) G. A. Olah *Angew. Chem. Int. Ed. Engl*. **1993**, *32*, 767.

(5) A. Corma *Chem. Rev*. **1995**, *95*, 559.

(6) D. A. Klumpp; R. Rendy; Y. Zhang; A. Gomez; A. McElrea *Org. Lett*. **2004**, *6*, 1789.

INDEX

ab initio calculations
  estimated activation energies, for
    superelectrophilic
    electrocyclizations 47–48
  of superelectrophiles
    energies of LUMOs, 49
    estimation of electronic chemical
      potential, 49–50
    estimation of electrophilicity indexes,
      49–50
    estimation of chemical hardness,
      49–50
    estimation of NBO charges, 49
    estimation of NMR chemical shifts,
      50–52
    superelectrophilic isodesmic reactions,
      48–49
Aceanthrenequinone
  condensation with deactivated arenes,
    141–142
  diprotonated in superacid, 141–142
  superelectrophilic, in polymer synthesis,
    142

Acenapthenequinone
  condensation with benzene, 141
  diprotonated in superacid, 141
Acetic acid
  diprotonated, calculated NMR chemical
    shifts, 51–52
  diprotonated, calculated structure and
    stability, 212
Acetoacetamide
  condensation with benzene, 199
  diprotonation, 199–200
Acetoacetic acid, diprotonation and NMR
  study, 199–200
Acetone, diprotonated, 86–88, 157, *see
  also* carboxonium ions
N-Acetonylpyridinium salts
  condensation with benzene, 205
  protonation in CF_3SO_3H, 205
Acetyl cation
  electrophilic solvation, 154
  superelectrophilic
    from $AlCl_3$, 154
    reactions with arenes, 155

Superelectrophiles and Their Chemistry, by George A. Olah and Douglas A. Klumpp
Copyright © 2008 John Wiley & Sons, Inc.